Advanced Mathematics

Advanced Mathematics

A Transitional Reference

Stanley J. Farlow
University of Maine
Orono, ME, USA

Registered Office
John Wiley & Sons, Inc., 111 River Street, Hoboken, NJ 07030, USA

Editorial Office
111 River Street, Hoboken, NJ 07030, USA

For details of our global editorial offices, customer services, and more information about Wiley products visit us at www.wiley.com.

Library of Congress Cataloging-in-Publication data applied for

ISBN: 9781119563518

Cover Design: Wiley
Cover Image: © agsandrew/Shutterstock

Set in 10/12pt Warnock by SPi Global, Pondicherry, India

Printed in the United States of America

V10013645_090519

Contents

Preface

This book is the culmination of a set of notes that has been growing and fermenting in a bottom drawer of my desk for over 30 years. The goal was that after the right amount of pedantic aging, they would be ready to uncork the bottle and share with the world. It is my hope that the finished product harbors the perfect blend of intuition and rigorous preciseness, theory versus pragmatism, hard-nosed mathematics, and an enjoyable learning experience.

I must confess that when the entrails of this book were taking form, the intended audience was the 25 or so students in my *Introduction to Higher Math* class, but now that they have been released into the wild, my hope is they will find a wider audience.

Possible Beneficial Audiences

- **College students** who have taken courses in calculus, differential equations, and perhaps linear algebra, may be unprepared for the more advanced courses of real analysis, abstract algebra, and number theory that lie ahead. (If you cannot negate a logical sentence, you are probably not ready for prime time.) Although the basic calculus et al. sequence is important for developing a rounded background in mathematics as well as developing problem-solving skills, most calculus courses are not intended to prepare the student for advanced mathematics. Few students, after going through the basic sequence, are even familiar with the basic language of mathematics, such as the sentential and predicate logic. If a student is to develop skills for reading, analyzing, and appreciating mathematical arguments and ideas, knowledge of the basic language of mathematics is a must.
- **Mathematics teachers and math education students** will hopefully find this book a valuable aid for teaching an inspiring and exciting introduction to mathematics that goes beyond the basics. The large and varied collection of historical notes and varied problems at the end of each section should be worth the "price of admission" alone.
- **Bright high school students** with good backgrounds and a strong interest in mathematics will hopefully fall in love with this book, all the way from the "Important Note" inserts to the exciting problem sets.
- **Scientists, engineers, and out-of-practice others** in the professional workforce, who are discovering the mathematics they learned long ago is not adequate for their current needs, will benefit by spending some time with this book. This book might even show self-study individuals just how uncomplicated and enjoyable mathematics can be, possibly at variance from their college days, and in fact might develop newly found mathematical confidence.

Wow Factors of the Book

Although this book is intended to raise the mathematical thought process of the reader, it is not intended for the reader to come away thinking that mathematics consists simply in reading and proving theorems. The goal is to blend a nuanced amalgam of inductive and deductive reasoning. Utmost in the goal of the book was to avoid the dogmatism of page-after-page of theorems and proofs, to prompt the reader in thinking about the ideas presented.

The book is loaded with informative sidebars, historical notes, and tons of graphics, which hopefully provides an enjoyable atmosphere for a constructive learning process.

Over the years, I have managed to track down, dig up, sniff out, and even make up a few fascinating problems myself. Each section of the book is packed with loads of problems for readers of all interests and abilities. Each section is filled with a broad range of examples intended to add clarity and insight to abstract concepts. The book takes time to explain the thinking and intuition behind many concepts.

Chapter by Chapter (the nitty-gritty)

Chapter 1: Logic and Proofs

This book is no *Principia Mathematica* by Alfred North Whitehead and Bertrand Russell, who famously prove $1 + 1 = 2$ after 378 arduous pages in their seminal 1910 work on the foundations of mathematics. That said, there are many mathematical proofs in this book, and each and every one of them is intended to act as a learning experience. The first and foremost, of course, being that before anything can be proven true or false, mathematics must be stated in a precise mathematical language, predicate logic. Chapter 1 introduces the reader to sentential and predicate logic and mathematical induction. After two introductory sections of sentential logic and the connectives: "and," "or," "not," "if then, if and only if," we move up the logical ladder to predicate logic and the universal and existential quantifiers and variables. Sections 1.4 and 1.5 are spent proving theorems in a variety of ways, including direct proofs, proofs by contrapositive, and proofs by contradiction. The chapter ends with the principle of mathematical induction.

Chapter 2: Sets and Counting

Sets are basic to mathematics, so it is natural that after a brief introduction to the language of mathematics, we follow with an introduction to sets. Section 2.1 gives a barebones introduction to sets, including the union, intersection, and complements of a set. Section 2.2 introduces the reader to the idea of families of sets and operations on families of sets, tools of the trade for more advanced mathematical subjects like analysis and topology. Section 2.3 is an introduction to counting, including permutations, combinations, and the pigeonhole principle. Sections 2.4, 2.5, and 2.6 introduce the reader to the basics of Cantor's discoveries on the cardinality of infinite sets. Brief discussions are included on the continuum hypothesis and the axiom of choice, as well as the Zermelo–Frankel axioms of set theory.

Chapter 3: Relations

As the English logician Bertrand Russell once said, mathematics is about relations. This chapter introduces order relations, e.g. partial, strict, and total order, followed by equivalence relations and the function relation. Section 3.5 introduces the idea of the image and inverse image of a set, concepts important in analysis and topology.

Chapter 4: The Real and Complex Number Systems

Sections 4.1 and 4.2 show how the real numbers can either be defined axiomatically, or constructed all the way from the natural numbers, to the integers, to the rational numbers, and finally to the real numbers using equivalence relations introduced in Chapter 3. The Dedekind cut is introduced in Section 4.2, which defines the real numbers in terms of the rational numbers. Finally, in Section 4.3, a brief tutorial on the complex numbers, a subject often overlooked in the undergraduate curricula.

Chapter 5: Topology

This chapter introduces a number of sides of topology, starting with Sections 5.1 and 5.2 on graph theory. Section 5.3 introduces some basic ideas of geometric topology, such as homeomorphisms, topological equivalence, Euler's equation for Platonic solids, and a hint at its verification. Finally, Section 5.4 introduces the basic ideas of point-set topology on the real line, including open and closed sets, limit points, interior, exterior and boundary of a set, and so on.

Chapter 6: Algebra

Chapter 6 provides a brief introduction to symmetries and abstract groups, rings, and fields. Section 6.1 finds the symmetries of geometric objects and shows who a composition of symmetries gives rise to the algebraic structure of a group. Sections 6.2, 6.3, and 6.4 introduce the symmetric and cyclic groups of permutations, along with the idea of a subgroup. Finally, Section 6.5 gives a brief summary of rings and finite fields.

Conclusion

May the ghosts of past students of this book be with you as you find your way to a mathematical utopia. In proper hands, this book can be used in a one-semester course, covering the entire book or portions of the book. For a more leisurely pace, an instructor can pick and choose depending on one's preferences. If rushed, one may pass over portions of the book, depending on one's preferences.

Enjoy.
Stanley Farlow
Professor Emeritus of Mathematics
University of Maine

Note to the Reader

And here let me insert a parenthesis to insist on the importance of written exercises. Compositions in writing are perhaps not given sufficient prominence in certain examinations. In *the École Polytechnique,* I am told that insistence on compositions would close the door to very good pupils who know the subject, yet are incapable of applying it in the smallest degree. The word "understand" has several meanings. Such pupils understand only in the first sense of the word, and this is not sufficient to make either an engineer or a geometrician.

—Henri Poincaré

Keeping a Scholarly Journal

One cannot help but be impressed with the huge number of important English naturalists who lived during the nineteenth century. In addition to Darwin, there were Wallace, Eddington, Thompson, Haldane, Galton, and others. One characteristic that permeated their work was the keeping of detailed journals where every observation and impression was recorded. In addition to recording data, a journal provides a way to organize one's thoughts, explore relationships, and formulate ideas. In fact, they stimulate learning through writing.

Journal keeping has declined in the twentieth and twenty-first centuries, but readers of this book have the ability to recapture that important tool of learning-through-writing.

Entries in your (leather-bound) journal can be entered daily or in conjunction with each section of the book. It is useful to date entries and give them short descriptive titles. While there are no specific rules on what to include in the journal or how to write them, you will eventually find your "voice" on what works best. From the habit of rereading old entries each time you write new ones, see how your grasp of material grows. After years have passed, you will be impressed on the value you give past journal entries.

About the Companion Website

This book is accompanied by a companion website:

www.wiley.com/go/farlow/advanced-mathematics

The famous mathematician, George Polya, once remarked that mathematics is not a spectator sport, and to learn mathematics, you must *do* mathematics. Otherwise, he argued, you are simply a *mathematical spectator*.

In the context of this book, Polya's observation means that while reading the book, the reader is a mathematical spectator, after which the reader develops serious mathematical skills and becomes a mathematician by solving, untangling, playing with, etc., yes, even enjoying the problems in the book.

Solving mathematical problems is similar to other endeavors in life, even activities far afield from mathematics, like swimming or playing a musical instrument. We acquire skills in these activities by trial and error, by practice, by making observations, throwing out what does not work, and keeping what does. After a while, we become adept, even proud of our skills, and more often than not, enjoy such activities, that earlier we might have disliked, even abhorred.

I seriously hope the reader will enjoy and develop mathematical skills from the problems in this book, as much as I enjoyed making them.

Enjoy,
Stanley Farlow

Chapter 1

Logic and Proofs

1.1

Sentential Logic

> **Purpose of Section**   To introduce **sentential (sen-TEN-shuhl)** or **propositional logic** and the fundamental idea of a **sentence** (or **proposition**) and show how simple sentences can be combined using the **logical connectives** "and," "or," and "**not**" to form **compound sentences**. We then analyze the meaning of these connectives by means of **truth tables** and then introduce the concepts of **logical equivalence**, **tautologies**, and **contradictions**. We close by introducing **conjunctive** and **disjunctive** normal forms of logical expressions.

1.1.1   Introduction

So what is mathematics? The word itself is derived from the Greek word *mathēmatikē*, meaning "knowledge" or "learning." To both the practitioners of mathematics as well as the general public, the definition of the *Queen of the Sciences* varies widely.

- One of the greatest mathematician of Greek antiquity, Aristotle (384–322 BCE) defined mathematics as follows:

    Mathematics is the science of quantity.

- Later, the Italian physicist Galileo (1564–1642), who was more interested in how it was applied, wrote:

    "Mathematics is the language of the Universe and its characters are triangles, circles, and other geometric figures, without which it is impossible to understand a single word of it."

- A more generic definition is given by the encyclopedia Britannica, which defines mathematics as follows:

Advanced Mathematics: A Transitional Reference, First Edition. Stanley J. Farlow.
© 2020 John Wiley & Sons, Inc. Published 2020 by John Wiley & Sons, Inc.
Companion website: www.wiley.com/go/farlow/advanced-mathematics

> *Mathematics is the science of numbers and shapes and the relations between them.*

- Then there is the beauty in mathematics as observed by the physicist Albert Einstein, who wrote

  > *Pure mathematics is, in its way, the poetry of logical ideas.*

- Iranian mathematician Maryam Mirzakhani once shared her thoughts about mathematics.

  > *There are times when I feel like I'm in a big forest and don't know where I'm going, but then I come to the top of a hill and can see everything more clearly.*

1.1.2 Getting into Sentential Logic

Although mathematics uses symbolic notation, its logical arguments are formulated in natural languages and so it becomes necessary to examine the truth value of natural language sentences. We begin by introducing the logical system behind reasoning called **sentential** (or **propositional**) **logic**, which is the most basic formal system of logic[1] and uses **symbols** and **rules of inference** such as

  if P is true and P implies Q, then Q is true[2]

Sentential logic is generally the first topic introduced in a formal study of logic, followed by more involved systems of logic, like predicate logic and modal logic.

> **Important Note**   The study of logic had its origins in many ancient cultures, but it was the writings of the Greek logician Aristotle (384–322 BCE) who most influenced Western culture in a collection of works known collectively as the *Oranon*.

English, as do all natural languages, contains various types of sentences, such as declarative, interrogative, exclamatory, and so on, which allow for the

---

1 There are other formal systems of logic other than sentential (or propositional) logic, such as **predicate** and **modal** logic. Because the rules of formal logic are precise, they can be programmed for a computer and are capable of analyzing mathematical proofs. Formal logical systems are important in artificial intelligence, where computers are programmed in the language of formal logic to carry out logical reasoning.
2 Greek logicians called this rule of inference *modus ponens*, meaning the *way that affirms*.

effective communication of thoughts and ideas. Some sentences are short and to the point, whereas others are long and rambling.

Some sentences can be classified as being either true or false, such as the sentence, "It will rain tomorrow." Although we may not know if it will rain or not rain, the sentence is nevertheless either true or false.

Other sentences, like the interrogatory sentence, "Why doesn't Burger King sell hotdogs?" or the exclamatory sentence, "Don't go there!" express thoughts, but have no truth or false value. Although statements like these are useful for effective communication, they are no concern in our study of logic. The sentences we study in this book are declarative and are intended to convey information. In formal logic, the word "sentence" is used in a technical sense as described in the following definition.

Definition   A **sentence** (or **proposition**) is a statement which is either true or false. If the sentence is true, we denote its truth value by the letter T and by F if it is false. In computer science they are often denoted by 1 and 0, respectively.

Example 1   Are the Following Statements Sentences?

a) $\dfrac{17}{231} - \dfrac{4}{10}$ is a positive number.

b) $\displaystyle\int_0^1 e^x dx = e$ is false.

c) $\sqrt{2}$ is a rational number.

d) Love is sharing your popcorn.

e) Come here!

f) N is an even integer.

g) Who first proved that π is a transcendental number?

h) Who was the greatest mathematician of the twentieth century?

i) This sentence is false.

a)–d) The statements are sentences. The reader can decide whether they are true or false.

e) The statement has no truth value so it is not a sentence.

f) Granted the statement is true or false, its truth value depends on the value of an unknown number N, so it is not considered a sentence. In Section 1.3, we will introduce quantifiers that will turn this statement into a sentence.

g) This is an interesting question, but questions are not sentences. The person who first proved π is a transcendental number was the German mathematician Ferdinand von Lindemann, who proved it in 1882. The number π is also irrational, which was proved by another German mathematician Johann Lambert in 1768.

h) This statement is not a sentence, but you can find candidates for the greatest mathematician by googling "famous mathematicians of the 20th century."

i) If you say the statement is *true*, then according to what the statement says, it is *false*. But on the other hand, if you say the statement is *false*, then the statement says it is *true*. In either case, we reach a contradiction. Hence, we conclude the statement is *neither* true nor false, and hence, it is not a *sentence*. This logical paradox that arises when a statement refers to *itself* in a *negative* way is one of the many forms of what is called the Russell paradox.

> **Liar Paradox** Consider the following statement
>
> *This sentence is not true.*
>
> This statement is paradoxical since if we say the sentence is true, the sentence itself says it is false, which yields a contradiction. On the other hand, if we say the sentence is false, the sentence itself says it is false, hence, it must be true. In either case, one is led to a contradiction, so one cannot say the sentence is true or false, and thus not what we call a *sentence*.

> **Russell's Barber Paradox** The Russell Barber Paradox is an example of a *self-referential* statement that refer to itself in a negative way. The paradox considers a barber in a small town that *shaves all men in the town, but only those men, who do not shave themselves.* The prophetic question then arises, does the barber shave himself? If you say the barber shaves himself, then barber does not shave himself since he only shaves those who do not shave themselves. On the other hand, if you say the barber does *not* shave himself, then he shaves himself since he shaves those who do not shave themselves. This paradox was formulated by the English logician Bertrand Russell (1872–1970) in 1901, and played a major role in the modern development set theory.

1.1.3 Compound Sentences ("AND," "OR," and "NOT")

In arithmetic, we combine numbers with operations $+$, $\times$, $-$, and so on. In logic, we combine sentences with logical expressions. The sentences discussed thus far are examples of **simple** (or **atomic**) **sentences** since they are made up of a single thought or idea. It is possible to combine these sentences to form **compound sentences** using **logical connectives**.[3]

---

3 We can also combine compound sentences to form more complex sentences.

> **Important Note** It has been said that love and the ability to reason are the two most important human traits. Readers interested in the former must go elsewhere for advice, but if the reader is interested in the human trait of reasoning, you are in the right place.

Definition Logical Connectives.

Given the sentences P and Q, we define:

Logical AND: The **conjunction** of P and Q, denoted $P \wedge Q$, is the sentence "P and Q" which is true when *both* P and Q are true, otherwise false.

Logical OR: The **disjunction** of P and Q, denoted $P \vee Q$, is the sentence "P or Q" which is true when *at least one* of P or Q are both true, otherwise false.

NOT operator: The **negation** of P, denoted $\sim P$, is the sentence "not P" and $\sim P$ is true when P is false, and $\sim P$ is false when P is true.

Example 2   Logical Conjunction

Let P and Q be the sentences

   P: "Jack went up the hill."
   Q: "Jill went up the hill."

The conjunction $P \wedge Q$ refers to the sentence

   "Jack and Jill went up the hill."

The truth of $P \wedge Q$ depends on the truth values of the two simple sentences P and Q. The conjunction $P \wedge Q$ is true if *both* Jack and Jill went up the hill, and false if *either* Jack or Jill (or *neither*) did not go up the hill. We summarize this symbolically by means of the **truth table** shown in Table 1.1, which examines the truth value of $P \wedge Q$ for the four possible truth values of P and Q. The two columns at the left list the four possible truth values for the sentences P and Q.

Table 1.1 Logical AND.

P	Q	$P \wedge Q$	
T	T	T	
T	F	F	$P \wedge Q$ is true if both P and Q
F	T	F	are true, otherwise false.
F	F	F	

> **Important Note** Because the rules of formal logical systems (like sentential logic, predicate logic, ...) are precisely defined, it is possible to write computer programs and sometimes have a computer evaluate mathematical proofs.

Example 3   Logical Disjunction
Again let

> P: "Jack went up the hill."
> Q: "Jill went up the hill."

The disjunction $P \vee Q$ refers to the sentence

> "Jack or Jill went up the hill."

The disjunction $P \vee Q$ is true if *either* Jack *or* Jill (or *both*) went up the hill. If *neither* Jack *nor* Jill went up the hill, then the disjunction is false. This is summarized in the truth table in Table 1.2.

Table 1.2 Logical OR.

P	Q	$P \vee Q$	
T	T	**T**	
T	F	**T**	$P \vee Q$ is false if both P and Q
F	T	**T**	are false, otherwise true
F	F	**F**	

Sometimes, in normal English discourse, the word "or" is used in an **exclusive** sense. For example, when someone says "for dessert I will have pie or cake," it is normally understood to mean the person will have one of the two desserts, but not both. In this case, we would say "or" is an **exclusive OR**. In sentential logic, unless otherwise stated, "or" means the **inclusive** OR as defined in Table 1.2.

Example 4   Negation of a Sentence
If

> P: "Jack went up the hill."

then $\sim P$ is the sentence "Jack did *not* go up the hill." In other words, if P is true, then $\sim P$ is false, and if P is false then $\sim P$ is true. This is summarized by Table 1.3.

Table 1.3 Negation.

P	$\sim P$	
T	F	$\sim P$ is true when P is false, and
F	T	$\sim P$ is false when P is true.

Important Note To check the accuracy of a truth table, there are webpages that carry out these computations.

1.1.4 Compound Sentences

We can also combine sentences to form more complex sentences as illustrated in Table 1.4.

Table 1.4 Forming new sentences.

$P \wedge \sim Q$	Jack went up the hill but Jill did not.
$\sim(P \vee Q)$	Neither Jack nor Jill went up the hill.
$\sim(P \wedge Q)$	It is not true both Jack and Jill went up the hill.
$\sim P \vee Q$	Either Jack did not go up the hill or Jill did.

The truth values of $\sim P \vee Q$ can be analyzed using the values in Table 1.5. The numbers above the columns give the order in which the truth values in the columns were filled.

Table 1.5 Truth table for $\sim P \vee Q$.

P	Q	(1) $\sim P$	(2) $\sim P \vee Q$
T	T	F	T
T	F	F	F
F	T	T	T
F	F	T	T

Note that $\sim P \vee Q$ is false only when P is true and Q is false, otherwise it is true.

The compound sentence $(P \vee Q) \wedge \sim R$ contains three sentences P, Q, and R, and its truth value is determined by enumerating all $2^3 = 8$ possible truth values for P, R, and R, followed by finding the truth values of the component parts of the sentence until arriving at the truth values of $(P \vee Q) \wedge \sim R$. These computations are shown in the truth table in Table 1.6.

Table 1.6 Truth table for $(P \vee Q) \wedge \sim R$.

			(1)	(2)	(3)
P	*Q*	*R*	*P* ∨ *Q*	~*R*	(*P* ∨ *Q*) ∧ ~*R*
T	T	T	T	F	**F**
T	T	F	T	T	**T**
T	F	T	T	F	**F**
T	F	F	T	T	**T**
F	T	T	T	F	**F**
F	T	F	T	T	**T**
F	F	T	F	F	**F**
F	F	F	F	T	**F**

> **Historical Note** In 1666, the philosopher and mathematician **Gottfried Wilhelm Leibniz** (1646–1716) laid out a plan in his work *De Arte Combinatorial* in which reasoning could be reduced to mental calculations. He wrote,
>
> > *The method should serve as a universal language whose symbols and special vocabulary can direct reasoning in such a way that errors, except for fact, will be mistakes in computation.*
>
> It is tragic that Leibniz' ideas were considered fantasy and his ideas sank into oblivion. It was not until the twentieth century, well after symbolic logic had been rediscovered by George Boole, that Leibniz' ideas became known to the general public.

An example of a sentence containing four component sentences is $(P \wedge Q) \vee (R \wedge \sim S)$, which depends on P, Q, R, and S. The truth value of this sentence is found by examining a truth table with $2^4 = 16$ rows listing the 16 possible truth values of the four components. Table 1.7 shows how this truth table is computed.

Table 1.7 Truth table for $(P \wedge Q) \vee (R \wedge \sim S)$.

				(1)	(2)	(3)	(4)
P	Q	R	S	$\sim S$	$P \wedge Q$	$R \wedge \sim S$	$(P \wedge Q) \vee (R \wedge \sim S)$
T	T	T	T	F	T	F	T
T	T	T	F	T	T	T	T
T	T	F	T	F	T	F	T
T	T	F	F	T	T	F	T
T	F	T	T	F	F	F	F
T	F	T	F	T	F	T	T
T	F	F	T	F	F	F	F
T	F	F	F	T	F	F	F
F	T	T	T	F	F	F	F
F	T	T	F	T	F	T	T
F	T	F	T	F	F	F	F
F	T	F	F	T	F	F	F
F	F	T	T	F	F	F	F
F	F	T	F	T	F	T	T
F	F	F	T	F	F	F	F
F	F	F	F	T	F	F	F

Historical Note After Leibniz, the next major development in sentential logic was due to the work of English logicians George Boole (1815–1854) and Augustus DeMorgan (1806–1871). Boole was interested in developing a logical algebra, whereby the symbols x, y represented sets[4] or classes, where the empty set was denoted by 0, the universal set denoted by 1, intersection of sets by xy, and union of sets by $x + y$. Boole interpreted $x = 1$ to mean "x is true" and $x = 0$ as "x is false." In this system, $xy = 1$ means x and y are both true, $x + y = 1$ means x or y is true, and so on. Boole's ideas sparked immediate interest among logicians and Boole's **Boolean algebras** are the basis for sentential logic and are used today in computer science and design of circuits.

---

4 We are getting a little ahead of ourselves, but we will get to sets in Chapter 2.

1.1.5 Equivalence, Tautology, and Contradiction

In mathematics and in particular logic, the statement "means the same thing" is often used and has a precise meaning, which brings us to the concept of logical equivalence.

Definition   Two sentences P, Q simple or compound, are **logically equivalent**, denoted by $P \equiv Q$, if they have the same truth values for all truth values of their component parts.

For example the sentences $1 + 1 = 3$ and $10 < 5$ are logically equivalent, the reason being they have the same truth value. The fact they have nothing to do with each other is irrelevant from the point of view of sentential logic, although the logical statements we study will be more useful. The following De Morgan's Laws are an example of two useful logical equivalences.

1.1.6 De Morgan's Laws

Two useful logical equivalences are **De Morgan's Laws**[5] stated in Table 1.8.

Table 1.8 De Morgan's laws.

De Morgan's laws
$\sim(P \vee Q) \equiv \sim P \wedge \sim Q$
$\sim(P \wedge Q) \equiv \sim P \vee \sim Q$

Can you speak these equivalences in the language of Jack and Jill?

We can verify De Morgan's laws by making truth tables for each side of the equivalences. The top De Morgan Law[6] in Table 1.8 is valid since columns (2) and (5) in Table 1.9 have T and F values for all truth/false values of P and Q.

---

5 Augustus De Morgan (1806–1871) was an Indian-born British mathematician and logician who formulated what we call De Morgan's Laws. He was the first person to make the idea of mathematical induction rigorous.

6 This De Morgan law says the negation of a disjunction is the conjunction of its negatives.

Table 1.9 Verification of De Morgan's law.

P	Q	(1) $P \vee Q$	(2) $\sim(P \vee Q)$	(3) $\sim P$	(4) $\sim Q$	(5) $\sim P \wedge \sim Q$
T	T	T	F	F	F	F
T	F	T	F	F	T	F
F	T	T	F	T	F	F
F	F	F	T	T	T	T

same truth values

> **Interesting Note** Although sentential logic captures the truth or falsity of simple sentences, it will never replace natural languages. Consider two lines from T.S. Eliot's Love Song of J. Alfred Prufrock:
>
> *In the room the women come and go*
> *Talking of Michelangelo.*
>
> As logicians, we might let
>
> - *WC* = women come into the room
> - *WL* = women leave the room
> - *WT* = women talk of Michelangelo
>
> and restate Eliot's poem as *WC* ∧ *WL* ∧ *WT*. This conjunction captures the underlying facts of the situation, but the essence of the poem is undoubtedly lost.

1.1.7 Tautology

Definition   A **tautology** is a sentence (normally compound) that is true for all truth values of its individual components.[7]

The sentence $P \vee \sim P$ is called the **Law of the Excluded Middle** and is a tautology since it is always true, regardless of the truth of its component parts. This is shown by the truth table in Table 1.10.

The **Law of the Excluded Middle** is a principle that says *everything* is either true or false and nothing else. This seems an obvious and trivial concept, but this principle is fundamental when we study proofs by contradiction.

---

7 In natural language, a tautology is often thought of as a sentence that says the same thing twice in a different way.

Table 1.10 Verification of a tautology.

P	~P	P ∨ ~P
T	F	T
F	T	T

Example 5 **Verifying a Tautology**

Show that

$$(P \wedge Q) \vee (\sim P \vee \sim Q)$$

is a tautology.

Solution

The truth value of $(P \wedge Q) \vee (\sim P \vee \sim Q)$ is T for all truth values of P and Q which can be seen in Table 1.11.

Table 1.11 Verification of a tautology.

P	Q	P ∧ Q	~P	~Q	~P ∨ ~Q	(P ∧ Q) ∨ (~P ∨ ~Q)
T	T	T	F	F	F	T
T	F	F	F	T	T	T
F	T	F	T	F	T	T
F	F	F	T	T	T	T

The opposite of a tautology is a contradiction.

Definition   A **contradiction** is a sentence that is false for all truth value of its components.

An example of a contradiction is $P \wedge \sim P$, as in "It is raining and it is not raining." Regardless of the truth value of P, the truth value of $P \wedge \sim P$ is false. Many contradictions are obvious, while others are not so obvious. For example is the sentence $(P \wedge \sim Q) \wedge (Q \wedge R)$ *always* false regardless of the truth values of P, Q, and R? The answer is yes, and you might try to convince yourself of this fact without resorting to Table 1.12.

Table 1.12 Verification of a contradiction.

P	Q	R	(1) ~Q	(2) P ∧ ~Q	(3) Q ∧ R	(4) (P ∧ ~Q) ∧ (Q ∧ R)
T	T	T	F	F	T	F
T	T	F	F	F	F	F
T	F	T	T	T	F	F
T	F	F	T	T	F	F
F	T	T	F	F	T	F
F	T	F	F	F	F	F
F	F	T	T	F	F	F
F	F	F	T	F	F	F

1.1.8 Logical Sentences from Truth Tables: DNF and CNF

Until now, we have found truth tables associated with various compound sentences. We now go backwards and find the logical expression that creates a given truth table. For example, what is the compound sentence that yields the truth table in Table 1.13.

Table 1.13 Three variable truth table.

P	Q	R	?	Row
T	T	T	T	1
T	T	T	F	2
T	T	F	F	3
T	T	F	T	4
T	F	T	T	5
T	F	T	F	6
T	F	F	T	7
T	F	F	F	8

1.1.9 Disjunctive and Conjunctive Normal Forms

To find a logical sentence that yields a given truth table, there are two equivalent logical expressions that can be found, one called the disjunctive normal form and the other called the conjunctive normal form. These forms are defined as follows:

Definition   Disjunctive and conjunctive normal forms are

a) **disjunctive normal** form (DNF) is a disjunction ($\vee$) of conjunctions ($\wedge$).
b) **conjunctive normal** form (CNF) is a conjunction ($\wedge$) of disjunctions ($\vee$).

Simple examples are the following:

A disjunctive normal form : $(P \wedge Q) \vee (R \wedge S)$

A conjunctive normal form : $(P \vee Q) \wedge (R \vee S)$

To find the DNF for the truth table in Table 1.13, we look at rows where the value of the truth table is **TRUE**, which are rows 1, 4, 5, and 7, then write the create formulas that yield a T. Doing this, we get

- row 1 true as a result that $P \wedge Q \wedge R$ is true
- row 4 true as a result that $P \wedge Q \wedge \sim R$ is true
- row 5 true as a result that $P \wedge \sim Q \wedge R$ is true
- row 7 true as a result that $P \wedge \sim Q \wedge \sim R$ is true

Hence, the truth table yields a T when at least one of the above conjunctions is true, which leads to the disjunctive normal form (DNF) for the truth table in Table 1.13.

$$\text{DNF} : (P \wedge Q \wedge R) \vee (P \wedge Q \wedge \sim R) \vee (P \wedge \sim Q \wedge R) \vee (P \wedge \sim Q \wedge \sim R).$$

To find the conjunctive normal form, we look at the rows where the truth table is **FALSE**, which are rows 2, 3, 6, and 8, which happens when the following are satisfied:

- row 2 is false when $\sim P \vee \sim Q \vee \sim R$
- row 3 is false when $\sim P \vee \sim Q \vee R$
- row 6 is false when $\sim P \vee Q \vee \sim R$
- row 8 is false when $\sim P \vee Q \vee R$

Hence, we have the conjunctive normal form for the truth table in Table 1.13:

$$\text{CNF} : (\sim P \vee \sim Q \vee \sim R) \wedge (\sim P \vee \sim Q \vee R) \wedge (\sim P \vee Q \vee \sim R)$$
$$\wedge (\sim P \vee Q \vee R)$$

You can verify that both the above disjunctive and conjunctive forms create the truth table in Table 1.13 and are thus logically equivalent. Conjunctive and disjunctive normal forms are important in circuit design where designs of integrated circuits are based on Boolean functions involving on/off switches.

> **Historical Note**   The first systematic discussion of sentential logic comes from the German logician Gottfried Frege (1848–1925) in the work *Begriffscrift. Many consider* him the greatest logician of the nineteenth century.

Problems

1. **Simple Sentences**
   Which of the following are sentences?
   a) WE TAKE YOUR BAGS AND SEND THEM IN ALL   DIRECTIONS (posted at an airline ticket counter)
   b) The Riemann hypothesis is still unsolved.
   c) The Battle of Hastings was fought in 1492.
   d) The constant π is an algebraic number.
   e) DROP YOUR TROUSERS HERE FOR THE BEST   RESULT. (sign posted at a dry cleaners)
   f) The constant π is a transcendental number.
   g) I am a monkey's uncle.
   h) Never again!
   i) ABSOLUTELY NO SWIMMING
   j) $2 + 5 = 1$
   k) e is an irrational number
   l) It is not true that $1 + 1 = 2$.
   m) This sentence is false.
   n) $\int_0^1 x^2 dx = 0$
   o) Digit 0 does not appear in π.
   p) We sit on the porch watching cows playing Scrabble.
   q) The Chicago Cubs will win the World Series this year
   r) To be or not to be, that is the question.
   s) A woman, without her man, is nothing.
   t) A woman, without her, man is nothing.
   u) I think, therefore I am.
   v) All generalizations are false, including this one.
   w) These pretzels are making me thirsty.
   x) The fifth order polynomial equation has not been solved.
   y) Who was Niels Abel?

2. **Truth Tables**

For simple sentences P, Q, R make the truth table for the following compound sentences.

a) $P \vee \sim P$

b) $P \wedge \sim P$

c) $P \vee \sim Q$

d) $\sim P \vee \sim Q$

e) $\sim(P \vee Q)$

f) $(P \vee Q) \wedge R$

g) $P \wedge (Q \vee R)$

h) $P \vee (Q \wedge R)$

i) $(P \vee Q) \vee (\sim P \vee \sim Q)$

j) $(P \vee Q) \wedge (Q \vee R)$

k) $(P \vee Q) \vee (Q \wedge R)$

3. **True or False**

Let

- P be the sentence "$4 > 2$,"
- Q be the sentence "$1 + 2 = 3$"
- R be the sentence "$5 + 2 = 9$"

What is the truth value of the following compound sentences?

a) $P \wedge \sim Q$

b) $\sim(P \wedge Q)$

c) $\sim P \wedge \sim Q$

d) $\sim(P \vee Q)$

e) $P \wedge Q$

f) $(\sim P \wedge Q) \vee P$

g) $(\sim P \wedge Q) \vee R$

h) $(\sim R \wedge Q) \vee R$

i) $(\sim P \vee Q) \wedge (P \wedge R)$

4. **Tautologies and Contradictions**

Suppose P and Q are sentences. Tell whether the following compound sentences are tautologies, contradictions, or neither. Verify your conclusion.

a) $P \wedge Q$

b) $P \vee Q$

c) $Q \vee \sim Q$

d) $P \wedge \sim P$

e) $P \wedge (Q \vee \sim P)$

f) $P \vee (Q \wedge \sim P)$

5. **In Plain English**
   In English, fill in the blanks in Table 1.14.

   Table 1.14 Fill in the blanks.

The sentence	Is TRUE when	Is FALSE when
P		
$\sim P$		
$P \vee Q$		
$P \wedge Q$		

6. **Denial of Sentences**
   State the negation of each of the following sentences.
   a) π is a rational number.
   b) 317 is a prime number.
   c) The function $f(x) = x^2 + 1$ has exactly one minimum.
   d) It will be cold and rainy tomorrow.
   e) It will be either cold or rainy tomorrow.
   f) It is not true that I am shiftless and lazy.
   g) It is not true that I am either lazy or shiftless.
   h) 2 is the only even prime number.

7. **Exclusive OR**
   In natural language the word "or" requires a certain amount of care. In this lesson we defined the word "or" in the inclusive sense, meaning one or the other or both. The **exclusive or** is slightly different; it means one or the other but *not both*
   a) Make a truth table for the exclusive or (denote it by $\oplus$)
   b) Verify that $P \oplus Q \equiv (P \vee Q) \wedge \sim (P \wedge Q)$.

8. **Prove or Disprove**
   Prove or disprove the statement

   $$P \oplus Q \equiv (P \wedge \sim Q) \vee (Q \wedge \sim P)$$

   where $P \oplus Q$ denotes the **exclusive or**, meaning P or Q are true but not both.

9. **Alternate Forms for Truth Tables**
   Truth tables for logical disjunction, conjunction and negation are analogous to addition, multiplication and negation in Boolean algebra, where 1 is defined as truth, 0 is taken as false, and minus as negation. Verify the following Boolean algebra statements in Figure 1.1 and find their logical equivalents in the sentential calculus.

	P	Q
×	0	1
P 0	0	0
Q 1	0	1

	P	Q
+	0	1
P 0	0	1
Q 1	1	1

P	0	1
~P	1	0

Figure 1.1 Numeric truth table.

a) $1 + (-1) = 1$
b) $1 \times (-1) = 0$
c) $-(1 + 0) = 0 \times 1$
d) $-(0 \times 1) = 1 + 0$
e) $1 \times (1 + 0) = (1 \times 1) + (1 \times 0)$
f) $1 + (1 \times 0) = (1 + 1) \times (1 + 0)$

10. **Logical AND for IP Addresses**
    An IP address is a numerical label assigned to each device connected to a computer network. A typical IP address is a four-dotted string of eight binary numbers (0s and 1s), which in decimal and binary form might be

    $$193.170.6.1 = 11000001.10101010.00000110.00000001$$

    The network might consist of 10 computers connected to a router, where each computer in the network is assigned a *subnet mask*. If the subnet mask of one of the computers is

    $$11111111.11111111.11111111.00000000$$

    then the IP of this computer is the logical AND of the network IP and this subnet mask, which is the same as logical binary multiplication. What is the IP address of this computer using binary arithmetic $0 \cdot 0 = 1, 0 \cdot 1 = 0,$ $1 \cdot 0 = 0, 1 \cdot 1 = 1$?
    Show that this sentence is a tautology.

11. **Disjunctive and Conjunctive Normal Forms**
    Find disjunctive and conjunctive normal forms that create the following truth tables.
    a)

Table 1.15 Find the CNF and DNF for the truth table.

P	Q	Logical function
T	T	T
T	F	T
F	T	T
F	F	F

b)

Table 1.16 Find the CNF and DNF for the truth table.

P	Q	Logical function
T	T	T
T	F	F
F	T	F
F	F	F

c)

Table 1.17 Find the CNF and DNF for the truth table.

P	Q	Logical function
T	T	T
T	F	T
F	T	T
F	F	T

d)

Table 1.18 Find the CNF and DNF for the truth table.

P	Q	R	Logical function
T	T	T	T
T	T	F	F
T	F	T	T
T	F	F	F
F	T	T	T
F	T	F	F
F	F	T	T
F	F	F	F

12. **Syllogisms**
    In Aristotelian logic,[8] a **syllogism** is an argument in which two premises (which are assumed true) lead to a valid conclusion. The most general form being

    Major premise : A general statement.
    Minor premise : A specific statement.
        Conclusion : based on the two premises.

    An example would be

    Major premise : All humans are mortal.
    Minor premise : Mary is human.
        Conclusion : Mary is mortal.

    In the language of sentential logic, this syllogism takes the form

    $$[(P \Rightarrow Q) \wedge P] \Rightarrow Q$$

    Show that this sentence is a tautology.

13. **Digital Logical Circuits I**
    Find the truth table for each of the following digital logical circuits in Figure 1.2 to prove they are equivalent. What logical law does the equivalence of these circuits represent? The individual electronic components are self-explanatory.

    (a)

    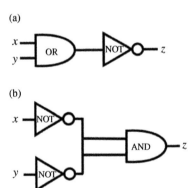

    (b)

    Figure 1.2 Logical circuit.

---

8 This form of a syllogism is but one of several devised by the Greek philosopher Aristotle (384 B.C. 322 BCE). Aristotle, along with Plato and Socrates, are three of the most important founding figures of Western philosophy and thought.

14. **Digital Logical Circuits II**

    Find the truth table for each of the following digital logical circuits in Figure 1.3 to prove they are equivalent. What logical law does the equivalence of these circuits represent? The individual electronic components are self-explanatory.

    (a)

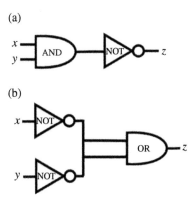

    (b)

    Figure 1.3 Logical circuit.

15. **Internet Research**

    There is a wealth of information related to topics introduced in this section just waiting for curious minds. Try aiming your favorite search engine toward *sentential logic, first-order-logic, liar paradox, de Morgan's laws,* and *disjunctive normal form*.

1.2

Conditional and Biconditional Connectives

> **Purpose of Lesson** To introduce the **conditional** and **biconditional sentences** along with three equivalent forms of the conditional sentence called the **converse**, **inverse**, and the **contrapositive**.

1.2.1   The Conditional Sentence

The **conditional sentence** (or **implication**), is a compound sentence of the form

"if P then Q"

From a purely logical point of view, conditional sentences do not necessarily imply a cause and effect between P and Q, although generally there is a definite cause and effect. For example the conditional sentence

If $1 + 1 = 3$, then pigs fly.

is a true conditional sentence, although the reader would have to think long and hard to find a cause and effect relation between $1 + 1 = 3$ and flying pigs. A more common implication in mathematics would be

If a positive integer n is composite, then n has a prime divisor less than or equal to $\sqrt{n}$.

which provides an important cause and effect between P and Q. No doubt the reader has seen conditional sentences in Euclidean geometry, where the subject is explained through cause and effect implications of this type. The sentence, "If a polygon has three sides, then it is a triangle," is a conditional sentence relating two important concepts in geometry.

Advanced Mathematics: A Transitional Reference, First Edition. Stanley J. Farlow.
© 2020 John Wiley & Sons, Inc. Published 2020 by John Wiley & Sons, Inc.
Companion website: www.wiley.com/go/farlow/advanced-mathematics

Historical Note The idea of enumerating all possible truth values in tables we now call "truth tables" seems to have been rediscovered several times through-out history. The philosopher and logician, Ludwig Wittgenstein and Emil Post, respectively, used them in the late 1800s where Wittgenstein labeled them "truth tables." Other early researchers who made contributions to symbolic logic are the English logician Bertrand Russell and the American logician Benjamin Peirce.

Conditional Sentence If P and Q are sentences, then the **conditional sentence** "if P then Q" is denoted symbolically by

$$P \Rightarrow Q$$

and whose truth values are defined by the truth table: See Table 1.20.

Table 1.20 Truth table.

P	Q	$P \Rightarrow Q$
T	T	T
T	F	F
F	T	T
F	F	T

Note that a conditional sentence is false when $T \Rightarrow F$, otherwise it is true. The sentence P is called the **assumption** (or **antecedent** or **premise**) of the conditional sentence (or implication) and Q is called the **conclusion** (or **consequent**[1]).

The conditional statement $P \Rightarrow Q$ can be visualized by the **Euler** (or **Venn**) diagram as drawn in Figure 1.4.

For example all polygons are triangles that we illustrate by the diagram in Figure 1.5.

---

1 In pure logical systems, P and Q are generally referred to as antecedent and consequent, respectively. In mathematics, they are more likely to be called the **assumption** and **conclusion**.

Points in this region
have property P

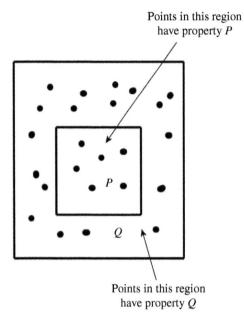

Points in this region
have property Q

Figure 1.4 Euler diagram for $P \Rightarrow Q$.

Polygons

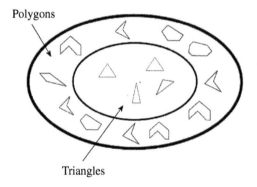

Triangles

Figure 1.5 Subsets of triangles.

Example 1   Conditional Sentences

The following sentences are conditional sentences. Are they true or false?

i) If f is a real-valued differentiable function, then f is continuous. TRUE
ii) If N is an even number greater than 2, then N is the sum of two primes.[2]

---

2 You get an A for the course if you can prove this statement. Just slide your solution under your professor's door. The truth value of this conjecture, called the Goldbach Conjecture, is unknown.

iii) If a and b are the lengths of the legs of a right triangle, and c is the length of the hypotenuse, then $c^2 = a^2 + b^2$. TRUE

1.2.2 Understanding the Conditional Sentence

The conditional sentence "if P, then Q" is best understood as a promise, where if the promise is kept, the conditional sentence is true, otherwise the sentence is false. As an illustration suppose your professor makes you the promise:

> *If pigs fly, then you will receive an A for the course.*

The proposition is true since your professor has only promised an A *if* pigs fly, but since they do not, all bets are off. However, if you see a flying pig outside your classroom and your professor gives you a C, then you have reason to complain to your professor since the promise was broken, hence the proposition false.

The conditional sentence $P \Rightarrow Q$ is often called an **inference**, and we say P **implies** Q. Another way of stating $P \Rightarrow Q$ is to say P is a **sufficient condition** for Q, which means the truth of P is sufficient for the truth of Q. We also say that Q is a **necessary condition** for P, meaning the truth of Q necessarily follows from the truth of P.

Example 2   Necessary Conditions and Sufficient Conditions
Table 1.21 illustrates some typical necessary and sufficient conditions.

Table 1.21 Necessary conditions and sufficient conditions.

P	Q	Condition
Being pregnant	Being female	Q is necessary for P
N is an integer	$2N$ is an integer	P is sufficient for Q
Life on earth	Air	Q is necessary for P
Run over by a truck	Squashed	P is sufficient for Q

Important Note   A famous conditional statement is due to the French philosopher/mathematician *Rene Decartes* (1591–1650) who once said "Cogito ergo sum," which means "I think therefore I am." As a conditional sentence it would be stated as "If I think, then I am."

> **Important Note** Normally, in mathematics, when one writes the implication $P \Rightarrow Q$ one assumes the assumption is true since it makes no sense to assume a false hypotheses. In fact, if one assumes a false hypothesis, the implication is true regardless of the truth value of Q.

1.2.3 Converse, Inverse, and the Contrapositive

The implication $P \Rightarrow Q$ gives rise to three related implications shown in Table 1.22, one equivalent to the implication, the others not.

Table 1.22 Converse, inverse, contrapositive.

Implication	Converse	Inverse	Contrapositive
$P \Rightarrow Q$	$Q \Rightarrow P$	$\sim P \Rightarrow \sim Q$	$\sim Q \Rightarrow \sim P$

It is easy to show by truth tables that

$$\text{converse}: \ P \Rightarrow Q \not\equiv Q \Rightarrow P$$

$$\text{inverse}: \ P \Rightarrow Q \not\equiv \sim P \Rightarrow \sim Q$$

$$\text{contrapositive}: \ P \Rightarrow Q \equiv \sim Q \Rightarrow \sim P$$

1.2.4 Law of the Syllogism

A fundamental principle of logic, called the **law of the syllogism**, states:

"if P implies Q, and Q implies R, then P implies R"

which is equivalent to the compound conditional sentence

$$[(P \Rightarrow Q) \wedge (Q \Rightarrow R)] \Rightarrow (P \Rightarrow R)$$

This sentence is a tautology since Table 1.23 shows all T's in column (5).

> **Well-Formed Sentences** The statement $P \vee Q \wedge R$ is not what is called a **well-formed** sentence since its meaning is unclear. One would have to include parentheses to tell which sentence $(P \vee Q) \wedge R$ or $P \vee (Q \wedge R)$ is intended. Well-formed sentences are similar to "well-formed sentences" in natural languages like English: a capital letter at the start, a period at the end, and all the other rules of grammar in between.

Table 1.23 Truth table verification of the syllogism.

			(1)	(2)	(3)	(4)	(5)
P	*Q*	*R*	$P \Rightarrow Q$	$Q \Rightarrow R$	$P \Rightarrow R$	$(P \Rightarrow Q) \wedge (Q \Rightarrow R)$	$[(P \Rightarrow Q) \wedge (Q \Rightarrow R)] \Rightarrow (P \Rightarrow R)$
T	T	T	T	T	T	T	**T**
T	T	F	T	F	F	F	**T**
T	F	T	F	T	T	F	**T**
T	F	F	F	T	F	F	**T**
F	T	T	T	T	T	T	**T**
F	T	F	T	F	T	F	**T**
F	F	T	T	T	T	T	**T**
F	F	F	T	T	T	T	**T**

1.2.5 A Useful Equivalence for the Implication

The implication $P \Rightarrow Q$ is true if either P is false or Q is true. Hence, we have the useful logical equivalence

$$P \Rightarrow Q \equiv \sim P \vee Q$$

which we verify by means of the truth table in Table 1.24.

Table 1.24 Equivalence of $P \Rightarrow Q \equiv \sim P \vee Q$.

P	*Q*	$P \Rightarrow Q$	$\sim P$	$\sim P \vee Q$
T	T	**T**	F	**T**
T	F	**F**	F	**F**
F	T	**T**	T	**T**
F	F	**T**	T	**T**

Also the negation of the implication $P \Rightarrow Q$ is another useful equivalence that we obtain by one of De Morgan's laws:

$$\sim (P \Rightarrow Q) \equiv \sim (\sim P \vee Q) \equiv P \wedge \sim Q$$

In other words, an implication is (only) false when the premise is true and the conclusion false.

1.2.6   The Biconditional

Compound sentences of the form

"*P* if and only if *Q*"

are fundamental in mathematics, which leads to the following definition.

Definition   Given that *P* and *Q* are sentences, the compound sentence

"*P* if and only if *Q*"

is called a **biconditinal sentence**, denoted by

$$P \Leftrightarrow Q$$

And whose truth values are defined by the truth table in Table 1.25.

Table 1.25 Biconditional sentence.

P	Q	$P \Leftrightarrow Q$
T	T	T
T	F	F
F	T	F
F	F	T

The biconditional $P \Leftrightarrow Q$ is generally read "*P* **if and only if** *Q*" or *P* iff *Q* for shorthand. Another common phrasing of $P \Leftrightarrow Q$ is *P* is a **necessary and sufficient condition** for Q. In other words, $P \Leftrightarrow Q$ is true if *P* and *Q* have the same truth value, otherwise false.

Example 3   Biconditional Equivalent to Two Implications
Show that the biconditional $P \Leftrightarrow Q$ is logically equivalent to

$$(P \Rightarrow Q) \wedge (Q \Rightarrow P).$$

Solution
The truth values in the following truth table under columns (3) and (4) in Table 1.26 are identical.   ∎

Table 1.26 Equivalence of $(P \Rightarrow Q) \wedge (Q \Leftarrow P) \equiv P \Leftrightarrow Q$.

		(1)	(2)	(3)	(4)
P	Q	$P \Rightarrow Q$	$Q \Rightarrow P$	$(P \Rightarrow Q) \wedge (Q \Rightarrow P)$	$P \Leftrightarrow Q$
T	T	T	T	**T**	**T**
T	F	F	T	**F**	**F**
F	T	T	F	**F**	**F**
F	F	T	T	**T**	**T**

Warning Be careful not to confuse the biconditional sentence $P \Leftrightarrow Q$ with logical equivalence $P \equiv Q$. Logical equivalence $P \equiv Q$ says that P and Q are *logical equivalent* sentences and always have the same truth values regardless of the truth values of component parts. On the other hand, the biconditional $P \Leftrightarrow Q$ does *not* necessarily mean P and Q have the same truth values because the biconditional sentence *can* be false that in case they do not.

Example 4 Truth Values of Biconditional Sentences

A few typical biconditional sentences are shown in Table 1.27.

Table 1.27 Biconditional sentences.

Biconditional	Truth value
$1 + 2 = 5$ if and only if $1 - 3 = 4$	True
$3 + 5 = 8$ if and only if $3 \times 4 = 12$	True
$1 + 2 = 3$ if and only if $(a + b)^2 = a^2 + b^2$	False
$\pi = 22/7$ if and only if $6/3 = 2$	False

Important Note Greek philosophers referred to **Modus Ponens** as the valid argument

"if P is true and if $P \Rightarrow Q$ is true, then Q is true"

which in sentential logic notation is

$$[P \wedge (P \Rightarrow Q)] \Rightarrow Q.$$

> **Important Note** The origin of the "iff" notation, meaning "if and only if" first appeared in print in 1955 in the text *General Topology* by John Kelly, although its invention is generally credited to the Hungarian/American Paul Halmos.

Problems

> **Working Definitions** The following definitions are needed in some problems in problem set as well as later ones.
>
> - An integer n **divides** an integer m, denoted $n|m$) if there exists an integer q satisfying $m = n \times q$. If n does **not** divide m, we write $n \nmid m$.
> - An integer n is **even** if there exists an integer k such that $n = 2k$.
> - An integer n is **odd** if there exists an integer k such that $n = 2k + 1$.
> - A natural number $\mathbb{N} = \{1, 2, 3, ...\}$ is **prime** if it is only divisible by 1 and itself.

1. **True or False**
   Identify the assumption and conclusion in the following conditional sentences and tell if the implication is true or false.
   a) If pigs fly, then I am richer than Bill Gates.
   b) If a person got the plague in the seventeenth century, they die.
   c) If you miss class over 75% of the time, you are in trouble.
   d) If x is a prime number, then x^2 is prime too.
   e) If x and y are prime numbers, then so is $x + y$.
   f) If the determinant of a matrix is nonzero, the matrix has an inverse.
   g) If f is a 1–1 function, then f has an inverse.

2. **Contrapositive**
   Write the contrapositive of the conditional sentences in Problem 1.

3. **True or False**
   Let P be the sentence "$4 > 6$," Q the sentence "$1 + 1 = 2$," and R the sentence "$1 + 1 = 3$." What is the truth value of the following sentences?
   a) $P \wedge \sim Q$
   b) $\sim(P \wedge Q)$
   c) $\sim(P \vee Q)$
   d) $\sim P \wedge \sim Q$
   e) $P \wedge Q$
   f) $P \Rightarrow Q$
   g) $Q \Leftrightarrow R$
   h) $P \Rightarrow (Q \Rightarrow R)$

i) $(P \Rightarrow Q) \Rightarrow R$
j) $(R \vee Q \vee R) \Leftrightarrow (P \wedge Q \wedge R)$

4. **True or False**
   Let P be the sentence "Jerry is richer than Mary," Q is the sentence "Jerry is taller than Mary," and R is the sentence "Mary is taller than Jerry." For the following sentences, what can you conclude about Jerry and Mary if the given sentence is true?
   a) $P \vee Q$
   b) $P \wedge Q$
   c) $\sim P \vee Q$
   d) $Q \wedge R$
   e) $\sim Q \wedge \sim R$
   f) $P \wedge (P \Rightarrow Q)$
   g) $P \Leftrightarrow (Q \vee R)$
   h) $Q \wedge (P \Rightarrow R)$
   i) $P \vee Q \vee R$
   j) $P \vee (Q \wedge R)$

5. **Truth Tables**
   Construct truth tables to verify the following logical equivalences.
   a) $(P \Leftrightarrow Q) \equiv (\sim P \Leftrightarrow \sim Q)$
   b) $[\sim(P \Leftrightarrow Q)] \equiv [(P \wedge \sim Q) \vee (\sim P \wedge Q)]$
   c) $(P \Rightarrow Q) \equiv (\sim P \vee Q)$

6. **Conditional Sentences**
   Translate the given English language sentences to the form $P \Rightarrow Q$.
   a) Unless you study, you will not get a good grade.
   b) "Do you like it? It is yours."
   c) Get out or I will call the cops.
   d) Anyone who does not study deserves to flunk.
   e) Criticize her and she will slap you.
   f) With his toupee on, the professor looks younger.

7. **In Plain English**
   Without making a truth table, say why the following implications are true.
   a) $[(P \vee Q) \wedge \sim P] \Rightarrow Q$
   b) $[P \wedge (Q \wedge \sim Q)] \Rightarrow \sim P$
   c) $(P \vee Q) \Rightarrow (\sim P \Rightarrow Q)$

8. **Distributive Laws for AND and OR**
   For P, Q, and R verify the distributive laws
   a) $P \wedge (Q \vee R) \equiv (P \wedge Q) \vee (P \wedge R)$
   b) $P \vee (Q \wedge R) \equiv (P \vee Q) \wedge (P \vee R)$

9. **Inverse, Converse, and Contrapositive**
   One of the following sentences is logically equivalent to the implication
   $P \Rightarrow Q$. Which one is it?

   $$inverse : \sim P \Rightarrow \sim Q$$
   $$converse : Q \Rightarrow P$$
   $$contrapositive : \sim Q \Rightarrow \sim P$$

   For the two sentences not equivalent to $P \Rightarrow Q$, find examples illustrating
   this fact.

10. **True or False**
    Is the following statement a tautology, a contradiction, or neither?

    $$[(P \Rightarrow Q) \wedge Q] \Rightarrow P$$

11. **Logical Equivalent Implications**
    Show that the following five implications are all logically equivalent.
    a) $P \Rightarrow Q$ (direct form of an implication)
    b) $\sim Q \Rightarrow \sim P$ (contrapositive form)
    c) $(P \wedge \sim Q) \Rightarrow \sim P$ (proof by contradiction)
    d) $(P \wedge \sim Q) \Rightarrow Q$ (proof by contradiction)
    e) $(P \wedge \sim Q) \Rightarrow R \wedge \sim R$ (*reduction ad absurdum*)

12. **Hmmmmmmmmmmmm**
    Is the statement

    $$(P \vee Q) \Leftrightarrow (P \vee \sim Q)$$

    true for all truth values of P and Q, or is it false for all values, or is it some-
    times true and sometimes false?

13. **Interesting Biconditional**
    Is the statement

    $$(P \vee Q) \Leftrightarrow (\sim P \vee \sim Q)$$

    true for all truth values of P and Q, or is it false for all values, or is it some-
    times true and sometimes false?

14. **Finding Negations**
    Find the negation of the following sentences.
    a) $(P \vee Q) \wedge R$
    b) $(P \vee Q) \wedge (R \vee S)$
    c) $(\sim P \vee Q) \wedge R$

15. If possible, find an example of a true conditional for which
    a) its contrapositive is true.
    b) its contrapositive is false.
    c) its converse is true.
    d) its converse is false.

16. The **inverse** of the implication $P \Rightarrow Q$ is $\sim P \Rightarrow \sim Q$.
    a) Prove or disprove that an implication and its inverse are equivalent.
    b) What are the truth values of P and Q for which an implication and its inverse are both true?
    c) What are the truth values of P and Q for which the implication and its inverse are both false?

17. For the sentence

    "If N is an integer, then 2N is an even integer."

    write the converse, contrapositive, and inverse sentences.

18. Let P, Q, and R be sentences. Show that
    a) $P \Rightarrow (Q \Leftrightarrow R)$ requires the given paranthesis
    b) $(P \wedge Q) \vee R$ requires the given paranthesis
    c) $(\sim P \vee Q) \Rightarrow R$ can not be written as $\sim P \vee (Q \Rightarrow R)$

19. **Challenge**
    Rewrite the sentence

    $$P \Rightarrow (Q \Rightarrow R)$$

    in an equivalent form in which the symbol "$\Rightarrow$" does not occur.

20. **Nonobvious Statement**
    The statement

    $$P \Rightarrow (Q \Rightarrow P)$$

    can be read "If P is true, then P follows from any Q." Is this a tautology, contradiction, or does its truth value depend on the truth or falsity of P and Q?

21. **Another Nonobvious Statement**
    The statement

    $$(Q \Rightarrow P) \vee (P \Rightarrow Q)$$

    can be read "For any two sentences P and Q, it is always true that P implies Q or Q implies P."

Is this a tautology, contradiction, or does its true value depend on the truth or falsity of P and Q?

22. **Three-Valued Logic**
Two-valued (T and F) truth tables were basic in logic until 1921 when the Polish logician Jan Lukasiewicz (1878–1956) and American logician Emil Post (1897–1954) introduced n-valued logical systems, where n is any integer greater than one. For example, sentences in a three-valued logic might have values True, False, and Unknown. Three-value logic is useful in computer science in database work. The truth tables for the AND, OR, and NOT connectives are given in Table 1.28.
From these connectives, derive the conditional $P \Rightarrow Q$ and biconditional $P \Leftrightarrow Q$ by drawing a truth table.

Table 1.28 Three-valued logic.

P	Q	P OR Q	P AND Q	NOT P
True	True	True	True	False
True	Unknown	True	Unknown	False
True	False	True	False	False
Unknown	True	True	Unknown	Unknown
Unknown	Unknown	Unknown	Unknown	Unknown
Unknown	False	Unknown	False	Unknown
False	True	True	False	True
False	Unknown	Unknown	False	True
False	False	False	False	True

23. **Modus Ponens**[3] and **Modus Tollens?**[4] are systematic ways of making logical arguments of the form:

If P, then Q If P, then Q

P $\sim Q$

Therefore, Q Therefore, $\sim P$
Modus Ponens   Modus Tollens

Write Modus Ponens and Modus Tollens as compound sentences, and show they are both tautologies.

3 Latin: *mode that affirms.*
4 Latin: *mode that denies.*

24. **Interesting**
   Are the following two statements equivalent?

   $P \wedge (Q \Rightarrow R)$

   $(P \wedge Q) \Rightarrow R$

25. **Sixteen Logical Functions of Two Variables**
   Figure 1.6 below shows the totality of 16 relations between 2 logical variables. One expression can be proven from another if it lies on an upward path from the first. For example

   $(P \wedge Q) \Rightarrow Q \Rightarrow (P \Rightarrow Q)$.

   Verify a few of these implications using truth tables. The compound sentence $P \Delta Q$ refers to the exclusive OR, which means either P or Q true but not both.

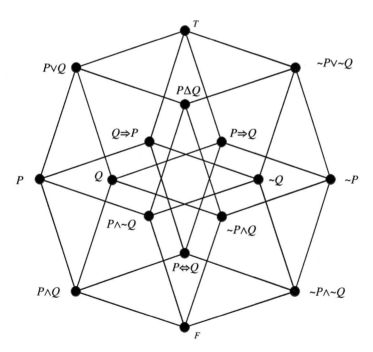

Figure 1.6 Sixteen logical functions.

26. **Internet Research**
   There is a wealth of information related to topics introduced in this section just waiting for curious minds. Try aiming your favorite search engine toward *conditional connectives, biconditional, truth tables,* and *necessary and sufficient conditions.*

1.3

Predicate Logic

> **Purpose of Section** This section introduces **predicate logic** (or **first-order logic**) that is the "language" of mathematics. We will see how predicate logic extends the language of sentential (or propositional) logic, introduced in the first two sections, by adding **universal** and **existential quantifiers, functions,** and **variables.**

1.3.1 Introduction

Although sentential logic studied in Sections 1.1 and 1.2 is probably sufficient to get you through your daily activities, it is not sufficient for higher mathematics. This was realized in the late 1800s by the German logician Gottlob Frege, who observed that mathematics requires a more extensive language than simple logical sentences connected by $\wedge, \vee, \sim, \Rightarrow, \Leftrightarrow$. Frege introduced what is called **predicate logic**[1] (or **first-order logic**), which are sentences which in addition to the logical connectives of sentential logic, includes **quantifiers, variables,** and functions called **predicates** (or **propositions**). Predicate logic allows one to express concepts we often hear in mathematics, like

for any real number x, there exists a real number y such that $x < y$

which is impossible to express in sentential logic.

1.3.2 Existential and Universal Quantifiers

Two phrases one hears again and again in mathematics are *for all*, and *there exists*. These expressions are called **quantifiers** and are necessary to

---

1 Predicate logic is also called **first-order logic** in contrast to sentential calculus which is sometimes called zero-order logic.

Advanced Mathematics: A Transitional Reference, First Edition. Stanley J. Farlow.
© 2020 John Wiley & Sons, Inc. Published 2020 by John Wiley & Sons, Inc.
Companion website: www.wiley.com/go/farlow/advanced-mathematics

describe mathematical concepts. The meaning of an expression like $x < y$ in itself is not clear until we describe the extent of x and y. This leads to the two basic quantifiers of predicate logic, the **universal quantifier**, meaning "for all" and denoted by $\forall$ (upside down A), and the **existential quantifier**, meaning "there exists," and denoted by $\exists$ (backwards E). Inherent in the use of quantifiers is the concept of a **universe** set, which is the collection or set of objects under discussion, often being the real numbers, integers, natural numbers, and so on.

Quantifiers of Predicate Logic   Let U be the universe, or a collection of all objects under consideration.

- **Universal Quantifier:** The statement $(\forall x \in U)\ P(x)$ means "for all (or any) x in U, the proposition $P(x)$ is true."
- **Existential Quantifier:** The statement $(\exists x \in U)\ P(x)$ means "there exists an x in U such that the proposition $P(x)$ is true."

Some common number universes in mathematics are the following.

$\mathbb{N} = \{1, 2, 3, \ldots\}$ (natural numbers)

$\mathbb{Z} = \{0, \pm 1, \pm 2, \ldots\}$ (integers)

$\mathbb{Q} = \{p/q : p \text{ and } q \text{ are integers with } q \neq 0\}$ (rational numbers)

$\mathbb{R} = \{\text{real numbers}\}$

$\mathbb{C} = \{\text{complex numbers}\}$

1.3.3   More than One Variable in a Proposition

Propositions (or predicates) in predicate logic often contain more than one variable, which means the proposition must contain more than one quantifier to put limits on the variables. The following Table 1.29 illustrates propositions with two variables and what it means for the propositions to be true or false. To simplify notation, we assume the universe is known, so we do not include it.

Example 1   Converting Predicate Logic to English
The following propositions in Table 1.30 are translated into natural language.[2]

---

2 The expression $x \in A$ is set notation meaning an element x belongs to a set A. This notation will be front and center when we arrive at sets in Chapter 2.

Table 1.29 Propositions with two variables.

Proposition	Proposition is true when	Proposition is false when
$(\forall x)(\forall y)P(x, y)$	For all x and y, $P(x, y)$ is true	There exists some x, y. such that $P(x, y)$ is false
$(\exists x)(\forall y)P(x, y)$	There is an x such that for all y, $P(x, y)$ is true	For all x, $P(x, y)$ is false for some y
$(\forall x)(\exists y)P(x, y)$	For all x, $P(x, y)$ is true for some y	There exists an x such that for all y $P(x, y)$ is false
$(\exists x)(\exists y)P(x, y)$	There exists x, y. such that $P(x, y)$ is true	For all x, y. $P(x, y)$ is false

Table 1.30 Sentences in predicate logic.

Proposition	English meaning	Truth value
$(\forall x \in \mathbb{R})\ (x^2 \geq 0)$	For any real number, its square is nonnegative.	True
$(\exists n \in \mathbb{N})\ (n$ is a prime number$)$	There exists a prime number.	True
$(\exists n \in \mathbb{N})\ (2 \nmid n)$	There exists at least one odd natural number.	True
$(\exists n \in \mathbb{N})\ (2 \mid n)$	There exists at least one even natural number.	True
$(\forall x \in \mathbb{N})\ (\exists y \in \mathbb{N})\ (x = y + 1)$	For any natural number x, there is a natural number y satisfying $x = y + 1$.	False
$(\forall x \in \mathbb{R})\ (\exists y \in \mathbb{R})\ (x < y)$	For any real number x, there is a real number y greater than x.	True
$(\exists x \in \mathbb{R})\ (\forall y \in \mathbb{R})\ (x < y)$	There exists a real number x such that all real numbers y are greater than x.	False
$(\forall x \in \mathbb{R})(\exists y \in \mathbb{R})\ (y = 2x)$	For any positive real number x, there is a real number y that satisfies $y = 2x$.	True

Historical Note   Frege's 1879 seminal work *Begriffsschrift* ("Conceptual Notation") marked the beginning of a new era in logic, which allowed for the quantification of mathematical variables, just in time for the more precise **arithmetization of analysis** of calculus, being carried out in the late 1800s by mathematicians like German Karl Weierstrass (1815–1897).

1.3.4 Order Matters

Here's a question to ponder. Do the two propositions $(\exists x)(\forall y)P(x, y)$ and $(\forall y)(\exists x)P(x, y)$ mean the same thing or does one imply the other or are they completely unrelated? For example do the following to two propositions:

- $(\exists x \in \mathbb{R})(\forall y \in \mathbb{R})(x < y)$
- $(\forall y \in \mathbb{R})(\exists x \in \mathbb{R})(x < y)$

mean the same thing? The answer is they do not since the first proposition is false, while the second proposition is true. Do you understand that?

To show how the order of the universal and existential quantifiers makes a difference in the meaning of a proposition, here's a simple example of a third-grade class consisting of three boys and three girls, members of the sets

$$B = \{Abe, Bob, Carl\}$$

$$G = \{Ann, Betty, Carol\}$$

and the two-variable predicate (or proposition)

$$P(b, g) \text{ means boy } b \text{ likes girl } g.$$

where Figure 1.7 illustrates the relationships between the quantifiers $\forall \forall$, $\exists \forall$, $\forall \exists$, $\exists \exists$. The dot at the intersection of a boy and girl indicates the boy likes the

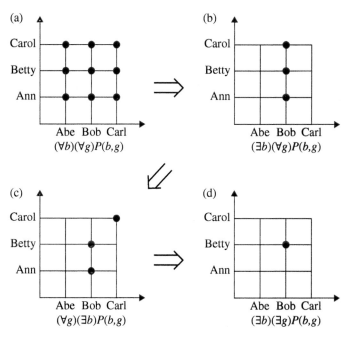

Figure 1.7 Universal and existential implications.

girl. For example, Figure 1.7a means every boy likes every girl, while Figure 1.7c means Bob likes Ann and Betty, and Carl likes Carol.

Figure 1.7 gives visual support for the implications

$$\forall g \forall b \equiv \forall b \forall g \Rightarrow \exists b \forall g \Rightarrow \forall g \exists b \Rightarrow \exists g \exists b \equiv \exists b \exists g$$

which is stated explicitly in Figure 1.8.

Predicate logic	Equivalent English
$(\forall b \in B)$ $(\forall g \in G)$ $P(b,g)$	Every boy likes every girl
$\Downarrow$	$\Downarrow$
$(\exists b \in B)$ $(\forall g \in G)$ $P(b,g)$	Some boy likes every girl
$\Downarrow$	$\Downarrow$
$(\forall g \in G)$ $(\exists b \in B)$ $P(b,g)$	Every girl is liked by some boy
$\Downarrow$	$\Downarrow$
$(\exists g \in B)$ $(\exists g \in G)$ $P(b,g)$	Some girl is like by some boy

Figure 1.8 $\forall g \forall b \equiv \forall b \forall g \Rightarrow \exists b \forall g \Rightarrow \forall g \exists b \Rightarrow \exists g \exists b \equiv \exists b \exists g.$

1.3.5 Negation of Quantified Propositions

In the next few sections, we will introduce strategies for proving theorems, and it will be necessary to know how quantified propositions are negated. The propositions in Table 1.31 show their negations in the language of predicated logic. The universe U, where the variables x and y are members might be any one of the common number systems $\mathbb{N}, \mathbb{Z}, \mathbb{Q}, \mathbb{R}, \mathbb{C}$ or a subset of one line the intervals $[0, 1], [0, \infty)$, and so on.

Table 1.31 Negation in predicate logic.

Proposition	Negation of proposition
$(\forall x \in U)$ $P(x)$	$(\exists x \in U)$ $[\sim P(x)]$
$(\exists x \in U)$ $P(x)$	$(\forall x \in U)$ $[\sim P(x)]$
$(\forall x \in U)$ $(\exists y \in U)$ $P(x, y)$	$(\exists x \in U)$ $(\forall y \in U)$ $[\sim P(x, y)]$
$(\exists x \in U)$ $(\forall y \in U)$ $P(x, y)$	$(\forall x \in U)$ $(\exists y \in U)$ $[\sim P(x, y)]$
$(\forall x \in U)(\forall y \in U)P(x, y)$	$(\exists x \in U)(\exists y \in U)[\sim P(x, y)]$
$(\exists x \in U)(\exists y \in U)P(x, y)$	$(\forall x \in U)(\forall y \in U)[\sim P(x, y)]$

Example 2   Negations

State the negation of the following propositions. In these cases, we do not bother to specify the universe for the variables. You might imagine for yourself some possible universes.

a) $(\forall x)[x > 0 \Rightarrow (\exists y)(x + y = 1)]$
b) $(\exists n)(n$ is a prime number)
c) $(\forall x)(\exists y)(xy = 10)$
d) $(\exists x)(\forall y)(xy \neq 10)$

Solution

a) $(\exists x)[(x > 0) \wedge (\forall y)(x + y \neq 1)]$
b) $(\forall n)(n$ is not a prime number)
c) $(\exists x)(\forall y)(xy \neq 10)$
d) $(\forall x)(\exists y)(xy = 10)$ ∎

Logicism   The development of predicate logic is generally attributed to the German logician Gottlob Frege (1848–1925), considered by many to be the most important logician of the nineteenth century. It was Frege's belief (misguided as it turned out) that all mathematics could be derived from logic. The philosophy that states mathematics is a branch of logic and mathematical principles are reducible to logical principles, is called **logicism**.

Predicate Logic in Analysis   Real analysis is an important area of mathematics. Here are a few definitions about sequences x_n, $n = 1, 2, \ldots$ of real numbers stated in the language of predicate logic.

- $\lim_{n \to \infty} x_n = 2 \Leftrightarrow (\forall \varepsilon > 0)(\exists N \in \mathbb{N})(\forall n > N)(|x_n - 2| < \varepsilon)$
- $\lim_{n \to \infty} x_n \neq 2 \Leftrightarrow (\exists \varepsilon > 0)(\forall N \in \mathbb{N})(\exists n > N)(|x_n - 2| \geq \varepsilon)$
- $\{x_n\}$ is Cauchy $\Leftrightarrow (\forall \varepsilon > 0)(\exists N \in \mathbb{N})(\forall m, n > N)(|x_m - x_n| < \varepsilon)$
- $\{x_n\}$ not Cauchy $\Leftrightarrow (\exists \varepsilon > 0)(\forall N \in \mathbb{N})(\exists m, n > N)(|x_m - x_n| \geq \varepsilon)$

1.3.6   Conjunctions and Disjunctions in Predicate Logic

As an aid to understanding and involving conjunctions and disjunctions, Professor Snarf has conducted a survey among his students about their liking for peanuts and qumquats.[3] If we let

---

[3] Author apologizes for the misspelling of kumquats, but the example desperately needs a fruit that starts with "q."

$P(s)=$ student s likes peanuts (p)
$Q(s)=$ student s likes qumquats (q)

we can write the following propositions about students likings of peanuts and qumquats.

1) $(\forall s)[P(s) \wedge Q(s)]$ means all students like p and q
2) $(\forall s)P(s) \wedge (\forall s)Q(s)$ means all students like p and all students like q
3) $(\exists s)[P(s) \wedge Q(s)]$ means some student likes p and q
4) $(\exists s)P(s) \wedge (\exists s)Q(s)$ means some student likes p and some student likes q
5) $(\exists s)[P(s) \vee Q(s)]$ means some student like p or q
6) $(\exists s)P(s) \vee (\exists s)Q(s)$ means some student likes p or some student likes q
7) $(\forall s)P(s) \vee (\forall s)Q(s)$ means all students like p or all students like q
8) $(\forall s)[P(s) \vee Q(s)]$ means all students like p or q

There are several important relations between these propositions, including

$$(\forall s)P(s) \vee (\forall s)Q(s) \Rightarrow (\forall s)[P(s) \vee Q(s)]$$

which means that if every student likes peanuts or every student likes qumquats, then it follows that every student likes peanuts or qumquats. But the converse does not follow since just because every student in the class likes either peanuts or qumquats that does not mean the class belongs to one of two distinct groups, peanut lovers or qumquat lovers.

Several important relationships between the previous propositions are as follows:

a) $(\forall s)[P(s) \wedge Q(s)] \Leftrightarrow (\forall s)P(s) \wedge (\forall s)Q(s)$
b) $(\exists s)[P(s) \wedge Q(s)] \Rightarrow (\exists s)P(s) \wedge (\exists s)Q(s)$
c) $(\forall s)P(s) \vee (\forall s)Q(s) \Rightarrow (\forall s)[P(s) \vee Q(s)]$
d) $(\exists s)[P(s) \vee Q(s)] \Leftrightarrow (\forall s)P(s) \vee (\forall s)Q(s)$

No Truth Tables in Predicate Logic Proofs in predicate logic cannot be verified with truth tables. It is not easy to prove $\forall \forall \Rightarrow \exists \vee \Rightarrow \forall \exists \Rightarrow \exists \exists$. For example although we know the statement $(\forall x \in \mathbb{R})(x^2 \geq 0)$ is true, to make a truth table to verify it would require a row for each real number x which would require an infinite number of rows.

Example 3   All Even or Odd versus All Even or All Odd

Consider the propositions involving an integer $n \in \mathbb{Z}$:

$O(n) = n$ is odd

$E(n) = n$ is even

and compare the following two statements.

- $(\forall n \in \mathbb{Z})(E(n)) \vee (\forall n \in \mathbb{Z})(O(n))$
- $(\forall n \in \mathbb{Z})(E(n) \vee O(n))$

Are the statements equivalent, if not does one imply the other?

Solution

The following implication

$$[(\forall n \in \mathbb{Z})(E(n)) \vee (\forall n \in \mathbb{Z})(O(n))] \Rightarrow (\forall n \in \mathbb{Z})(E(n) \vee O(n))$$

is true since the assumption states "all integers are even or all integers are odd" is false. The assumption is clearly false since it states that $\mathbb{Z}$ consists of all even integers or all odd integers, but we know that $\mathbb{Z}$ contains both even and odd numbers, and hence the implication is true.

On the other hand, the converse of the implication states that if every integer is even or odd, which we know to be true, then we conclude a fact we know as false. Hence, we have a true assumption and a false conclusion, thus the converse implication

$$(\forall n \in \mathbb{Z})(E(n) \vee O(n)) \not\Rightarrow [(\forall n \in \mathbb{Z})(E(n)) \vee (\forall n \in \mathbb{Z})(O(n))]$$

is false. ∎

Historical Note In the early 1900s, German mathematician David Hilbert (1862–1943) attempted to *formalize* all mathematics within the language of predicate logic by proposing a series of axioms from which all of mathematics would follow. If Hilbert's grand plan had succeeded, it would put mathematicians out of business and turn them into "deduction robots" and turn mathematics into a "turn-the-crank" predicate-logic machines. Unfortunately, Hilbert's grand scheme failed due to Gödel's **Incompleteness Theorem** of 1931, which proved that no matter what axiom system[4] is chosen, there are always theorems that can never be proven or disproven.[5]

---

4 Hilbert's plan to formalize mathematics can be found online by interested readers.
5 Then too, mathematics is more than a pure deductive discipline. It relies a great deal on intuition and creativity.

Problems

1. Write the following quotes in the symbolic language of predicate logic.
   a) Learn from yesterday, live for today, hope for tomorrow.

   ...Albert Einstein
   b) A woman can say more in a sigh than a man in a sermon.

   ...Arnold Haultain
   c) "We all go a little mad sometimes."

   ...Norman Bates
   d) Cowards die many times before their deaths, the valiant never taste of death but once.

   ...William Shakespeare
   e) All people are mortal.
   f) All that glitters is not gold.

2. **Translate to Predicate Logic**
   Write the following sentences in the symbolic language of predicate logic. The universe of each variable is given in parentheses. For these problems, we use the notation

   $\mathbb{Z}$ = integers

   $\mathbb{R}$ = real numbers

   a) If $a|b$ and $b|c$, then $a|c$, where a, b, c are integers. (Integers)
   b) 4 does not divide $n^2 + 2$ for any integer. (Integers)
   c) $x^3 + x + 1 = 0$ for some real number x. (Real numbers)
   d) Everybody loves mathematics. (All people)
   e) For every positive real number a, there exists a real number x that satisfies $e^x = a$. (Real numbers)
   f) For every positive real number $\varepsilon > 0$, there exists a real number $\delta > 0$ such that $|x - a| < \delta \Rightarrow |f(x) - f(a)| < \varepsilon$, where a, x are arbitrary real numbers.
   g) Everyone always attends class. (All students)
   h) The equation $x^2 + 1 = 0$ has no solution. (Real numbers)
   i) The equation $x^2 - 2 = 0$ has no solution. (Rational numbers)

3. **True or False**
   Which of the following propositions are true in the given universe? The universe is given in parentheses.
   a) $(\forall x)(x \leq x)$ (Real numbers)
   b) $(\exists x)(x^2 = 2)$ (Real numbers)
   c) $(\exists x)(x^2 = 2)$ (Rational numbers)

d) $(\exists x)(x^2 + x + 1 = 0)$ (Complex numbers)
e) $(\forall x)[x \equiv 1 \,(\mathrm{mod}\,5)]$ (Integers)
f) $(\exists x)(e^x = 1)$ (Real numbers)
g) $(\forall x)(x \leq x)$ (Integers)

4. **True or False**
   Very quickly, true or false.
   a) $\sim(\forall x)(P(x)) \equiv (\exists x)[\sim P(x)]$
   b) $\sim(\exists x)(P(x)) \equiv (\forall x)[\sim P(x)]$
   c) $\sim(\forall x)[\sim P(x)] \equiv (\exists x)[P(x)]$
   d) $\sim(\exists x)[\sim P(x)] \equiv (\forall x)[P(x)]$
   e) $(\forall x)[P(x) \Rightarrow Q(x)] \equiv \sim(\exists x)[P(x) \wedge \sim Q(x)]$
   f) $\sim(\exists x)[P(x) \wedge Q(x)] \equiv (\forall x)[P(x) \Rightarrow \sim Q(x)]$

5. **Expanding Universes**
   In which of the universes $\mathbb{N}, \mathbb{Z}, \mathbb{Q}, \mathbb{R}, \mathbb{C}$ are the following sentences true for x and y.
   a) $(\forall x)(\exists y)(y = 1 - x)$
   b) $(\forall x \neq 0)(\exists y)(y = 1/x)$
   c) $(\exists x)(x^2 - 2 = 0)$
   d) $(\exists x)(x^2 + 2 = 0)$

6. **Not as Easy as It Looks**
   Tell if the sentence

   $$(\exists x \in U)[x \text{ is even} \Rightarrow 5 \leq x \leq 10]$$

   is true or false in the following universe U.
   a) $U = \{4\}$
   b) $U = \{3\}$
   c) $U = \{6, 8, 10\}$
   d) $U = \{6, 8, 10, 12\}$
   e) $U = \{6, 7, 8, 10, 12\}$

7. **Small Universe**
   Which statements are true in the universe $U = \{1, 2, 3\}$.
   a) $1 < 0 \Rightarrow (\exists x)(x < 0)$
   b) $(\exists x)(\forall y)(x \leq y)$
   c) $(\forall x)(\exists y)(x \leq y)$
   d) $(\exists x)(\exists y)(y = x + 1)$
   e) $(\forall x)(\forall y)(xy = yx)$
   f) $(\forall x)(\exists y)(y \leq x + 1)$

8. **Well-Known Universe**
   Given the statements

   $R(x)$: x is a rational number

   $I(x)$: x is an irrational number

   which of the following sentences are true in the universe of real numbers.
   a) $(\forall x)[I(x) \vee R(x)]$
   b) $(\forall x)[I(x) \wedge R(x)]$
   c) $(\forall x)R(x) \vee (\forall x)I(x)$
   d) $(\forall x)[R(x) \vee I(x)] \Rightarrow [(\forall x)R(x) \vee (\forall x)I(x)]$
   e) $[(\forall x)R(x) \vee (\forall x)I(x)] \Rightarrow (\forall x)[R(x) \vee I(x)]$

9. **Famous Theorems**
   State the following famous theorems in the language of predicate logic.
   a) **Intermediate Value Theorem** Let f be a continuous function on the closed interval $[a, b]$. If f changes sign from negative to positive on $[a, b]$, then there exists a number c between a and b, such that $f(c) = 0$.
   b) **Fermat's Last Theorem** If n is a natural number greater than 2, then there are no natural numbers a, b, and c that satisfy $a^n + b^n = c^n$.
   c) **Euler's Theorem** If P is any regular polyhedron and if v, e, f are the number of vertices, edges, and faces, respectively, then $v - e + f = 2$.
   d) **Binomial Theorem** If a, b are real numbers and n is a positive integer, then

   $$(a+b)^n = \sum_{k=0}^{n} \frac{n!}{k!(n-k)!} a^k b^{n-k}$$

10. **Negation**
    Negate the following sentences in words.
    a) All women are moral.
    b) Every player on the team was over 6 feet tall.
    c) For any real number y, there exists a real number x that satisfies $y = \tan x$.
    d) There exists a real number x that satisfies $0 < x < 5$ and $x^3 - 8 = 0$.
    e) The equation $a^n + b^n = c^n$ does not have positive integer solutions a, b, c for n a natural number $n > 2$.

11. **Negation in Predicate Logic**
    Negate the following sentences in symbolic form.
    a) $(\forall x)[P(x) \Rightarrow Q(x)]$
    b) $(\forall x)[x > 0 \Rightarrow (\exists y)(x^2 = y)]$
    c) $(\forall x)(\exists y)(\forall z)(xyz = 1)$
    d) $(\exists x)(\exists y)(\forall z)(xy < z)$

12. **Convergence and Nonconvergence**

A sequence $\{x_n\}_{n=1}^{\infty}$ of real numbers converges to a limit L if and only if

$$(\forall \varepsilon > 0)(\exists N \in \mathbb{N})(\forall n > N)(|x_n - L| < \varepsilon)$$

a) State the negation of this sentence.
b) Using this negation show the sequence $1/2^n$, $n = 1, 2, \ldots$ does not converge to $1/4$.

13. **Graph to the Rescue**

If $P(x, y) : y \le x^2 + 1$, where (x, y) are points in the plane, determine which of the following is true. Hint: A picture (i.e. graph) is worth a thousand words.

a) $(\forall x)(\forall y)P(x, y)$
b) $(\forall x)(\exists y)P(x, y)$
c) $(\exists x)(\forall y)P(x, y)$
d) $(\exists x)(\exists y)P(x, y)$
e) $(\exists y)(\forall x)P(x, y)$
f) $(\forall y)(\exists x)P(x, y)$

14. **Order Counts**

Which of the following statements are true and which are false for real numbers x, y?

a) $(\forall x)(\forall y)(x < y)$
b) $(\exists x)(\forall y)(x < y)$
c) $(\forall y)(\exists x)(x < y)$
d) $(\exists x)(\exists y)(x < y)$

15. **Fun Time**

State the denial of words of wisdom attributed to Abraham Lincoln: "You can fool some of the people all the time and all the people some of the time, but you can't fool all the people all of the time."

16. **In Plain English**

Restate the following sentences in plain English.

a) $(\forall x \in \mathbb{R})(\forall y \in \mathbb{R})[(x < y) \Rightarrow (\exists z \in \mathbb{R})[(x < z) \wedge (z < y)]]$
b) $(\forall x \in \mathbb{R})[(x > 0) \Rightarrow (\exists y \in \mathbb{R})(x = y^2)]$
c) $(\forall m, n \in \mathbb{N})[(n > 1) \Rightarrow [m \mid n \Rightarrow (m = 1) \vee (m = n)]]$

17. **True Here, False There**

Tell if the following sentences are true or false in each universe $\mathbb{N}, \mathbb{Z}, \mathbb{Q}, \mathbb{R}$.

a) $(\forall x)(\exists y)(x < y)$
b) $(\forall y)(\exists x)(x < y)$
c) $(\exists x)(\forall y)(x < y)$

d) $(\exists y)(\forall x)(x < y)$

e) $(\forall x)[(x > 0) \Rightarrow (\exists y)(y = x^2)]$

18. **Satisfiable in Predicate Logic**

A **satisfiable** sentence is a sentence that is true in at least *some* universe. For example, the sentence $(\exists x \in U)(x > 0)$ is satisfiable since it is true in the universe of real numbers. Tell if the following sentences are satisfiable, and if so, give a universe.

a) $(\exists x \in U)[(x > 0) \wedge (x < 0)]$

b) $(\exists x \in U)(x^2 = -3)$

c) $(\exists x \in U)[P(x) \wedge \sim P(x)]$

d) $(\exists x \in U)[(x > 0) \vee (x < 0)]$

19. **Translation into Predicate Logic**

Letting

$E(x) = x$ is even

$O(x) = x$ is odd

translate the following sentence to predicate logic.

a) Not every integer is even.

b) Some integers are odd.

c) Some integers are even, and some integers are odd.

d) If an integer is even, then it is not odd.

e) If an integer is even, then the integer two larger is even.

20. **Internet Research**

There is a wealth of information related to topics introduced in this section just waiting for curious minds. Try aiming your favorite search engine toward *predicate logic, negation of logical statements, first-order-logic,* and *universal quantifier.*

1.4

Mathematical Proofs

Purpose of Section Most theorems in mathematics take the form of a conditional $P \Rightarrow Q$ or biconditional statement $P \Leftrightarrow Q$, where the biconditional can be verified by proving both $P \Rightarrow Q$ and $P \Leftarrow Q$. In this section, we describe several different ways for proving $P \Rightarrow Q$ including a **direct proof**, proof by **contrapositive**, and three variations of **proof by contradiction**, one of which is called proof by **reductio ad absurdum**.

1.4.1 Introduction

Proof is basic to mathematics; we do not know if a proposition is true or false until we have proved or disproved it, which raises the obvious question, what is a mathematical proof? The precise definition of mathematical proof varies from mathematician to mathematician. The famed mathematician GianCarlo Rota once remarked: "Everybody knows what a mathematical proof is, it's a series of steps which leads to the desired conclusion." A more rigid viewpoint of proof might be manipulating definitions and accepted rules of logic in a valid way, going from the assumption to the conclusion. But regardless of its definition, the history of what constitutes a mathematical proof has gone through several refinements over the years, each refinement attaining a higher level of "rigor" from its predecessors.[1] Some mathematical proofs proposed by such greats as Newton and Euler[2] do not hold up to today's scrutiny.

---

1 Although a mathematical argument must be logically precise, the American mathematician George Simmons once said, "Mathematical rigor is like clothing, it ought to suit the occasion."
2 Euler often manipulated infinite series without regard to convergence of the series.

Advanced Mathematics: A Transitional Reference, First Edition. Stanley J. Farlow.
© 2020 John Wiley & Sons, Inc. Published 2020 by John Wiley & Sons, Inc.
Companion website: www.wiley.com/go/farlow/advanced-mathematics

As an example of a mathematical proof, consider the proposition:

Proposition: If a, b, c are real numbers, $a \neq 0$ that satisfy $ax^2 + bx + c = 0$, then

$$x = \frac{-b \pm \sqrt{b^2 - 4ac}}{2a}.$$

This proposition has the form $P \Rightarrow Q$ and says if P is true, then so is Q. A proof of the proposition consists of the following algebraic steps:

Assume x satisfies $ax^2 + bc + c = 0$ $(a \neq 0)$

a) $x^2 + \frac{b}{a}x = -\frac{c}{a}$ (divide by a and transpose a term)

b) $x^2 + \frac{b}{a}x + \left(\frac{b}{2a}\right)^2 = \left(\frac{b}{2a}\right)^2 - \frac{c}{a}$ $\left(\text{add } \left(\frac{b}{2a}\right)^2 \text{ to each side}\right)$

c) $\left(x + \frac{b}{2a}\right)^2 = \frac{b^2 - 4ac}{4a^2}$ (complete the square on the left)

d) $x + \frac{b}{2a} = \pm \frac{\sqrt{b^2 - 4ac}}{2a}$ (take the square root)

e) $x = \frac{-b \pm \sqrt{b^2 - 4ac}}{2a}$ (isolate x and simplify)

Is this proof convincing or is there something about the argument that is lacking? You might also ask if the converse holds. Is it possible to go backwards by starting with the conclusion and reproducing the quadratic equation? In this case, the answer is yes, so we would say that the quadratic equation holds if and only if the given solution holds.

The previous statement is an example of a mathematical theorem.

> A **theorem**[3] is a mathematical statement that can be demonstrated to be true by accepted mathematical operations and arguments. The chain of reasoning used to convince one of the truth of the theorem is called a **proof** of the theorem.[4]

Theorems are ultimately based on a collection of principles considered so self-evident and obvious that their truth value is taken as fact. Such accepted maxims are called **axioms**, and in every area of mathematics, be it real or complex analysis, algebra, geometry, topology, and even arithmetic is based on a collection of self-evident truths or axioms.

---

3 The Hungarian mathematician Paul Erdos (1913–1996) once said that a mathematician is a machine for converting coffee into theorems.

4 Keep in mind that there may be more than one proof of a theorem. For example, there are over 200 different proofs of the Pythagorean Theorem.

1.4.2 Types of Proofs

Many theorems in mathematics have the form of a conditional statement or implication $P \Rightarrow Q$ where one assumes the validity of P, then with the aid of existing mathematical truths and accepted rules of inference, arrives at Q. Although the goal is always to "go from P to Q," there is more than one way of achieving this goal, which are displayed in Table 1.32.

Table 1.32 Five ways to prove $P \Rightarrow Q$.

Five ways to prove $P \Rightarrow Q$	
$P \Rightarrow Q$	Direct proof
$\sim Q \Rightarrow \sim P$	Proof by contrapositive
$(P \wedge \sim Q) \Rightarrow Q$	Proof by contradicting the conclusion
$(P \wedge \sim Q) \Rightarrow \sim P$	Proof by contradicting the hypothesis
$(P \wedge \sim Q) \Rightarrow (R \wedge \sim R)$	Proof by *reductio ad absurdum*

The five statements in Table 1.34 are equivalent inasmuch the five columns numbered (4) through (8) in Table 1.33 that have the same truth values of TFTT. Hence, there are five forms of the basic implication $P \Rightarrow Q$, each providing a different approach for proving the basic implication.

Table 1.33 Five equivalent forms for the basic implication.

		(1)	(2)	(3)
P	*Q*	$\sim P$	$\sim Q$	$P \wedge \sim Q$
T	T	F	F	F
T	F	F	T	T
F	T	T	F	F
F	F	T	T	F

(4)	(5)	(6)	(7)	(8)
$P \Rightarrow Q$	$\sim Q \Rightarrow \sim P$	$(P \wedge \sim Q) \Rightarrow Q$	$(P \wedge \sim Q) \Rightarrow \sim P$	$(P \wedge \sim Q) \Rightarrow (R \wedge \sim R)$
T	T	T	T	T
F	F	F	F	F
T	T	T	T	T
T	T	T	T	T

In this and the next section, we demonstrate different methods of proof. Before presenting some theorems and proofs, we begin by stating a few definitions that are important since they give precise meaning to mathematical concepts.

Primes, Composites, Even, and Odd Integers
Prime number: A prime number is a natural number greater than 1 that is divisible only by 1 and itself. In the language of predicate logic, a natural number $p \in \mathbb{N}$ greater than 1 is prime if and only if

$$(\forall n \in \mathbb{N})[n \mid p \Rightarrow [(n = 1) \vee (n = p)]]$$

An integer $n \in \mathbb{Z}$ is composite if it is not a prime. In the language of predicate logic, an integer n is composite if and only if

$$(\exists m \in \mathbb{Z})[[(m \neq 1) \wedge (m \neq n)] \Rightarrow m \mid n]$$

Odd integer: An integer $n \in \mathbb{Z}$ is an **odd integer** if it is not divisible by 2. In the language of predicate logic, an integer $n \in \mathbb{Z}$ is odd if and only if

$$(\exists k \in \mathbb{Z})(n = 2k + 1)$$

Even integer: An integer $n \in \mathbb{Z}$ is an **even integer** if it is divisible by 2. In the language of predicate logic, an integer $n \in \mathbb{Z}$ is even if

$$(\exists k \in \mathbb{Z})(n = 2k)$$

Proof in Experimental Sciences   Experimental sciences (as biology, physics, chemistry, ...) use laboratory experiments to *proof* results, which are then verified by repeated experimentation. Most experimental sciences are not based on fundamental axioms as they are in mathematics, the net result being that over time more sophisticated experiments may change the accepted beliefs. That is in contrast with mathematics, which is based on fundamental axioms and definitions, what was proven true 2000 years ago is just as valid today.

1.4.3   Analysis of Proof Techniques

- **Direct Proofs** $[P \Rightarrow Q]$ A direct proof starts with an assumption P, then uses existing accepted mathematical truths and rules of inference to establish the truth of the conclusion Q.

  Indirect proofs[5] refer to proof by contrapositive or some proof by contradiction.

---

5 The formal name for an indirect proof is *modus tollens* (Latin for "mode that denies").

- **Proof by Contrapositive** $[\sim Q \Rightarrow \sim P]$ Here, one assumes the conclusion Q false and then proves the hypothesis P false.
- **Proofs by Contradiction** The two proofs by contradiction have the form

$$(P \wedge \sim Q) \Rightarrow \sim P$$
$$(P \wedge \sim Q) \Rightarrow Q$$

In each case, one assumes the hypotheses P as true and the conclusion Q false and then arrives at a contradiction either contradicting the assumption P in the first case, or contradicting the assumed denial $\sim Q$ in the second case. Both of these proofs by contradiction[6] are powerful methods of proof.

- *Reductio ad absurdum*

$$[(P \wedge \sim Q) \Rightarrow (R \wedge \sim R)]$$

This is another form of proof by contradiction. Here, one assumes P true and Q false, then seeks to prove some type of internal contradiction[7] (like $1 = 0$ or $x^2 < 0$), which is denoted by $R \wedge \sim R$.

1.4.4 Modus Operandi for Proving Theorems

Before getting into the nitty-gritty of proving theorems, the following steps are always useful, maybe crucial.

1) Be sure you understand the terms and expressions in the theorem.
2) Ask yourself if you believe the theorem is true or false.
3) Write the theorem in the language of first-order logic so that you understand its logical structure.
4) Determine how the proof might proceed, i.e. a direct proof, proof by contrapositive, or a proof by contradiction.
5) Start the proof.

Theorem 1   Direct Proof
If n is an odd natural number, then n^2 is odd.

Direct Proof   Since $n \in \mathbb{N}$ is odd, it can be written in the form $n = 2k + 1$ for some integer $k \in \mathbb{Z}$. Squaring both sides of this equation yields

---

6 The English mathematician, G. H. Hardy, said proof by contradiction is one of the finest weapons in the mathematician's arsenal.

7 Hence, the name *reductio ad absurdum* (reduction to the absurd).

$$n^2 = (2k+1)^2$$

$$= 4k^2 + 4k + 1$$

$$= 2(2k^2 + 2k) + 1$$

Since $k \in \mathbb{Z}$ is an integer so is $2k^2 + 2k$. Hence, n^2 has the form of an odd integer. ∎

Lemmas and Corollaries In addition to theorems, there are lemmas and corollaries.[8] Although theorems, lemmas, and corollaries are similar logically, it is how they are used and their importance that distinguishes them.

A **lemma** is a statement that is proven as an aid in proving a theorem. Often, unimportant details are included in a lemma so as not to clutter a the proof of a theorem. Occasionally, lemmas take on a life of their own and become as important or more important than the theorem they support. (i.e. Zorn's lemma, Burnside's lemma, Urysohn's lemma, ...)

A **corollary** is a statement that easily follows from a theorem and whose results are generally secondary to that of the theorem. The statement:

if a, b are the sides of a right isosceles triangle, then the hypotenuse has length $c = \sqrt{2}\,a$

is a simple corollary of the Pythagorean theorem.

Historical Note The American logician Charles Saunders Pierce (1839–1914) introduced what is called **second-order logic**, which in addition to quantifying variables like x, y, ... also quantifies *functions* and *entire sets* of variables. Although Pierce also made contributions to the development of first-order logic, it was Frege who carried out first-order logic earlier and whose name is associated with its development. It was Pierce who coined the word "first-order" logic.

Theorem 2 Proof by Contrapositive

$$(\forall n \in \mathbb{N})(n^2 \text{ even} \Rightarrow n \text{ even})$$

Proof
Since it is more natural to start with n rather than n^2, we prove this result by proving its contrapositive, where we assume the conclusion false and prove

---

8 Someone once remarked, we plant the lemmas, grow the theorems, and harvest the corollaries.

the assumption false. That is, we assume n odd and prove n^2 odd. But this is the result we proved in Theorem 1. Hence, the proof is complete. ∎

Important Note   Note the important equivalences of the conditional

$$\sim(P \Rightarrow Q) \equiv \sim(\sim P \vee Q) \equiv (P \wedge \sim Q).$$

Theorem 3   Proof by Contrapositive
If $a, b \in \mathbb{N}$, then

$$(a + b \geq 15) \Rightarrow [(a \geq 8) \vee (b \geq 8)].$$

Proof
The contrapositive form of this implication is

$$\sim[(a \geq 8) \vee (b \geq 8)] \Rightarrow \sim(a + b \geq 15)$$

Using one of De Morgan's laws, this implication becomes

$$[(a < 8) \wedge (b < 8)] \Rightarrow (a + b < 15)$$

But $a < 8$ and $b < 8$ implies $a \leq 7$ and $b \leq 7$, which gives the desired result $a + b \leq 14 < 15$. ∎

There is no magic bullet for proving theorems. Sometimes the result to be proven provides the starting point, and the theorem can be proven by working backwards. The proof of the following theorem shows how this technique is carried out.

Theorem 4   Backwards Proof
Prove for any two positive real numbers x and y, the algebraic mean is greater than or equal to the geometric mean. That is

$$\frac{x + y}{2} \geq \sqrt{xy}$$

Proof
This result is a prime example of "working backwards." We begin by rewriting the conclusion as

$$x + y \geq 2\sqrt{xy}$$

or

$$x + y - 2\sqrt{x}\sqrt{y} \geq 0$$

and factoring gives

$$\left(\sqrt{x}-\sqrt{y}\right)^2 \geq 0.$$

But this statement is true so to prove the desired result, we simply carry out the steps in reverse order. ∎

Important Note A **conjecture** is a mathematical statement which is believed to be true but has not been proven. Once proven to be true, it is called a theorem.[9] The **Goldbach conjecture** is one of the oldest unsolved problems in mathematics, which claims that every even integer greater than 2 can be written as the sum of two (not necessarily distinct) primes. For example

$$4 = 2 + 2, \quad 6 = 3 + 3, \quad 8 = 3 + 5, \quad 10 = 3 + 7 = 5 + 5, \ldots$$

and so on. There are many conjectures in number theory, including Legendre's Conjecture (unsolved as of 2019) that claims there exists a prime number between n^2 and $(n+1)^2$ for all natural numbers n (check out some yourself). The Poincare Conjecture,[10] was proved by Russian mathematician Grigori Perelman in 2002, and should be retired from conjecture status and be called Poincare's Theorem, but will probably keep its original name as a conjecture.

Important Note *Not* Mathematical Proofs:

- The proof is so easy we will skip it.
- Do not be stupid, of course, it's true!
- It's true because I *said* it's true!
- Scribble, scribble QED
- God let it be true!
- I have this gut feeling.
- I did it last night.
- It works for 2 and 3.
- I *define* it to be true!
- Sounds good to me.
- All in favor ...?
- My boyfriend said it's true.

Although we suspect readers of this book would never be guilty of applying one of the aforementioned proof techniques, there is one habit, however, almost equally heinous that is often used, and that is the overuse of the word "obvious." If something is "obvious," then go ahead and prove it.

---

9 Sometimes, however, a conjecture is also proven false, as in the case of the Polya Conjecture. Interested readers can read about this (false) conjecture online.
10 Every simply-connected, closed 3-manifold is homeomorphic to the 3-sphere.

Fundamental Theorem of Arithmetic   The **Fundamental Theorem of Arithmetic** states that any natural number greater than 1 can be uniquely factored as the product of prime numbers. For example

$$21 = 3 \times 7$$
$$40 = 2^3 \times 5$$
$$180 = 2^2 \times 3^2 \times 5$$
$$235 = 5 \times 47$$
$$453,569,345 = 5 \times 773 \times 117,353$$

You might think of the prime numbers as building blocks for the natural numbers.

The Prime Number Theorem (PNT): We now prove that there are an infinite number of prime numbers, a proof that goes back to 300 BCE to the Greek mathematician Euclid of Alexandra (present-day Egypt). Although Euclid proved the theorem by *reductio ad absurdum,* the German mathematician Dirichet (DEER-a-shlay) later developed a direct proof using analytic function theory.

Theorem 5   Infinite Number of Prime Numbers
There are an infinite number of prime numbers.[11]

Proof
We assume that there are only a finite number of prime numbers, which we enumerate in the increasing order $p_1 = 2, p_2 = 3, p_3 = 5, \dots, p_n$, where p_n is the largest prime number. We now construct the product of these primes plus 1, or

$$M = p_1, p_2, \dots, p_n + 1$$

We now show one of the assumed prime numbers $p_1, p_2, \dots, p_n$ both divides M and does *not* divide M, which is a contradiction, and contradicts the assumption that there are only finitely many primes.

Since M is larger than the largest prime number p_n, it must be a composite number, and thus is divisible by one of our finite number of primes,[12] say p_m (with $1 \le m \le n$). But when we divide M by p_m, we get a remainder of 1 as seen by

$$\frac{p_1, p_2, \dots, p_m, \dots, p_n + 1}{p_m} = p_1, p_2, \dots, p_n + \frac{1}{p_m}$$

---

11 You might ask what is the assumption in this theorem? A more explicit statement of the theorem might be if a prime number is a natural number divisible only by 1 and itself, then there are an infinite number of such numbers.
12 We have used the Fundamental Theorem of Arithmetic here, which states that every natural number greater than 1 is either a prime number or can be written as the product of primes.

Hence, p_m does not divide M, thus showing that p_m divides M and does not divide M, contradicting the assumption of a finite number of primes. Hence, we are left to conclude that there are an infinite number of prime numbers. ∎

Euler's Proof of the PNT: Another proof that there are an infinite number of prime numbers is obtained from the identity

$$\left(\frac{2}{2-1}\right)\left(\frac{3}{3-1}\right)\left(\frac{5}{5-1}\right)\cdots\left(\frac{p}{p-1}\right)\cdots = \sum_{n=1}^{\infty}\frac{1}{n} = \infty$$

proved by **Leonard Euler** (1707–1783), which uses the fact that since the harmonic series on the right diverges, then the product on the left, taken over prime numbers p = 2, 3, 5, 7, ... is also infinite, giving rise to an infinite number of prime numbers.

The Prime Number Theorem The next question to ask after knowing there are an infinite number of prime numbers is what proportion of the natural numbers are prime? It was observed by Karl Friedrich Gauss (1777–1856) and A. M. Legendre that although prime numbers do not occur in any regularity, the proportion of prime numbers among the first n natural numbers is approximately $1/\ln(n)$. For example, of the first million numbers the fraction of primes is $1/\ln(1,000,000) \doteq 0.07$. It took almost a hundred years after Gauss and Legendre made their conjecture for French and Belgian mathematicians Jacques Hadamard (1865–1963) and de la Vallee Poussin (1866–1962) to simultaneously and independently prove the **Prime Number Theorem** correct.

Important Note Indirect arguments or proofs by contradiction are not foreign to our psyche. When a parent tells a child not to do something, the child thinks the contrapositive, "If I do it, what will they do to me?"

We now come to one of the most famous theorem of antiquity, the proof of the irrationality of $\sqrt{2}$.

Theorem 6 $\sqrt{2}$ Is Irrational: Proof by Contradiction

$\sqrt{2}$ is an irrational number.[13]

Proof
Assume the contrary, which says $\sqrt{2}$ is rational. Hence, we can write $\sqrt{2} = p/q$, where $p, q \in \mathbb{Z}$ with $q \neq 0$ and p and q have no common factors. If they have common factors, we cancel the common factors.

---

13 It may not be obvious, but this theorem is of the form $P \Rightarrow Q$. The theorem simply says $\sqrt{2}$ is irrational, so where is the "if" in the theorem? The "if" would define the square root. We generally do not say it here in the theorem since it is understood by everyone.

We will now show that even after p/q is reduced to lowest form, both p and q must be even, thus contradicting the fact that we have reduced p/q to lowest form.

We begin by squaring both sides of the above equation, getting

$$\left(\sqrt{2}\right)^2 = \left(\frac{p}{q}\right)^2 \Rightarrow 2 = \frac{p^2}{q^2} \Rightarrow p^2 = 2q^2$$

which means p^2 is even, but we have seen in Theorem 2 that if p^2 is even so is p. But if p is even, it can be written as $p = 2k$, which implies $p^2 = 4k^2$, where k is an integer, hence $p^2 = 2q^2$ can be written as

$$4k^2 = 2q^2 \Rightarrow 2k^2 = q^2$$

which in turn means q^2 is even and thus q is even. Thus, we have shown that both p and q are even, which contradicts the fact we reduced p/q to lowest terms. Hence, our assumption of the rationality $\sqrt{2} = p/q$ leads to a contradiction, and hence, by *reductio ad absurdum*, we conclude that $\sqrt{2}$ is irrational. ∎

Ugh   There is the story about a student who was asked to prove a given theorem or find a counterexample. The student asked the teacher if extra credit was given for doing both.

Historical Note   The reader should know the story of Hippasus, supposedly a Pythagorean who first proved that $\sqrt{2}$ is irrational. The Pythagoreans were a religious sect that flourished in Samos, Greece, around 500 BCE and founded by the Greek philosopher and mathematician Pythagoras. They believed that all numbers were either natural numbers 1, 2, 3, ... or fractions. So when Hippasus proved $\sqrt{2}$ was *irrational*, which according to legend was made at sea, the Pythagoreans considered the proof an act of heresy and threw him overboard. So much for making one of the greatest mathematical discoveries of all time.

Important Note   Many important theorems in mathematics are proven by contradiction. Three of the most famous are

- Cantor's seminal theorem: the real numbers are uncountable.
- Euclid's proof: there are an infinite number of primes.
- Pythagoras' proof: $\sqrt{2}$ is irrational.

> **Tips for Proving Theorems** Here are a few guidelines that might be useful for proving theorems.
>
> - Draw figures to visualize the concepts.
> - Construct examples that illustrate the general principles.
> - Try working backwards
> - For "⇒" theorems, ask if the converse "⇐" is true.
> - Modify the theorem to make it easier.
> - Generalize, does the theorem hold in more general cases?
> - Did you actually *use* all the assumptions?

1.4.5 Necessary and Sufficient Conditions (NASC)

We are all familiar with the concept of a *necessary condition*. For example, air is *necessary* for human survival, since without it there *is* no life. But on the other hand, air is not *sufficient* for human life inasmuch as we also need food to survive. Thus, we would say air is a necessary condition for human life, but not a sufficient condition.

On the other hand, there are conditions that are *sufficient*, but not *necessary*. For example, being a resident of California is sufficient for being a resident of the United States, but not necessary. In other words, sufficient conditions are more restrictive than necessary conditions.

And finally, there are conditions that are both *necessary* and *sufficient*.[14] For example, a differential function has a maximum value at $x = a$ is a *necessary* and *sufficient* condition for the function to have a zero first derivative and negative second derivative at $x = a$.

Table 1.34 summarizes these ideas.

Table 1.34 Necessary and sufficient conditions.

(A is sufficient for B) $\equiv (A \Rightarrow B)$
(A is necessary for B) $\equiv (B \Rightarrow A)$
(A is necessary and sufficient for B) $\equiv (A \Leftrightarrow B)$

> **Historical Note** When Newton found the derivative (which he called the **flux-ion**) of x^2 (which he called the **fluent**), he arrived at the expression $2x + \Delta x$. The Δx was referred to as an "infinitely small" quantity and thus was omitted, giving the

---

14 Finally, there are conditions that are *neither* sufficient nor necessary. For example being the smartest student in the class is neither necessary nor sufficient for achieving the highest grade in the class.

derivative (or fluxion) of 2x. It was not until a hundred years later when mathematicians in the nineteenth century, such as Cauchy, Dedekind, Cantor, and Weierstrass, put mathematics on a more solid logical footing, and the old mathematical expressions like *infinitely small* were laid to rest, replaced by more precise "limits" and "$\varepsilon - \delta$ arguments."

The earlier discussion motivates logical statements of the form $P \Leftrightarrow Q$, which means "P is true if and only if Q is true," which is often stated as P is a necessary and sufficient condition for Q, due to the logical equivalence

$$P \Leftrightarrow Q \equiv (P \Rightarrow Q) \wedge (P \Leftarrow Q).$$

The methodology for proving theorems of this type is to prove both $P \Rightarrow Q$ and $Q \Rightarrow P$. The following theorem illustrates this idea:

Theorem 7 If and Only If
Let n be any natural number, then 3 divides $n^2 - 1 \Leftrightarrow$ 3 does not divide n.

Proof
($\Rightarrow$) First we prove

3 divides $n^2 - 1 \Rightarrow$ 3 does not divide n.

Since 3 is a prime number and divides $n^2 - 1 = (n-1)(n+1)$, it must divide either $n-1$ or $n+1$. If 3 divides $n-1$, it cannot divide n (it will have a remainder of 1), and if 3 divides $n+1$, it cannot divide n (it will have a remainder of 2). Hence, 3 does not divide n.

($\Leftarrow$) We now prove the other way that 3 divides $n^2 - 1 \Leftarrow$ 3 does not divide n: If 3 does not divide n, then we can write

$$\frac{n}{3} = q + \frac{r}{3}$$

or $n = 3q + r$, where the remainder r is either 1 or 2. If $r = 1$, then $n - 1 = 3q$, which means 3 divides $(n-1)(n+1) = n^2 - 1$. Finally, if $r = 2$, then $n - 2 = 3q$ or $n + 1 = 3q + 3 = 3(q+1)$, which also means 3 divides $n^2 - 1 = (n-1)(n+1)$. ∎

Who Has Proven the Stronger Theorem?   Jerry and Susan have each proven an important theorem with the same conclusion C, and each hopes to win a *Field's Medal*.[15] However, although their conclusions are the same, their hypotheses are different. Jerry has assumed a hypothesis J, so his theorem has the form $J \Rightarrow C$, whereas Susan has assumed a hypothesis S, so her theorem has the form $S \Rightarrow C$. In their battle to see who has the stronger theorem, or which

15 The Fields Medal is regarded as the "Nobel Prize" of mathematics, awarded every four years to one or more outstanding mathematicians under the age of 40.

theorem implies the other, Jerry makes the discovery that his hypothesis J is *sufficient* for Susan's hypothesis S. That is, $J \Rightarrow S$ and so he claims he has the stronger theorem. Is Jerry correct? The answer is no! Susan's weaker hypothesis means she has the stronger theorem as can be seen by the implication

$$(J \Rightarrow S) \Rightarrow [(S \Rightarrow C) \Rightarrow (J \Rightarrow C)]$$

The validity of this tautology is left to the reader. Hint: truth table.

The following Theorem 8 is best proved with the aid of the following Lemma 1.

Lemma 1   Every natural number n can be written in the form $n = s + 3m$, where s is the sum of the digits of n and m is some natural number. For example, $675 = 18 + 3(219)$.

Proof
Writing n as

$$n = a_k 10^k + a_{k-1} 10^{k-1} + \cdots + 10a_1 + a_0$$

and summing of its digits, getting $s = a_0 + a_1 + \cdots + a_k$, we compute the difference

$$n - s = \left(a_k 10^k + a_{k-1} 10^{k-1} + \cdots + 10a_1 + a_0\right) - \left(a_k + a_{k-1} + \cdots + a_0\right)$$
$$= 999...9a_k + \cdots + 99a_2 + 9a_1$$
$$= 3(333...3a_k + \cdots + 33a_2 + 3a_1)$$
$$= 3m$$

which proves the lemma.   ∎

We now use this lemma to prove the following interesting result.

Theorem 8   If and Only if

$$(\forall n \in \mathbb{N})[(3 \mid n) \Leftrightarrow (3 \mid \text{sum of the digits of } n)]$$

Proof
($\Rightarrow$) If 3 divides n, we can write $n = 3k$, where k is a natural number. Then using Lemma 1, we have $3k = s + 3m$ and solving for the sum s gives

$$s = 3k - 3m = 3(k - m)$$

which proves that the sum of the digits of n is divisible by 3.

($\Leftarrow$) Assuming 3 divides the sum of the digits of n, we write $s = 3k$, where k is a natural number. Appealing again to Lemma 1, we have

$$n = s + 3m = 3k + 3m = 3(k + m)$$

which proves n is divisible by 3.

■

Corollary: 3 divides 9031827540918.

Problems

1. **Direct Proof**
   Prove the following by a direct proof:
   a) The sum of two even integers is even.
   b) The sum of an even and an odd integer is odd.
   c) If a divides b, and b divides c, then a divides c.
   d) The product of two consecutive natural numbers plus the larger number is a perfect square.
   e) Every odd integer n greater than 1 can be written as the difference between two perfect squares. Give examples.
   f) If n is an even positive integer, then n is the difference of two positive integer squares if and only if $n = 4k$ for some integer $k > 1$.
   g) If a, b are real numbers, then $a^2 + b^2 \geq 2ab$.
   h) The sum of two rational numbers is rational.
   i) Let $p(x)$ be a polynomial, where E is the sum of the coefficients of the even powers, and O is the sum of the coefficients of the odd powers. Show that $E + O = p(1)$ and $E - O = p(-1)$, where by $p(1)$ and $p(-1)$ we mean $p(x)$ evaluated at $x = 1$ and $x = -1$, respectively.

2. **Divisibility by 4**
   Show that a natural number is divisible by 4 if and only if its last two digits are divisible by 4. For example both 256 and 56 are divisible by 4. The same holds for 64 and 34,595,678,206,754,964.

3. **Divisibility by 3**
   Show that a natural number is divisible by 3 if and only if the sum of its digits is divisible by 3. For example, the number 9,003,186 is divisible by 3.

4. **Proof by Contradiction**
   Prove the following by contradiction:
   a) If n is an integer and $5n + 2$ is even, then n is even.
   b) If I is an irrational number and R is a rational number, then $I + R$ is irrational.

5. **Divisibility Problem**
   Prove the following for any natural number n.
   a) $5 \mid n^2 \Rightarrow 5 \mid n$.
   b) $9 \mid n$ if $\Leftrightarrow 9$ divides the sum of the digits of n.
   c) $3n + 1$ even $\Rightarrow n$ odd.

6. **Counterexamples**
   A counterexample[16] is an exception to a rule. Counterexamples are used in mathematics to probe the boundaries of a result. Find counterexamples for the following faulty statement and tell how you could add new hypothesis to make the claim a valid theorem.
   a) If $a > b$, then $|a| > |b|$.
   b) If $(a - b)^2 = (m - n)^2$, then $a - b = m - n$.
   c) If x and y are real numbers, then $\sqrt{xy} = \sqrt{x}\sqrt{y}$.
   d) If f is a continuous function defined on $[a, b]$, then there exists a $c \in (a, b)$ such that

   $$f'(c) = \frac{f(b) - f(a)}{b - a}$$

7. **Valid Proof: Invalid Conclusion**
   If the assumption of a theorem is false, then the conclusion can be false even if the proof of the theorem is valid. For example if you assume there is a largest positive integer N, it is possible to prove $N = 1$. Can you find such a proof?

8. **Comparing Theorems**
   Verify the statement

   $$(J \Rightarrow S) \Rightarrow [(S \Rightarrow C) \Rightarrow (J \Rightarrow C)]$$

   showing if $J \Rightarrow S$, then $J \Rightarrow C$ is the weaker theorem.

9. **Comparing Theorems**
   Suppose a hypothesis J implies two different results C_1 and C_2, i.e. $J \Rightarrow C_1$ and $J \Rightarrow C_2$, and that C_2 is the weaker of the results, i.e. $C_1 \Rightarrow C_2$. Show that $J \Rightarrow C_1$ is the stronger theorem, or

   $$(C_1 \Rightarrow C_2) \Rightarrow [(J \Rightarrow C_1) \Rightarrow (J \Rightarrow C_2)]$$

---

16 A nice reference book for any mathematician is *Counterexamples in Mathematics* by Bernard Geldbaum and John Olmsted, Springer-Verlag (1990).

10. **Hmmmmmmmmm**

Infinite decimal expansions are sometimes needed to represent certain fractions. Prove that $1/3 = 0.333. \ldots$ by writing the decimal form $0.333\ldots$ as the infinite series

$$0.333\ldots = \frac{3}{10} + \frac{3}{100} + \frac{3}{1000} + \cdots$$

and showing the sum of this series is $1/3$.

11. **Another Irrational Number**

Prove that $\log_{10} 3$ is irrational.

12. **Not Proofs**

The following statements are not considered as valid proofs by most of the mathematicians. Maybe the reader knows of a few others.

a) Proof by obviousness: *Too trivial to prove.*
b) Proof by plausibility: *It sounds good, so it must be true.*
c) Proof by intimidation: *Do not be stupid; of course, it's true!*
d) Proof by definition: *I define it to be true.*
e) Proof by tautology: *It's true because it's true.*
f) Proof by majority rule: *Everyone I know says it's true.*
g) Proof by divine words: *And the Lord said, "Let it be true," and it was true.*
h) Proof by generalization: *It works for me, that's enough.*
i) Proof by hope: *Please, let it be true.*
j) Proof by intuition: *I got this gut feeling.*

13. **Just a Little Common Sense**

You are given a column of 100 ten-digit numbers by adding them, you get a sum of 2,437,507,464,567. Is your answer correct?

14. **Syllogisms**

The Greek philosopher Plato is recognized as the first person associated with the concept a logical argument. His arguments took the form of two premises followed by a conclusion. This basic logical form is called a **syllogism**, the most famous being the "Socrates syllogism":

- **First premise:** *All men are mortal.*
- **Second premise:** *Socrates is a man.*
- **Conclusion:** *Therefore, Socrates is mortal.*

which has the general form

- **First premise:** $M \sim R$
- **Second premise:** $S \sim M$
- **Conclusion:** $S \sim R$

where
M = "being a man"
S = "being Socrates"
R = "being mortal"
Plato classified each premise and conclusion as one of the four basic types:[17]

- **E** Every A is B (example: Every dog has a tail.)
- **S** Some A is B (example: Some dogs have black hair.)
- **N** No A is B (example: No dog has orange hair.)
- **SN** Some A are not B (example: Some dogs are not poodles.)

which means there are a total of $4 \times 4 \times 4 = 64$ possible syllogisms, some logically true, some false. For example, a syllogism of type NEN means the first premise is of basic type N, the second premise of type E, and the conclusion of type N. Which of the following syllogism types are valid and which are invalid? Draw Venn diagrams to support your argument.

a) NEN   Ans: true
b) ESS   Ans: true
c) NSSN   Ans: true
d) SSS   Ans: false
e) EES   Ans: false
f) SES   Ans: false

15. **Euler's Totient Function**
Euler's totient function, denoted by $\phi(n)$, gives the number of natural numbers less than a given number n, including 1, that are relatively prime to n, where two numbers are relatively prime if their greatest common divisor is 1.
For example $\phi(p) = p - 1$ for any prime number since $1, 2, 3, \ldots, p - 1$ are all relatively prime with p. On the other hand, $\phi(12) = 4$ since 1, 5, 7, and 11 are the numbers less than 12 relatively prime with 12. Prove that for a power of a prime number p^k, $k = 1, 2, \ldots$ the Euler totient function is $\phi(p^k) = p^{k-1}(p - 1)$ by proving the following results.
a) Find the number of natural numbers strictly between 1 and p^k that are not relatively prime with p^k, i.e. divide p^k.
b) Subtract the result a) from $p^k - 1$ to obtain $\phi(p^k)$.

16. **Pick's Amazing Formula**
In 1899, an Austrian mathematician, Georg Pick, devised a fascinating formula for finding the area A inside a simple polygon whose vertices lie on grid points (m, n) in the plane, where m and n are integers. The formula he came up with was

---

17 We use A and B to denote the properties S, P, and M.

$$A = \frac{B}{2} + I - 1$$

where B is the number of vertices that lie on the boundary of the polygon, and I is the number of vertices that lie interior to the polygon.

a) Verify that Pick's formula yields an area of 1 for a simple square bounded by four adjacent vertices.
b) Use Pick's formula to find the area inside the polygons in Figure 1.9.
c) Prove that Pick's formula yields the correct area of mn inside a rectangle with m rows and n columns. We draw a $m = 8$ by $n = 11$ rectangle in Figure 1.10 for illustration

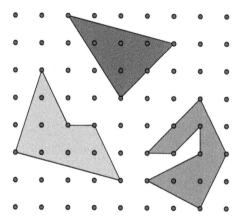

Figure 1.9 Applying Pick's formula.

17. **Twin Prime Conjecture**

Twin primes are pairs of prime numbers of the form $(p, p + 2)$ of which $(17, 19)$ and $(197, 199)$ are examples. It is not known that if there are

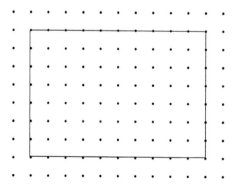

Figure 1.10 The area of mn inside an $m \times n$ rectangle.

an infinite number of such pairs, although it is currently known that there are an infinite number of prime pairs that are at most 246 apart. This leads one to the question about *triple* primes of the form $(p, p+2, p+4)$ of which $(3, 5, 7)$ is an example. Prove this is the only triple prime sequence of this form.

18. **Internet Research**

There is a wealth of information related to topics introduced in this section just waiting for curious minds. Try aiming your favorite search engine toward *types of mathematical proofs, proof by contradiction, reductio ad absurdum*, and *prime number theorem*.

1.5

Proofs in Predicate Logic

> **Purpose of Section** We continue our discussion of mathematical proofs by focusing on theorems of the form $(\forall x \in U)P(x)$, $(\exists x \in U)P(x)$ as well as theorems with multiple quantifiers, such as $(\forall x \in U)(\exists y \in V)\, P(x, y)$. To solve theorems with a quantified variable, like $\exists x, \forall x$, it is important we know how to negate such theorems in order to prove them by contradiction.

1.5.1 Introduction

Most theorems in mathematics begin with quantifiers such as "for all" or "there exists," or some variation of these such as "for all x, there exists a y," although often not stated explicitly inasmuch the quantification is understood. Euclid's famous theorem about prime numbers is often stated simply as "there are an infinite number of prime numbers," which begs the question, where is the "if" in the theorem? The answer is that the assumption is the definition of a prime number. In this section, we always include the all-important quantifiers and prove theorems stated in the language of predicate logic.

1.5.2 Proofs Involving Quantifiers

Since many theorems in mathematics are stated in the form $(\forall x \in U)P(x)$ or $(\exists x \in U)P(x)$, the question we ask is how do we go about proving them. We begin by proving theorems that include a single universal quantifier.

> **Universal Quantifier** To prove a theorem of the form
>
> $$(\forall x \in U)\, P(x)$$
>
> one selects an *arbitrary* $x \in U$, then proves that the proposition $P(x)$ is true.

Advanced Mathematics: A Transitional Reference, First Edition. Stanley J. Farlow.
© 2020 John Wiley & Sons, Inc. Published 2020 by John Wiley & Sons, Inc.
Companion website: www.wiley.com/go/farlow/advanced-mathematics

Theorem 1   Universal Direct Proof
All integers divisible by 6 are even.

Proof
Stated in the language of predicate logic, we have

$$(\forall n \in \mathbb{Z})(6|n \Rightarrow n \text{ is even})$$

Since n is assumed divisible by 6, there exists an integer m that satisfies $n = 6m$, which can be rewritten as $n = 2(3m) = 2k$, where $k = 3m$. Hence, n is an even integer, which completes the proof. We could streamline this argument symbolically as

$$(\forall n \in \mathbb{Z})[(6 \mid n) \Rightarrow (\exists m \in \mathbb{Z})(n = 6m = 2(3m))]$$

∎

Note: We could also prove Theorem 1 by writing the theorem in a contrapositive form as

$$(\forall n \in \mathbb{Z})[n \text{ is odd} \Rightarrow (6 \nmid n)]$$

where $6 \nmid n$ means 6 does *not* divide n.

Counterexample   To prove a theorem that contains a universal quantifier false, one only needs to find a counterexample, i.e. an example where the theorem is invalid. At one time, there was a conjecture in number theory by Fermat that stated

$$(\forall n \in \mathbb{N})(2^{2^n} + 1 \text{ is a prime number})$$

until Leonard Euler proved the conjecture was false with the embarrassing observation that for $n = 5$:

$$2^{2^5} + 1 = 2^{32} + 1 = 4\,294\,967\,297 = 641 \times 6\,700\,417$$

In other words, Euler proved the *negation* of the theorem

$$(\exists n \in \mathbb{N})(2^{2^n} \text{ is not prime})$$

Proofs involving the existential quantifier $\exists$ are often easier than ones involving the universal quantifier $\forall$ since it is only necessary to find *one* element in the universe that satisfies the given proposition.

> **Existential Sentences** To prove a theorem of the form
> $(\exists x \in U)\, P(x)$
> find one (or more) element $x \in U$ that satisfies $P(x)$.

Simply because we only have to find one object that satisfies the given condition, does not automatically mean the theorem is easy to prove. There are many unsolved conjectures related to finding just one thing. For example, a perfect number is a natural number that is equal to the sum of its proper divisors,[1] such as $6 = 1 + 2 + 3$, $28 = 1 + 2 + 4 + 7 + 14$, ... Currently, it is unknown if there are any *odd* perfect numbers. The largest perfect number currently known is

$$\left(2^{74\,207\,280}\right)\left(2^{7\,420\,728} - 1\right)$$

and contains $44\,677\,235$ digits.

Theorem 2   Proof by Demonstration
Show that there exists an even prime number.

Proof
One should never use the word trivial in mathematics, but in this case it is. The number 2 is both even and prime.

> **Intuitionism** In the philosophy of mathematics, there is a school of thought, called *Intuitionism* introduced by the Dutch mathematician L. E. J. Brouwer (1881–1961) with advocates like the German mathematician Leopold Kronecker (1823–1891). Intuitionists (or constructionists) feel that mathematics is purely the result of the constructive mental activity of humans, contrary to the belief that mathematical concepts exist outside of human existence. Kronecker once said, *God made the integers, all else is the work of man.* In the late 1800s and early 1900s, several intuitionists felt that new mathematical theories of the time such as Cantor's infinite sets, imaginary numbers, proof by contradiction and noneuclidean geometries were taking mathematics down the road to mysticism. Even the great French mathematician Henri Poincare (1854–1912) felt that Cantor's theory of infinite sets and transfinite arithmetic should be excluded from mathematics. Today, the intuitionist school of mathematics is not held in favor among many mathematicians.

---

1 The proper divisors of a number are the numbers that divide the number other than the number itself. For example the proper divisors of 6 are 1, 2, and 3. The proper divisors of 28 are 1, 2, 4, 7, and 14.

Theorem 3   Use of the Law of the Excluded Middle

There exists irrational numbers a, b such that a^b is rational.[2] In the language of predicate logic, if we call I the set of irrational numbers, we would write

$$(\exists a, b \in I)\left(a^b \in \mathbb{Q}\right)$$

Proof

One hesitates to call a proof "cute," but this is one strange theorem to say the least. Consider the number

$$\sqrt{2}^{\sqrt{2}}$$

which is either rational or irrational. We consider each case.

Case 1: If $\sqrt{2}^{\sqrt{2}}$ is *rational*, then the proof is complete since we simply pick the irrational numbers as $a = b = \sqrt{2}$.

Case 2: If $\sqrt{2}^{\sqrt{2}}$ is *irrational*, then since

$$\left(\sqrt{2}^{\sqrt{2}}\right)^{\sqrt{2}} = \left(\sqrt{2}\right)^2 = 2$$

we pick the two irrational numbers

$$a = \sqrt{2}^{\sqrt{2}}, \quad b = \sqrt{2}$$

∎

Note that we have not determined which of the two powers

$$\sqrt{2}^{\sqrt{2}} \text{ or } \left(\sqrt{2}^{\sqrt{2}}\right)^{\sqrt{2}}$$

is rational, but we have shown *one* of them is rational.

1.5.3   Proofs by Contradiction for Quantifiers

Proofs by contradiction are important tools in a mathematician's toolkit.

Proof of $(\forall x)P(x)$ by Contradiction   To prove a theorem of the form $(\forall x)P(x)$ by contradiction, assume the theorem false

$$\sim (\forall x)P(x) \equiv (\exists x)[\sim P(x)]$$

and arrive at a contradiction, thus proving the theorem cannot be negated.

---

2 The *Law of the Excluded Middle* states the accepted logical principle that every proposition is either true or false, unlike the weatherperson who says, rain, no rain, or maybe.

Theorem 4   Proof by Contradiction

If m, n are integers, then $14m + 21n \neq 1$, or in the language of predicate logic:

$$(\forall m, n \in \mathbb{Z})(14m + 21n \neq 1)$$

Proof

Assume the proposition is false, so taking the negative of the above proposition, we have

$$(\exists m,\ n \in \mathbb{Z})(14m + 21n = 1)$$

But this equation obviously[3] cannot hold since seven divides the left side of the equation but not the right side. Hence, the denial of the theorem is false, so the theorem is true. ∎

Important Note   If a theorem is true for thousands of cases, that does not prove the theorem. The equation

$$(n-1)(n-2)\cdots(n-1\,000\,000) = 0$$

has a solution for $n = 1, 2, \ldots, 1\,000\,000$, but has no solution for $n = 1\,000\,001$.

Historical Note   Frege's 1879 seminal work *Begriffsschrift* ("Conceptual Notation") marked the beginning of a new era in logic, which allowed the quantification of mathematical variables, just in time for the more precise **arithmetization of analysis** of calculus, being carried out in the late 1800s by mathematicians like the German Karl Weierstrass (1815–1897) and others.

Some theorems contain both universal and existential quantifiers. The following theorem is an example.

1.5.4   Unending Interesting Properties of Numbers

Is the number $10\,008\,036\,000\,540$ a multiple of 9? The answer is yes, and if you know a certain theorem, you can answer the question in about five seconds. To prove this result, one must make the observation that $10 = 9 + 1$. Here is the theorem stated in the language of predicate logic.

---

3 There is the story about a professor who pointed to an ominous looking equation on the board and after scratching his head for 15 minutes, said at last, "Aha, it's obvious!"

Theorem 5   Interesting Property of Numbers (Casting Out 9s)

$$(\forall n \in \mathbb{N})(9 \mid n \Leftrightarrow 9 \mid \text{sum of the digits of } n)$$

Semi-Proof

Sometimes proofs are easy conceptually, but the arithmetic or algebra of the theorem becomes messy, this problem being such an example. We prove the result for a two-digit number and let you prove it for a three-digit number. See Problem 9.

($\Rightarrow$) We assume $9 \mid d_1 d_2$, which implies $d_1 d_2 = 9k$ for some $k \in \mathbb{N}$, which further implies:

> starting with $d_1 d_2 = 9k$
>
> hence $10 \times d_1 + d_2 = 9k$
>
> hence $(9 + 1)d_1 + d_2 = 9k$
>
> hence $d_1 + d_2 = 9k - 9d_1 = 9(k - d_1) = 9k_1$ where $k_1 = k - d_1 \in \mathbb{N}$

Hence, the sum of the digits, $d_1 + d_2$, is a multiple of 9.

($\Leftarrow$) Assuming the sum of the digits a multiple of 9, we can write

> starting with $d_1 + d_2 = 9k$
>
> hence $d_1 + d_2 = 9k + (9d_1 - 9d_1)$
>
> hence $10d_1 + d_2 = 9k + 9d_1 = 9(k + d_1)$
>
> hence $d_1 d_2 = 9k_1$

where $k_1 = k + d_1$, which proves the result.   ∎

In general, to determine if a large number is a multiple of 9, one sums the digits to get a new number. If it is not immediately known if the new number is a multiple of 9, sum the digits again, and again. If the end product of all this is 9, then the original number is divisible by 9, otherwise no. Try a few numbers yourself.

It is also possible to prove theorems involving the existential quantifier by contradiction.

> **Proof of ($\exists x$) P(x) by Contradiction**   To prove a proposition of the form $(\exists x)P(x)$ by contradiction, assume the proposition false, or
>
> $$\sim (\exists x)\, P(x) \equiv (\forall x)[\sim P(x)]$$
>
> and then reach some kind of contradiction.
>
> Another proposition involving the existential quantifier $\exists$ is of the form $(\sim \exists x)P(x)$. To prove a proposition of this form by contradiction, assume the contrary, i.e.$(\exists x)$ $P(x)$ and then reach a contradiction.

Theorem 6   No Largest Even Integer

There is no largest even integer.

Proof

Assume there is a largest even integer we call N, which we can write $N = 2k$ for some $k \in \mathbb{Z}$. Now consider $N + 2$, which we can write as

$$N + 2 = 2k + 2 = 2(k + 1)$$

But this says $N + 2$ is an even integer greater than N, which contradicts the assumption that N is the largest even integer. Hence, we cannot claim there is a largest even integer. ∎

Historical Note   There are theorems and then there are *theorems*. In the *Classification Theorem for Simple Groups* (known lovingly at the *enormous* theorem). The proof required the work of hundreds of mathematicians and consists of an aggregate of hundreds of papers. If the theorem were to be written out, it is estimated it would take between 10 000 and 15 000 pages.

1.5.5   Unique Existential Quantification ∃!

A special type of existential quantifier is the **unique existential quantification**

Proving Unique Existential Theorems   A theorem of the form

$$(\exists! x \in U)\, P(x)$$

with an *exclamation point* ! after the existential quantifier ∃ states *there exists a unique element x such that P(x) is true*, the emphases being on the word "unique." To prove a theorem of this form, we must show P(x) is true for *exactly one* element of the given universe U. A common strategy is to first show P(x) is true for *some* x ∈ U, and then if P(x) is true for *another* element y ∈ U, then x = y.

The concept of uniqueness is important in mathematics. For many problems, the first step is to first show existence, and the second step is to show uniqueness.

Theorem 7   Uniqueness: Diophantine Equation

There exist unique natural numbers m and n that satisfy

$$m^2 - n^2 - 12$$

or $(\exists!\, m \in \mathbb{N})(\exists!\, n \in \mathbb{N})(m^2 - n^2 = 12).$

Proof

An equation that allows only integer solutions is called a **Diophantine equation**. We begin by factoring

$$m^2 - n^2 = (m + n)(m - n) = 12$$

and note that the difference between the two factors $(m + n)(m - n)$ is $2n$, which means that both factors must be even or both must be odd. But the only factor of 12 that meets this requirement is $12 = 2 \times 6$. Hence, we are left with

$$m + n = 6$$

$$m - n = 2$$

which only has the solution $m = 4$, $n = 2$. ∎

Historical Note   In the late 1800s and early 1900s, there was a shift in the philosophy of mathematics, from thinking that logic was simply the tool for mathematics to thinking that logic was the foundation or precursor of mathematical thought. This thesis, called the "logistic thesis" (or "Frege–Russell thesis"), contends that mathematics is an extension of logic, as described in Russell and Whitehead's seminal work, *Principia Mathematica*. For others, like Giuseppe Peano, symbolic logic is only a tool for mathematics, which is the philosophy of many mathematicians today.

Historical Note   The American logician Charles Saunders Pierce (pronounced "purse") introduced **second-order logic**, which in addition to quantifying variables like x, y, ... also quantifies functions and entire sets of variables. For most mathematics, first-order logic is adequate. Pierce also developed first-order logic, but Frege carried out his research earlier and is generally given credit for its development. However, it was Pierce who coined the term "first-order" logic.

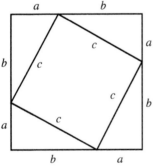

Figure 1.11 Visual proof of the Pythagorean theorem.

Theorem 8   Proof by Pictures

Does the drawing in Figure 1.11 constitute a "legitimate" proof of the Pythagorean theorem $a^2 + b^2 = c^2$, with legs having lengths a, b and a hypotenuse having length c? Some people say yes, others say no, but the best way to think about "visual" proofs is that they provide an idea that can be turned into a valid logical proofs.

Proof by Picture

Setting the area of the large square equal to the sum of the areas of the four triangles plus the area of the smaller square yields.

$$(a + b)^2 = 2ab + c^2 \Rightarrow a^2 + b^2 = c^2$$ ∎

> **Important Note** A great mathematical proof is one that is distinguished by beauty and economy. There are some proofs that get the job done but do not lift ones' intellectual spirit. On the other hand, some proofs overwhelm one with creative and novel insights. You might grade the proofs in this section according to those principles.

Problems

1. **True or False**
   Which of the following are true?
   a) $(\forall x \in \mathbb{R})(x^2 + x + 1 > 0)$
   b) $(\forall x \in \mathbb{R})[x^2 > 0 \vee x^2 < 0]$
   c) $(\forall x \in \mathbb{Z})(x^2 > x)$
   d) $(\forall x \in \mathbb{R})(\exists y \in \mathbb{R})(y = \sin x)$
   e) $(\forall x \in \mathbb{R})(\exists y \in \mathbb{R})(y = \tan x)$
   f) $(\exists x \in \mathbb{R})(\exists y \in \mathbb{R})(y = \sin x)$
   g) $(\exists x, y \in \mathbb{N})(\exists n > 2)(x^n + y^n = 1)$
   h) $(\exists x \in \mathbb{R})(\forall a, b, c \in \mathbb{R})(ax^2 + bx + c = 0)$
   i) $(\exists x \in \mathbb{C})(\forall a, b, c \in \mathbb{R})(ax^2 + bx + c = 0)$
   j) $(\forall \varepsilon > 0)(\exists N \in \mathbb{N})(\forall n > N)\left(\frac{1}{n} < \varepsilon\right)$
   k) $(\forall \varepsilon > 0)(\exists \delta > 0)(|x - 2| < \delta \Rightarrow |x^2 - 4| < \varepsilon)$

2. **Predicate Logic Form**
   Write the following theorems in the language of predicate logic.
   a) A number is divisible by 4 if and only if its last two digits are divisible by four.
   b) A natural number is divisible by 2^n if and only if its last n digits are divisible by 2^n.
   c) There exists irrational numbers a, b such that a^b is rational.
   d) $\sum_{k=1}^{n} k = \frac{n(n+1)}{2}$
   e) For positive real numbers a, b, we have $\sqrt{ab} \leq \frac{a+b}{2}$.
   f) If a, b are integers and $b \neq 0$, then there exists unique integers q, r such that $a = qb + r$ where $0 \leq r < |b|$.
   g) If p is a prime number that does not divide an integer a, then p divides $a^p - a$.
   h) The square of any natural number has each of its prime factor occurring an even number of times in its prime factorization.
   i) All prime numbers p greater than 2 are of the form either $p = 2n + 1$ or $p = 2n + 3$ for some natural number n.

j) Every even number greater than 4 can be expressed as the sum of two prime numbers.

k) **Euler's Conjecture:** There are no natural numbers a, b, c, d that satisfy $a^4 + b^4 + c^4 = d^4$.

3. Negation

Negate the following propositions and tell whether the original statement or the negation is true. Let x and y be real variables.

a) $(\exists x)(\forall y)(xy < 1)$

b) $(\forall x)(\forall y)(\exists z)(xyz = 1)$

c) $(\forall x)(\forall y)(\forall z)(\exists w)(x^2 + y^2 + z^2 + w^2 = 0)$

d) $\sim(\exists x)(\forall y)(x < y)$

e) $(\forall x)(\exists y)(xy < 1 \wedge xy > 1)$

f) $(\exists x)(\exists y)(xy = 0 \vee xy \neq 0)$

4. Counterexamples

All the following statements are wrong. Prove them wrong by finding a counterexample.[4]

a) All mathematicians make tons of money writing textbooks.

b) For all positive integers n, $n^2 + n + 41$ is prime.[5]

c) Every continuous function defined on the interval $(0,1)$ has a maximum and minimum value.

d) Every continuous function is differentiable.

e) If the terms of an infinite series approach zero, then the series converges.

f) The perimeter of a rectangle can never be an odd integer.

g) $(\forall x \in \mathbb{R})(\exists y \in \mathbb{R})(x = y^2)$

h) Every even natural number is the sum of two primes.

i) If m and n are positive integers such that m divides $n^2 - 1$, then m divides $n - 1$ or m divides $n + 1$.

j) If $\{f_n : n = 1, 2, ...\}$ is a sequence of continuous functions defined on $(0, 1)$ that converge to a function f, then f is continuous.

5. If and Only If Theorems

State each of the following theorems in the language of predicate logic. Take the function f as a real-valued function of a real variable.

a) A function f is even if and only if for every real number x we have $f(x) = f(-x)$.

b) A function f is odd if and only if for every real number x we have $f(x) = -f(-x)$.

---

4 A nice book outlining many counterexamples in mathematics is *Counterexamples in Mathematics* by Bernard R. Gelbaum and John Olmsted (Holden Day, Inc.), 1964.

5 A famous mathematician once said that he was once x^2 years old in the year x. Can you determine the year this person was born?

c) A function f is periodic if and only if there exists a real number p such that $f(x) = f(x + p)$ for all real numbers x.

d) A function is increasing if and only if for every real numbers x and y we have $x \le y \Rightarrow f(x) \le f(y)$.

e) A function f is continuous at x_0 if and only if for any $\varepsilon > 0$ there exists a $\delta > 0$ such that $|x - x_0| < \delta \Rightarrow |f(x) - f(x_0)| < \varepsilon$.

f) A function f is uniformly continuous on a set E if and only if for any $\varepsilon > 0$ there exists a $\delta > 0$ such that $|f(x) - f(y)| < \varepsilon$ for any x, y in E that satisfy $|x - y| < \delta$.

6. **Hard or Easy?**
Prove that if there is a real number $a \in \mathbb{R}$ that satisfies $a^3 + a + 1 = 0$, then there is a real number $b \in \mathbb{R}$ that satisfies $b^3 + b - 1 = 0$. This problem is either very hard or very easy, depending how it is approached. It is your job to determine which is true.

7. **Predicate Logic in Analysis**
The so-called $\varepsilon - \delta$ proofs in analysis were originated by the German mathematician Karl Weierstrass in the 1800s and involve inequalities and universal and existential quantifiers. They often start with $(\forall \varepsilon > 0)$ followed by $(\exists \delta > 0)$. The idea is that your adversary can pick $\varepsilon > 0$ as small as one pleases, but you have the advantage of picking the δ second. Of course, your choice of δ will most likely depend on ε.

a) Show that for every real number $\varepsilon > 0$ there exists a real number $\delta > 0$ such that $2\delta < \varepsilon$. In the language of predicate logic, prove $(\forall \varepsilon > 0)$ $(\exists \delta > 0)(2\delta < \varepsilon)$.

b) For every real number $\varepsilon > 0$, there exists an integer $N > 0$ such that for $n > N$ one has $1/n < \varepsilon$. In the language of predicate logic, prove $(\forall \varepsilon > 0)$ $(\exists N > 0)(\forall n > N)(1/n < \varepsilon)$.

c) Show that for every real number $\varepsilon > 0$, there exists a real number $\delta > 0$ such that if $|x| < \delta$ then $x^2 < \varepsilon$. In the language of predicate logic, prove $(\forall \varepsilon > 0)(\exists \delta > 0)(\forall x \in \mathbb{R})(|x| < \delta \Rightarrow x^2 < \varepsilon)$

d) Show that for every positive integer M there is a positive integer N such that $x > N \Rightarrow \sqrt{x} > M$.

In the language of predicate logic, this says

$$(\forall M > 0)(\exists N > 0)(\forall x \in \mathbb{R})(x > N \Rightarrow \sqrt{x} > M).$$

8. **Doing Mathematics**
Textbooks sometimes lead one to believe that theorems appear out of thin air for mathematicians to prove. If this were so, mathematics would be a purely deductive science, but in fact "doing mathematics" and mathematical research is as much an inductive science as deductive. Table 1.35 below

Table 1.35 Divisors of a few natural numbers.

n	$\tau(n)$	n	$\tau(n)$
1	1	13	2
2	2	14	4
3	2	15	4
4	3	16	5
5	2	17	2
6	4	18	6
7	2	19	2
8	4	20	6
9	3	21	4
10	4	22	4
11	2	23	2
12	6	24	8

lists the number of divisors $\tau(n)$ of n for the first 24 natural numbers. Looking at the table, can you think of any question to ask about the number of divisors of a natural number? A couple of candidates are

Theorem 1: $\tau(n)$ is odd if and only if n is a square, like 4, 9, 16, ...

Theorem 2: If m and n have no common factor, then $\tau(m)\tau(n) = \tau(mn)$.

9. **Casting Out 9s**
   Show that a number $d_1d_2d_3$ is a multiple of 9 if and only if $d_1 + d_2 + d_3$ is a multiple of 9.

10. **Internet Research**
    There is a wealth of information related to topics introduced in this section just waiting for curious minds. Try aiming your search engine toward *philosophy of intuitionism, Gottlob Frege, casting out nines, proofs in predicate logic,* and *proofs by pictures.*

1.6

Proof by Mathematical Induction

> **Purpose of Section**   To introduce the **Principle of Mathematical Induction,** both the weak and strong forms, and show how a certain class of theorems can be proven by this technique.

1.6.1   Introduction

An important technique for verifying proofs in combinatorics and number theory is the Principle of Mathematical Induction. The technique was used implicitly in Euclid's Elements in the "descent proof" that states that every natural number has a prime divisor. The term "Mathematical Induction" was first coined in 1828 by the English logician Augustus DeMorgan (1806–1871) in an article called *Induction.*

Mathematical Induction is generally not used in deriving new formulas, but is an effective tool to verify formulas and facts you suspect are true. That said, it is part of the repertoire of any mathematician.

The beauty of mathematical induction is it allows a theorem to be proven in cases when there are an infinite number of cases to explore without having to examine each case. Induction is the mathematical analogue of an infinite row of dominoes, where if you tip over the first domino, it tips over the next one, and so on, until they all are tipped over. The nice thing about induction is you do not have to prove that it works. It is an axiom[1] in the foundations of mathematics.

---

1 In 1889 Italian mathematician Giuseppe Peano (1858–1932) published a list of five axioms which define the natural numbers. Peano's fifth axiom is called the induction axiom, which states that *"any property which belongs to* 1 *and also to the successor of any number which has the property belongs to all numbers."*

Advanced Mathematics: A Transitional Reference, First Edition. Stanley J. Farlow.
© 2020 John Wiley & Sons, Inc. Published 2020 by John Wiley & Sons, Inc.
Companion website: www.wiley.com/go/farlow/advanced-mathematics

Mathematical induction provides a convenient way to establish that a statement is true for all natural numbers 1, 2, 3, ... The following statements are prime candidates for proof by mathematical induction.

- For all natural numbers n, $1 + 3 + 5 + \cdots + (2n - 1) = n^2$
- If a set A contains n elements, then the collection of subsets of A contains 2^n elements.
- $\dfrac{1}{2} \cdot \dfrac{3}{4} \cdot \dfrac{5}{6} \cdots \cdots \dfrac{2n-1}{2n} \leq \dfrac{1}{\sqrt{3n+1}}$ for all natural numbers n.

Here, then is how induction works.

1.6.2 Mathematical Induction

The Principle of Mathematical Induction is a method of proof for verifying that a proposition $P(n)$ is true for all natural numbers $n = 1, 2, \ldots$. The methodology for proving theorems by induction is as follows:

Methodology of Mathematical Induction

To verify that a proposition $P(n)$ holds for all natural numbers n, the **Principle of Mathematical Induction** consists of carrying out two steps.

- **Base Case:** Prove $P(1)$ is true.
- **Induction Step:** Assume $P(n)$ is true for an arbitrary n, then prove $P(n + 1)$ is true.

If the above two steps are proven, then the Principle of Mathematical Induction states that $P(n)$ is true for all natural numbers n. In the language of predicate logic, this states

$$\left. \begin{array}{l} P(1) \text{ is true} \\ (\forall n \in \mathbb{N})[P(n) \text{ true} \Rightarrow P(n+1) \text{ true}] \end{array} \right\} \Rightarrow (\forall n \in \mathbb{N})P(n) \text{ is true}$$

Important Note Do not confuse *mathematical induction* with *inductive reasoning* often associated with the natural sciences. Inductive reasoning in science is a scientific method whereby one *induces* general principles from specific observations. Mathematical induction is not the same thing: it is a *deductive* form of reasoning used to establish the validity of a proposition for all natural numbers.

Important Note There are many variations of the basic induction proof. For example, there is no reason the base case starts with $P(1)$. If the base case is replaced by proving $P(a)$, where "a" is any integer (positive or negative), mathematical induction would then conclude that $P(n)$ is true for all $n \geq a$. Also, if the induction step is replaced by $P(n) \Rightarrow P(n+2)$, then mathematical induction would conclude that $P(n)$ is true for $P(1), P(3), \ldots, P(2n+1), \ldots$

Theorem 1   Famous Identity

If n is a positive integer, then

$$1 + 2 + \cdots + n = \frac{n(n+1)}{2}.$$

Proof

Denote $P(n)$ as

$$P(n): \ 1 + 2 + \cdots + n = \frac{n(n+1)}{2}$$

Base Case: We can show that $P(1)$ is true since[2] $P(1)$ says

$$1 = \frac{1 \cdot (2)}{2}$$

Induction Step: Assuming $P(n)$ true for arbitrary n, this says

$$P(n): \ 1 + 2 + \cdots + n = \frac{n(n+1)}{2}$$

Now, adding $n+1$ to each side of this equation, we get

$$1 + 2 + \cdots + n + (n+1) = \frac{n(n+1)}{2} + (n+1)$$

$$= \frac{n(n+1) + 2(n+1)}{2}$$

$$= \frac{(n+1)(n+2)}{2}$$

---

2 The reader can verify that $P(2)$ and $P(3)$ are also true, but that is not relevant to proof by induction.

which is statement $P(n + 1)$. Hence, we have proven $P(n) \Rightarrow P(n + 1)$ and so by induction the result is proven. ∎

The result from Theorem 1 can also be found with pictures.

Visual Proof. The $n \times n$ array[3] drawn in Figure 1.12 has n^2 boxes where

- number of boxes containing xs is $1 + 2 + 3 + \cdots + n$
- number of unmarked boxes is $1 + 2 + \cdots + (n - 1)$,

Hence,

$$n^2 = (1 + 2 + \cdots + n) + (1 + 2 + \cdots + (n - 1))$$

Adding n to each side of this equation gives

$$n^2 + n = 2(1 + 2 + \cdots + n)$$

and solving for $1 + 2 + 3 + \cdots + n$ gives the desired result

$$1 + 2 + 3 + \cdots + n = \frac{n^2 + n}{2} = \frac{n(n + 1)}{2}$$

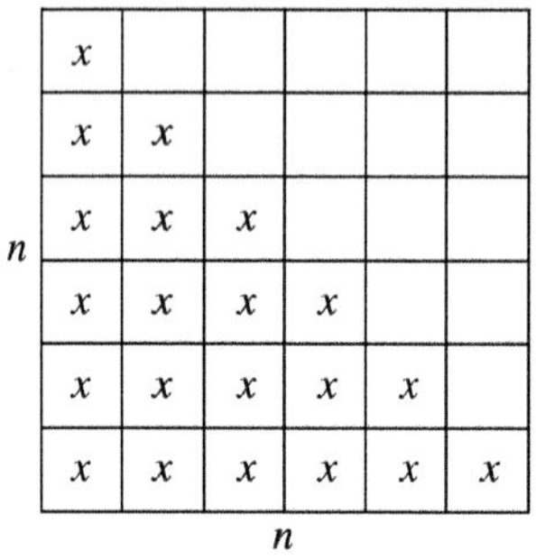

Figure 1.12 Visual proof.

Although a direct proof of the following theorem is fairly difficult, an induction proof is easy.

Theorem 2 Induction in Calculus
Prove that for every natural number n

$$P(n): \quad \frac{d^n(xe^x)}{dx^n} = (x + n)e^x$$

---

3 The array is really a 6×6 array but we imagine it is an $n \times n$ array.

Proof

Using mathematical induction, we have

Base Step: When $n = 1$ and using the product rule for differentiation, we have

$$\frac{d(xe^x)}{dx} = x\frac{d}{dx}e^x + e^x = (x+1)e^x.$$

Induction Step: Assuming

$$P(n): \frac{d^n(xe^x)}{dx^n} = (x+n)e^x$$

is true for arbitrary natural number $n > 1$, we compute

$$P(n+1): \frac{d^{n+1}(xe^x)}{dx^{n+1}} = \frac{d}{dx}\left(\frac{d^n(xe^x)}{dx^n}\right)$$

$$= \frac{d}{dx}[(x+n)e^x] \quad \text{(induction assumption)}$$

$$= (x+n)e^x + e^x \quad \text{(product rule)}$$

$$= [x+(n+1)]e^x$$

which proves $P(n+1)$, and so by induction the theorem is proven. ∎

Theorem 3   Inequality by Induction

If $n \geq 5$, then $2^n > n^2$.

Proof

Defining $P(n): 2^n > n^2$ we prove:

Base Case: $P(5): 2^5 = 32 > 25 = 5^2$.

Induction Step: $P(n) \Rightarrow P(n+1)$ for $n \geq 5$.

Assuming the statement is true for an arbitrary $n > 5$, the goal is to prove

$$2^n > n^2 \Rightarrow 2^{n+1} = (n+1)^2, \; n \geq 5$$

which we do in the following steps:

$$2^{n+1} = 2 \cdot 2^n$$

$$> 2n^2 \quad \text{(induction hypothesis)}$$

$$= n^2 + n^2$$

$$\geq n^2 + 5n \quad \text{(we are assuming } n > 5)$$

$$= n^2 + 2n + 3n$$

$$> n^2 + 2n + 1$$

$$= (n+1)^2$$

Hence, we have proven $P(n+1)$, so by induction $P(n)$ true for all $n \geq 5$. ∎

You may now ask, how did we ever arrive at the nonintuitive inequalities in the proof that made everything turn out so nice? The answer is we worked out the inequalities *backwards* starting with the conclusion.

Sometimes a result can be proven by induction or with a direct proof. The following problem is such an example. You can decide if you have a preference.

Theorem 4   Direct Proof or Proof by Induction?
For any $n \in \mathbb{N}$, the number $n(n + 1)$ is even.

Direct Proof
The idea is to show that for any natural number n, the number $n(n + 1)$ contains a factor 2. We consider both cases when n is even or odd.

If n is even, we have

$$(\exists k \in \mathbb{N})(n = 2k) \Rightarrow n(n + 1) = 2k(n + 1)$$

If n is odd, we have

$$(\exists k \in \mathbb{N})(n = 2k + 1) \Rightarrow n(n + 1) = (2k + 1)(2k + 2) = 2(k + 1)(k + 2) \quad \blacksquare$$

Proof by Induction
The basic proposition states $P(n) : n(n + 1)$ is even. Using mathematical induction, we show
Base Step: Show $P(1) : 1(1 + 1) = 2$ is even
Induction Step: Show $P(n) \Rightarrow P(n + 1)$. Assuming $n(n + 1)$ is even, we have $n(n + 1) = 2k$ for $k \in \mathbb{N}$. Hence, we can write

$$P(n + 1) : (n + 1)(n + 2) = n(n + 1) + 2(n + 1)$$

$$= 2k + 2(n + 1)$$

$$= 2(k + n + 1)$$

which proves $P(n + 1)$ and so by induction the result is proven. $\quad \blacksquare$

Important Note   Someone once said mathematical induction is the formal way of saying "and so on."?

The type of induction discussed thus far is sometimes called **weak induction**. We now introduce another version of induction called **strong induction**. Although the two versions are logically equivalent, there are problems where strong induction is more convenient.

1.6.3   Strong Induction

Surprising as it might seem, both weak and strong induction are logically equivalent, the difference being practical, sometimes strong induction is more convenient and sometimes weak induction is more convenient. The methodology of strong induction consists in carrying out the following steps.

Methodology of (Strong) Mathematical Induction   To verify a proposition P (n) holds for all natural numbers n, the **Principle of (Strong) Mathematical Induction** consists of carrying out the following steps.

1) **Base Case:** Prove $P(1)$ is true.
2) **Induction Step:** Show

$$(\forall n \in \mathbb{N})[P(1) \wedge P(2) \wedge \cdots \wedge P(n) \Rightarrow P(n+1)].$$

or equivalently

$$(\forall n \in \mathbb{N})[P(1) \wedge P(2) \wedge \cdots \wedge P(n-1) \Rightarrow P(n)]$$

The following theorem shows how strong can be useful in proving certain results.

Theorem 5   Every natural number n greater than 1 is divisible by a prime number.

Proof
Base Case: The result is true for $n = 2$ inasmuch as 2 is prime and 2 divides 2.
Induction Step: Assume all natural numbers 2 through $n - 1$ are divisible by a prime number, where $n - 1$ is an arbitrary natural number. The goal is to show n is divisible by a prime number. If n is prime, then it is divisible by a prime number, itself. If n is not prime, then n is a product of numbers less than n, and one (or any) of these numbers m, by hypothesis, is divisible by a prime number p. Hence, we have p divides m, and m divides n, and so, as we have proven earlier, p divides n, or symbolically

$$[(p \mid m) \wedge (m \mid n)] \Rightarrow (p \mid n).$$

Hence, the induction step in proven, so by the Principle of Strong Induction, the result is proven.   ∎

A fundamental result in number theory is the **Fundamental Theorem of Arithmetic**, which can be proven by strong induction.

Theorem 6   Fundamental Theorem of Arithmetic
Every natural number greater than 1 can be written as the product of prime numbers.[4] For example $350 = 2 \times 5^2 \times 7$, $1911 = 3 \times 7^2 \times 13$.

Proof
Using the principle of strong induction, we prove
Base Case: $P(2)$ holds since 2 is prime.
Induction Step: Using strong induction, we prove that for an arbitrary natural number $n \geq 2$:

$$P(2) \wedge P(3) \wedge P(4) \wedge \cdots \wedge P(n) \Rightarrow P(n+1)$$

Assuming $P(2), P(3), \ldots, P(n)$ true implies every natural number 2, 3, ..., n can be written as the product of primes. To prove that $n + 1$ can be written as a product of prime numbers, consider two cases:

Case 1: If $n + 1$ is a prime number, the result is proven since we can write $n + 1 = n + 1$.
Case 2: If $n + 1$ is not prime, it can be written as a product $n + 1 = pq$, where both factors p and q are less than $n + 1$ and greater than or equal to 2. By the induction hypothesis both p and q can be written as the product of primes:

$$p = p_1 p_2 \cdots p_m \quad q = q_1 q_2 \cdots q_n$$

Hence, we have

$$n + 1 = pq = (p_1 p_1 \cdots p_m)(q_1 q_2 \cdots q_n).$$

which shows $P(n + 1)$ is true. Thus, by the principle of strong induction $P(n)$ is true for all $n \geq 2$. ∎

History of Mathematical Induction   Although some elements of mathematical induction have been hinted at the time of Euclid, one of the oldest argument using induction goes back to the Italian mathematician *Francesco Maurolico*, who used induction in 1575 to prove that the sum of the first n odd natural numbers is n^2. The method was later discovered independently by the Swiss mathematician *John Bernoulli*, and French mathematicians *Blaise Pascal* (1623–1662) and *Pierre de Fermat* (1601–1665). Finally, in 1889, the Italian logician Giuseppe Peano (1858–1932) introduced five axioms, called Peano's axioms, for logically deducing the natural numbers in which the fifth axiom was the Principle of Mathematical Induction. Hence, the Principle of Mathematical Induction is an axiom of arithmetic.

---

4 Moreover, this representation is unique up to the order of the factors. However, we will not prove the uniqueness of the representation here.

The next example shows a variation of the base step in induction. Each problem is different, and the induction proof must be modified accordingly.

Problems

1. **Proof by Induction**
    Prove the following propositions, either by weak or strong induction.

    a) $1^2 + 2^2 + 3^2 + \cdots + n^2 = \dfrac{n(n+1)(2n+1)}{6}$

    b) $P(n): 1^3 + 2^3 + 3^3 + \cdots + n^3 = \dfrac{n^2(n+1)^2}{4}$

    c) $1 + 3 + 5 + \cdots + (2n - 1) = n^2$.

    d) $9^n - 1$ is divisible by 8 for all natural numbers n.

    e) For $n \geq 1$, $1 + 2^2 + 2^3 + 2^4 + \cdots + 2^n = 2^{n+1} - 1$

    f) For $n \geq 5$, $4n < 2^n$,

    g) $n^3 - n$ is divisible by 3 for $n \geq 1$.

    h) $2^{n-1} \leq n!$, $n \in \mathbb{N}$

    i) For all positive integers n, $n^2 + n$ is even.

    j) For any real numbers a, b and natural number n, we have $(ab)^n = a^n b^n$.

2. **Something Fishy**
    Let us prove by induction $n^2 + 7n + 3$ is even for all natural numbers $n = 1$, 2, ... What is wrong with the following induction argument? Letting $P(n)$ denote

    $$P(n): n^2 + 7n + 3 \text{ is an even integer}$$

    we prove $P(n) \Rightarrow P(n + 1)$. Assuming $P(n)$ true, we have

    $$P(n + 1): (n + 1)^2 + 7(n + 1) + 3 = \left(n^2 + 7n + 1\right) + 2n + 10$$

    $$= 2k + 2(n + 5) \text{(induction hypothesis)}$$

    $$= 2(k + n + 5)$$

    Hence, $P(n + 1)$ is true which by induction proves that $n^2 + 7n + 3$ is even for all natural numbers n. Note: Check the result for $n = 2$.

3. **Clever Mary**
    To prove the identity

    $$1 + 2 + 3 + \cdots + n = \frac{n(n+1)}{2}$$

Mary evaluates the left-hand side of the equation for $n = 0, 1, 2$ getting

n	0	1	2
$p(n)$	0	1	3

and then finds the quadratic polynomial $p(n) = an^2 + bn + c$ that passes through those points, getting

$$p(n) = \frac{1}{2}n^2 + \frac{1}{2}n = \frac{n(n+1)}{2}.$$

Mary turns in this proof to her professor. Is her proof[5] valid?

4. **Hmmmmmmmmm**

   Is there something fishy with this argument that Mary can carry a 50-ton load of straw on her back. Clearly, she can carry one straw on her back, and if she can carry n straws on her back, she can certainly carry one more. Hence, she can carry any number of straws on her back, which can amount to a 50-ton load.

5. **Geometric Principle by Induction**

   Show that every convex polygon[6] can be divided into triangles. An example illustrating a triangulation of an eight-sided convex polygon is drawn in Figure 1.13.

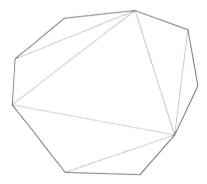

Figure 1.13 Triangulation of a polygon.

---

5 This problem is based on a problem in the book $A = B$ by Marko Petkovšek, Doron Zeilberger, and Herbert Wilf.

6 A convex polygon is a simple polygon (sides do not cross) whose interior is a convex set. (i.e. the line segment connecting any two points in the set also belongs to the set.

6. **Fibonacci Sequence**
   The Fibonacci sequence

   $$\{F_n, n = 1, 2, \ldots\}$$

   is defined for $n \geq 1$ by the equations

   $$F_1 = F_2 = 1, \quad F_{n+1} = F_n + F_{n-1}, \quad n \geq 2.$$

   A few terms of the sequence are 1,1,2,3,5,8,13. Show the nth term of the sequence is given by

   $$F_n = \frac{\alpha^n - \beta^n}{\sqrt{5}}$$

   where $\alpha = (1 + \sqrt{2})/2$, $\beta = (1 - \sqrt{2})/2$.

   Parting Note   Just because something is true for the first million numbers does not mean it is true for the millionth and one. For example, the equation

   $$(n-1)(n-2)\cdots(n-1\,000\,000) = 0$$

   is true for n from 1 to a million, but not true when $n = 1\,000\,001$.

   ---

7. **Peano's Axioms**
   The Principle of Mathematical Induction is generally taken as an axiom for the natural numbers and is in fact the fifth axiom for **Peano's axioms**. Google Peano's axioms and read about them on the Internet.

8. **Internet Research**
   There is a wealth of information related to topics introduced in this section just waiting for curious minds. Try aiming your favorite search engine toward *mathematical induction*, and *Giuseppe Peano*.

Chapter 2

Sets and Counting

2.1

Basic Operations of Sets

> **Purpose of Section**   We now present an informal discussion of some of the basic ideas related to sets, including membership in sets, subsets, the empty set, power set, union and intersection of sets, and other fundamental concepts. The material presented in this section will provide a foundation for study in all areas of mathematics.

2.1.1   Sets and Membership

Sets are (arguably) the most fundamental of all mathematical objects. Quite simply, a set is

> *a collection things, the things belonging to the set are called members*

or elements of the set.[1] Other synonyms for the word "set," are *collection, class, family*, and *ensemble*. We refer to a collection of people, family of functions, an ensemble of voters, and so on. We even consider sets whose members themselves are sets, such as the set of classes at a university, where each class consists of students. In mathematics, we might consider the set of all open intervals on the real line or the set of solutions of an equation.

If a set does not contain too many members, we can specify the set by simply writing down the members of the inside a pair of brackets, such as {3, 7, 31},

---

1 However, we will see in Section 2.6 when it comes to infinite sets, more care must be taken in its definition.

Advanced Mathematics: A Transitional Reference, First Edition. Stanley J. Farlow.
© 2020 John Wiley & Sons, Inc. Published 2020 by John Wiley & Sons, Inc.
Companion website: www.wiley.com/go/farlow/advanced-mathematics

which denotes the first three Mersenne primes.[2] Sometimes sets contain an infinite number of elements, like the natural numbers, where we might specify them by {1, 2, 3, ...}, where the three dots after the 3 signify "and so on."

We generally denote sets by capital letters such as $A, B, C, ...$ and members of sets by small letters, like $a, b, c, ...$ If a is a member of a set A, we denote this by writing $a \in A$, and we say "a is a member of A" or "a belongs to A." If an element does not belong to a set, we denote this by $a \notin A$.

One can also specify a set by specifying defining properties of the member of the set, such as illustrated in Figure 2.1.

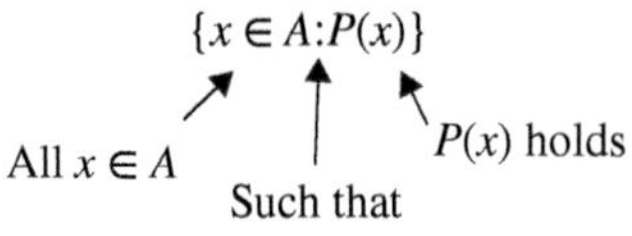

$$\{x \in A : P(x)\}$$

All $x \in A$ Such that $P(x)$ holds

Figure 2.1 Set notation.

which we read as "the set of all x in a set A that satisfies the proposition $P(x)$." The set of even integers could be denoted by

Even integers = $\{n \in \mathbb{Z} : n$ is an even integer$\}$.

Some common sets in mathematics are listed in Table 2.1.

Table 2.1 Some common sets.

Common sets in mathematics
$\mathbb{N} = \{1,2,3,...\}$ (the natural numbers or positive integers)
$\mathbb{Z} = \{...,-3,-2,-1,0,1,2,3,...\}$ (integers)
$\mathbb{Q} = \{n : n = p/q, p \text{ and } q \neq 0 \text{ integers}\}$ (rational numbers)
$\mathbb{R}$ = the set of real numbers
$\mathbb{C}$ = the set of complex numbers
$(a,b) = \{x \in \mathbb{R} : a < x < b\}$ (open interval from a to b)
$[a,b] = \{x \in \mathbb{R} : a \leq x \leq b\}$ (closed interval from a to b)
$(a, \infty) = \{x \in \mathbb{R} : x > a\}$ and $(-\infty, a) = \{x \in \mathbb{R} : x < a\}$ (open rays)
$[a, \infty) = \{x \in \mathbb{R} : x \geq a\}$ and $(-\infty, a] = \{x \in \mathbb{R} : x \leq a\}$ (closed rays)

The half-open intervals $(a, b]$ and $[a, b)$ are defined similarly.

Some examples illustrating membership and nonmembership in sets are the following.

---

2 Mersenne primes are prime numbers of the form $2^n - 1$, $n = 1, 2, ...$ At the present time 2019, there are 51 known Mersenne primes, the largest being $2^{82\,589\,993} - 1$, which has 24 862 048 digits.

$\pi \in \mathbb{R}$

$\pi \notin \mathbb{Q}$

$\dfrac{7}{12} \in \mathbb{Q}$

$e \in \{x \in \mathbb{R} : x \text{ is transcendental}\}$

$0 \notin \{x \in \mathbb{R} : 0 < x < 1\}$

> **Rigor in Mathematics**   Although in mathematics, there should always be a sufficient degree of precision or rigor, which generally refers to the degree in which mathematical arguments are logically valid, it is also important to create intuition about mathematical concepts. The great Swiss mathematician Leonhard Euler had an uncanny intuition about concepts and often did not prove results he believed to be true. That said, no mathematician would ever say he wasn't one of the greatest mathematicians who ever lived.

2.1.2   Universe, Subset, Equality, Complement, Empty Set

- **Universe:** The universe U is the set consisting of the **totality of elements** under consideration. A common universe in number theory is the natural numbers $\mathbb{N}$, whereas in calculus, a common universe is the real numbers $\mathbb{R}$, or intervals of real numbers such as $[0, 1]$ or $[0, \infty)$.
- **Subset:** One is often interested in a set A which is part of a larger set B. We say that a set A is a **subset** of a set B if every member of A is also a member of B. Symbolically, we write this as $A \subseteq B$ and is read "A is contained in B." If $A \subseteq B$ but $A \neq B$, we say that A is a **proper subset** of a set B and denote this by $A \subset B$.

  Sets are often illustrated visually by **Venn diagrams**, where sets are represented by circles or ovals and elements of a set as points inside the circle. Figure 2.2 shows a Venn diagram that might ring a bell for most students and illustrates $A \subseteq B$. The diagram also illustrates that A is not a subset of C.

> **Important Sets:** $\mathbb{N} \subset \mathbb{Z} \subset \mathbb{Q} \subset \mathbb{R} \subset \mathbb{C}$

- **Equality of Sets:** Two sets A and B are **equal** if they consist of the same members. In other words,

$$A = B \Leftrightarrow A \subseteq B \text{ and } B \subseteq A.$$

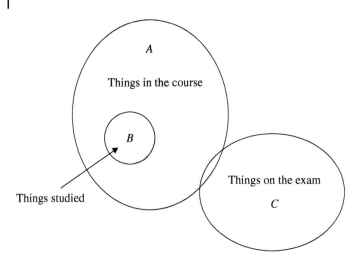

Figure 2.2 Venn diagram you may be familiar.

- **Complement of a Set:** If a set A belongs to a universe U, the **complement** of A, written $\bar{A}$ consists of all members of U that do not belong to A. In other words,

$$\bar{A} = \{x \in U : x \notin A\}$$

- **Empty Set**: The set that does not contain any members is called the **empty set**[3] (or **null set**) and plays an important role in set theory. It is denoted by the Greek letter Ø or by the empty bracket { }.

  The empty set is not is *not nothing*, it is *something*, it is just that it contains *nothing*. You might think of the empty set as a bag that has nothing in it. In this regard, it might be better to denote the empty set by { } rather than Ø. For example

$$\{\text{all people over } 500 \text{ years old}\} = \{x \in \mathbb{R} : x^2 + 1 = 0\}$$

which may seem a bit strange, but it makes a point. There is only one empty set regardless of how differently it is expressed.

> **Important Note** $\in$ versus $\subseteq$: When you ask if $a \in A$, you are asking the question is "a" a *member* of A, but when you ask if $B \subseteq A$, you are asking is every member *of B* is also a member *of A*. For example
>
> $$\{a, \{a\}\} \notin \{a, \{a\}\} \text{ but } \{a, \{a\}\} \subseteq \{a, \{a\}\}.$$

---

3 If you object to a set with no elements, you are like the persons in the past who objected to the number 0, since it stood for nothing. The number 0 was resisted for centuries as a legitimate number.

- **Power Set**: For every set A, the collection of *all* subsets of A is called the **power set** of A and denoted by P(A) or sometimes 2^A. For example the power set of the set A = {a, b, c} consists of the following set (or collection) of eight subsets of A:

$$P(A) = \{\varnothing, \{a\}, \{b\}, \{c\}, \{a,b\}, \{a,c\}, \{b,c\}, \{a,b,c\}\}.$$

A few power sets of some other sets are given in Table 2.2

Table 2.2 Typical power sets.

Set	Power set
$\varnothing$	$\{\varnothing\}$
$\{a\}$	$\{\varnothing, \{a\}\}$
$\{a, b\}$	$\{\varnothing, \{a\}, \{b\}, \{a, b\}\}$
$\{a, \{b\}\}$	$\{\varnothing, \{a\}, \{\{b\}\}, \{a, \{b\}\}\}$

Later, we will prove that for any set of n elements, the power set contains 2^n elements, which we will prove by induction.[4]

> **Important Note** Often, the power set of a set A is denoted by 2^A since a set with n members has 2^n subsets, hence, the power set has this many members.

Theorem 1 Guaranteed Subset
Every set contains at least one subset since for any set A, we have $\varnothing \subseteq A$.

Proof
We must show $x \in \varnothing \Rightarrow x \in A$. But our job is finished before we even begin since the hypothesis $x \in \varnothing$ is false inasmuch as the empty set contains *no* members, hence, the implication is true. ∎

Theorem 2 Transitive Subsets
For sets A, B, and C, the following property holds:

$$[(A \subseteq B) \wedge (B \subseteq C)] \Rightarrow A \subseteq C.$$

Proof
The strategy is to assume $x \in A$ and prove $x \in C$ with the help of the assumption $(A \subseteq B) \wedge (B \subseteq C)$. Letting $x \in A$ and since $A \subseteq B$, we have $x \in B$, but $B \subseteq C$

---

4 Things get a lot more interesting when we consider the family of all subsets of an infinite set like the natural numbers. It turns out that ... well, we do not want to ruin the fun for you.

and, hence $x \in C$. Hence, $A \subseteq C$ which proves the result. Figure 2.3 illustrates this result with a Venn diagram. ∎

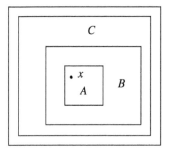

Figure 2.3 Venn diagram illustrating subsets.

Important Note   One reason the concept of a set is so powerful is the fact that its elements can be anything, even sets themselves. In the area of mathematics called *real analysis*, sets normally consist of sets of numbers, like intervals on the real line, and so on. In geometry, they are geometric objects, in probability, they are sample spaces and events, and so on. In topology, one studies certain *families of subsets* of a given set, called the *open sets* of the set.

Note: $5 \neq \{5\}$, $\{a, b\} \notin \{a, b\}$, $\varnothing \notin \varnothing$, $\varnothing \neq \{\varnothing\}$

Example 1   Membership and Subset
Do you understand why the following are correct?

a) $\varnothing \subseteq \{x \in \mathbb{R} : x^2 = -1\}$

b) $3 \in \mathbb{N}$

c) $-1 \notin \mathbb{N}$

d) $\pi \notin \mathbb{Q}$

e) $e \in \mathbb{R}$

f) $3.5 \in \mathbb{C}$

g) $3 + 2i \notin \mathbb{R}$

h) $\{x : x^2 - 1 = 0\} \not\subseteq \{x : x^3 - 1 = 0\}$

Example 2   Membership versus Subsets

Yes or no, do the following expressions make correct usage of $\in$ and $\subseteq$?

a) $\emptyset \in \{\emptyset, \{\emptyset\}\}$ **Ans:** Yes
b) $\emptyset \subseteq \{\emptyset, \{\emptyset\}\}$ **Ans:** Yes
c) $\emptyset \in \{\{\emptyset\}\}$ **Ans:** No
d) $a \in \{\{a\}, \{a, \{a\}\}\}$ **Ans:** No
e) $\{a\} \in \{\{a\}, \{a, \{b\}\}\}$ **Ans:** Yes
f) $\{\emptyset\} \subseteq \{\{\emptyset\}, \{\emptyset, \{\emptyset\}\}\}$ **Ans:** No
g) $\{a, \{b\}\} \in \{a, \{a, \{b\}\}\}$ **Ans:** Yes
h) $\{a, \{b\}\} \in \{\{b\}, a\}$ **Ans:** No

Theorem 3   Power Set

For any two sets A and B

$$A \subseteq B \Leftrightarrow P(A) \subseteq P(B).$$

Proof

There are two steps to the proof.

$(A \subseteq B) \Rightarrow (P(A) \subseteq P(B))$: If $X \in P(A)$ that means $X \subseteq A$. But $A \subseteq B$ so $X \subseteq B$, which in turn means $X \in P(B)$. Hence, $P(A) \subseteq P(B)$.

$(A \subseteq B) \Leftarrow (P(A) \subseteq P(B))$: If $x \in A$, then $\{x\} \in P(A)$ and by assumption $\{x\} \in P(B)$, which in turn means $x \in B$. Hence, we have $A \subseteq B$. ∎

2.1.3   Union, Intersection, and Difference of Sets

In arithmetic, we have the binary operations + and ×, whereas in logic, we have analogous binary operations of $\lor$ and $\land$, and now for sets, we have the binary operations of union $\cup$ and intersection $\cap$.

Definition   Let A, B be subsets of some universe U.

- **Union:** The **union** of two sets A and B, denoted $A \cup B$, is the set of elements that belong to A *or* B or both.[5]

$$A \cup B = \{x \in U : x \in A \ \text{or} \ x \in B\}$$

---

5 This "or" is the inclusive "or" in contrast to the exclusive "or" which means one or the other but not both.

See Figure 2.4

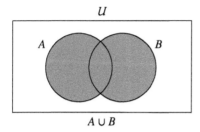

$$U$$

$$A \cup B$$

Figure 2.4 Set union.

- **Intersection:** The **intersection** of two sets A and B, denoted $A \cap B$, is the set of elements that belong to both A *and* B.
  See Figure 2.5.

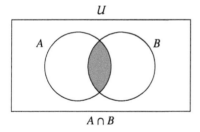

$$U$$

$$A \cap B$$

Figure 2.5 Set intersection.

$$A \cap B = \{x \in U : x \in A \text{ and } x \in B\}$$

If $A \cap B = \emptyset$, the sets A and B are called **disjoint**.

- The **difference** of two sets, denoted $A - B$, is defined to be the set of elements that belong to A but not B.

$$A - B = \{x \in U : x \in A \text{ and } x \notin B\}$$

Important Note The Italian mathematician Giuseppe Peano introduced the notation for set inclusion ($\in$), set union ($\cup$), and set intersection ($\cap$) in 1889 in a treatise on axioms for the natural numbers.

> **Historical Note**   Although the origin of the idea of a set is vague, Greek mathematicians defined a circle as
>
> > *A circle is a plane figure contained by one line such that all the straight-lines falling upon it from one point among those lying within the figure equal one another.*
>
> When the English mathematician George Boole referred to sets in 1854 in his seminal treatise, *An Investigation of the Laws of Thought*, the concept of a "set" per se was well established. That said, the distinction between "finite sets" and "infinite sets" eluded mathematicians over the centuries until the late 1800s in the work of the German mathematician George Cantor, whose works we study in Sections 2.4–2.6.

2.1.4   Venn Diagrams of Various Sets

Venn diagrams for subsets A, B of a universe U are shown in Figure 2.6. The universe is represented by the rectangle containing the sets.

> **Historical Note**   George Venn (1834–1923) was an English mathematician who further developed George Boole's symbolic logic but is known mostly for his pictorial representations of the relations between sets.

Figure 2.7 illustrates Venn diagrams for three subsets of a universe U.

> **Naive versus Axiomatic Set Theory**   **Naive set theory**, as we introduce in this section, studies basic properties of sets, such as complements, union, intersection, De Morgan's laws, and so on, using intuition. Unfortunately, when it comes to infinite sets, unless care is taken on exactly what kinds of collections of objects can be "accepted" as a set, it is possible to arrive at contradictions (i.e. Russell's paradox). **Axiomatic** set theory was created to place set theory on a firm axiomatic foundation where the axioms are **consistent** (one cannot prove contradictions) and **independent** (no one axiom can be proven from the others). The most accepted axioms of set theory are the Zermelo–Fraenkel (ZF) axioms, named after logicians Ernst Zermelo (1871–1953) and Abraham Fraenkel (1891–1965).

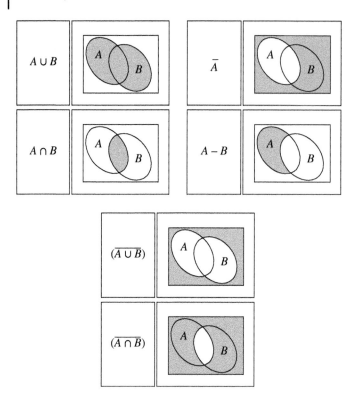

Figure 2.6 Venn diagrams for two sets.

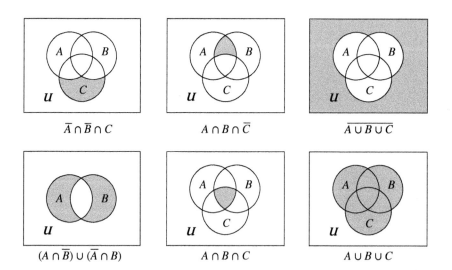

Figure 2.7 Venn diagrams for three sets.

2.1.5 Relation Between Sets and Logic

The properties of "∪" and "∩" in set theory have their counterparts in the properties of "∨" and "∧" in sentential logic. Table 2.3 illustrates these counterparts.

Table 2.3 Equivalence between some laws of logic and laws of sets.

Tautology	Set equivalence
$P \vee Q \equiv Q \vee P$	$A \cup B = B \cup A$
$P \wedge Q \equiv Q \wedge P$	$A \cap B = B \cap A$
$P \vee (Q \vee R) \equiv (P \vee Q) \vee R$	$A \cup (B \cup C) = (A \cup B) \cup C$
$P \wedge (Q \wedge R) \equiv (P \wedge Q) \wedge R$	$A \cap (B \cap C) = (A \cap B) \cap C$
$P \vee (Q \wedge R) \equiv (P \vee Q) \wedge (P \vee R)$	$A \cup (B \cap C) = (A \cup B) \cap (A \cup C)$
$P \wedge (Q \vee R) \equiv (P \wedge Q) \vee (P \wedge R)$	$A \cap (B \cup C) = (A \cap B) \cup (A \cap C)$
$P \wedge P \equiv P$	$A \cap A = A$
$P \vee P \equiv P$	$A \cup A = A$

Theorem 4   Complement Relation
For sets A and B, we have

$$A \subseteq B \Rightarrow \bar{B} \subseteq \bar{A}.$$

Proof
The goal is to show $\bar{B} \subseteq \bar{A}$ with the aid of $A \subseteq B$.

Letting $x \in \bar{B}$,

- we have $x \notin B$ from the definition of the compliment of B,
- hence, $x \notin A$ by the hypotheses $A \subseteq B$,
- hence, $x \in \bar{A}$ from the definition of the compliment of A. ∎

Theorem 5   Distributivity
Let A, B, and C be sets. Then "∩" distributes over "∪." That is

$$A \cap (B \cup C) = (A \cap B) \cup (A \cap C)$$

Proof
(⊆) We prove $A \cap (B \cup C) \subseteq (A \cap B) \cup (A \cap C)$
Since

$$B \subseteq B \cup C$$
$$C \subseteq B \cup C$$

we intersect each side with A, getting[6]

$$A \cap B \subseteq A \cap (B \cup C)$$
$$A \cap C \subseteq A \cap (B \cup C)$$

and so

$$(A \cap B) \cup (A \cap C) \subseteq A \cap (B \cup C).$$

($\supseteq$) We now prove $A \cap (B \cup C) \supseteq (A \cap B) \cup (A \cap C)$
If $x \in (A \cap B) \cup (A \cap C)$ we have $x \in A \cap B$ or $x \in A \cap C$.
We now use the logic indicated in the sentential logic equivalence

$$[(p \wedge q) \vee (p \wedge r)] \equiv [p \wedge (q \vee r)]$$

translated to sets to show that if $x \in A$ and $x \in B \cup C$, we have

$$x \in A \cap (B \cup C). \qquad \blacksquare$$

2.1.6 De Morgan's Laws for Sets

In sentential logic, we were introduced to the important tautologies
$$\sim(P \wedge Q) \equiv \sim P \vee \sim Q \text{ and } \sim(P \vee Q) \equiv \sim P \wedge \sim Q$$
called De Morgan's laws. We now prove the set versions of these laws.

Theorem 6   De Morgan's Laws
For sets A and B, prove De Morgan's laws:
a) $\overline{A \cup B} = \bar{A} \cap \bar{B}$
b) $\overline{A \cap B} = \bar{A} \cup \bar{B}$

Proof

a) We prove the first De Morgan law by first proving:

($\subseteq$) To show $\overline{A \cup B} \subseteq \bar{A} \cap \bar{B}$, we let

$$x \in \overline{A \cup B} \Rightarrow x \notin A \cup B$$
$$\Rightarrow x \notin A \text{ and } x \notin B$$
$$\Rightarrow x \in \bar{A} \text{ and } x \in \bar{B}$$
$$\Rightarrow x \in \bar{A} \cap \bar{B}$$

---

6 We have used the fact that if $B \subseteq C$ then $A \cap B \subseteq A \cap C$, but we will assume that fact as a lemma.

Hence, $\overline{A \cup B} \subseteq \bar{A} \cap \bar{B}$

($\supseteq$). To show $\overline{A \cup B} \supseteq \bar{A} \cap \bar{B}$, we let

$$x \in \bar{A} \cap \bar{B} \Rightarrow x \in \bar{A} \text{ and } x \in \bar{B}$$

$$\Rightarrow x \notin A \text{ and } x \notin B$$

$$\Rightarrow x \notin A \cup B$$

$$\Rightarrow x \in \overline{A \cup B}$$

Combining the results of a) and b) we have $\overline{A \cup B} = \bar{A} \cap \bar{B}$. ∎

The second De Morgan law is left to the reader. See Problem 21.

2.1.7 Sets, Logic, and Arithmetic

We compare the set operations of union and intersection with the logical operations of "and" and "or," and the arithmetic operations of + and + and × in Table 2.4.

Table 2.4 Sets, logic, and arithmetic.

Sets	Sentences	Arithmetic
$\cup$	$\vee$	$+$
$\cap$	$\wedge$	$\times$
$\subseteq$	$\Rightarrow$	$\leq$
$\bar{A}$	$\sim P$	$-$
$=$	$\equiv$	$=$
$\varnothing$	F	0
U	T	1

Problems

1. **Set Notation**
   Write the following sets in notation $\{x : P(x)\}$.
   a) The real numbers between 0 and 1.
   b) The natural numbers between 2 and 5.
   c) The set of prime numbers.
   d) $\{1, 2, 3, ...\}$

e) $\{5, 6, 7\}$

f) The solutions of the equation $x^2 - 1 = 0$.

2. **True or False**

If $A = \{\{a\}, \{b, c\}, \{d, e, f\}\}$, tell if the following are true or false:

a) $a \in A$

b) $a \subseteq A$

c) $c \in A$

d) $\{b, c\} \in A$

e) $\emptyset \in A$

f) $\emptyset \subseteq A$

3. **Checking Subsets: True or False**

a) $\mathbb{Z} \subseteq \mathbb{R}$

b) $\mathbb{R} \subseteq \mathbb{C}$

c) $(0, 1) \subseteq [0, 1]$

d) $(0, 1) \subseteq \mathbb{R}$

e) $(2, 5) \subseteq \mathbb{Q}$

f) $\mathbb{Q} \subseteq (2, 5)$

g) $[1, 3] \subseteq \{1, 3\}$

h) $\{1, 3\} \subseteq [1, 3]$

i) $\{3, 15\} \subseteq \{3, 5, 7, 15\}$

4. **The Empty Set: True or False**

a) $\emptyset = \{\emptyset\}$

b) $\emptyset \in \{\emptyset\}$

c) $\emptyset \subseteq \{\emptyset\}$

d) $A \cup \emptyset = A$

e) $\{\emptyset\} \subseteq \emptyset$

f) $\{\emptyset\} \in \{\{\emptyset\}\}$

g) $\{\{\emptyset\}\} \in \{\emptyset, \{\emptyset\}\}$

5. **True or False**

a) $A \in A$

b) If $A \subseteq B$ and $x \notin B$ then $x \notin A$

c) $A \subseteq B$ then $A \in B$.

d) $A \in B$ then $A \subseteq B$

e) $A \in B$ and $B \in C$ then $A \in C$

f) $A \in B$ and $B \in C$ then $A \subseteq C$

6. **Sets, Members, and Subsets**
   Which pair $(\in, \subseteq)$, $(\in, \not\subseteq)$, $(\notin, \subseteq)$, $(\notin, \not\subseteq)$ of connectives describe the relationship between the quantity on the left with the set on the right. The answer to a) is $(\in, \not\subseteq)$ since a is a member of $\{c, a, t\}$ but not a subset.
   a) a___$\{c, a, t\}$ **Ans:** $\in, \not\subseteq$
   b) a___$\{c, a, \{a\}, t\}$ **Ans:** $\in, \not\subseteq$
   c) $\{a, t\}$___$\{c, a, t\}$ **Ans:** $\notin, \subseteq$
   d) $\{a\}$___$\{c, \{a\}, t\}$
   e) $\{a, \{t\}\}$___$\{c, a, t, \{t\}\}$
   f) $\{a, \{t\}\}$___$\{c, a, t, \{t\}, \{a, \{t\}\}\}$
   g) $\{c, a, \{t\}\}$___$\{a, t, \{t\}\}$
   h) $\{a, \{t\}\}$___$\{c, a, t, \{t\}\}$
   i) $\varnothing$___$\varnothing$
   j) $\{\varnothing\}$___$\{\varnothing\}$
   k) $\{\varnothing\}$___$\{\varnothing, \{\varnothing\}\}$
   l) $\varnothing$___$\{\varnothing\}$
   m) $\{\varnothing\}$___$\varnothing$

7. **Power Sets**
   Find the power set of the given sets.
   a) $A = \{4, 5, 6\}$
   b) $A = \{\oplus, \odot, \otimes\}$
   c) $A = \{a, \{b\}\}$
   d) $A = \{a, \{b, \{c\}\}\}$
   e) $A = \{a, \{a\}\}$
   f) $A = \{\varnothing, \{\varnothing\}\}$

8. **Find the Set**
   Let $A = \{a_1, a_2, ...\}$, where a_n is the remainder when n is divided by 5. List the elements of the set A.

9. **Interesting**
   If $A = \{a, b, c\}$ are the following relations true or false?
   a) $A \in P(A)$
   b) $A \subseteq P(A)$

10. **Power Set as a Collection of Functions**
    The power set of a set can be interpreted as the set of all functions[7] defined on the set whose values are 0 and 1. For example the functions defined on the set $A = \{a, b\}$ with values 0 and 1 are

---

7 Although we have not introduced functions yet in this book, we are confident most readers have familiarity with the subject.

- $f(a) = 0, f(b) = 0$ corresponds to $\emptyset$
- $f(a) = 1, f(b) = 0$ corresponds to $\{a\}$
- $f(a) = 0, f(b) = 1$ corresponds to $\{b\}$
- $f(a) = 1, f(b) = 1$ corresponds to $\{a, b\}$

Show that the elements of the power set of $A = \{a, b, c\}$ can be placed in this "one-to-one" correspondence with the functions on A whose values are either 0 or 1.

11. **Second Power Set**
The second power set of a set A is the set of subsets of the set, or $P(P(A))$. What is the second power set of $A = \{a\}$?

12. **Power Set of the Empty Set**
Prove $P(\emptyset) = \{\emptyset\}$

13. **Identities**
Let A, B, and C be arbitrary subsets of a universe U. Prove the following.
a) $A \subseteq A$
b) $A \cup \emptyset = A$
c) $A \cap \emptyset = \emptyset$
d) $\emptyset = \bar{U}$
e) $A \cap U = A$
f) $A \cap \bar{A} = \emptyset$
g) $A \subseteq B \Rightarrow A \cup B = B$
h) $A \cup A = A \cap A$
i) $\bar{\bar{A}} = A$

14. **Difference Between Sets**
The formula $A - B = A \cap \bar{B}$ defines the difference between two sets in terms of the operations of intersection and complement. Can you find a formula for the union $A \cup B$ in terms of intersections and complements?

15. **NASC for Disjoint Sets**
Prove $A \cap B = \emptyset \Leftrightarrow A - B = A$.

16. **Distributive Law**
Prove that if A, B, and C are sets, then "$\cup$" distributes over "$\cap$." That is $A \cup (B \cap C) = (A \cup B) \cap (A \cup C)$.

17. **Set Identity**
Prove $A \subseteq B \Leftrightarrow A \cup B = B$.

18. **Proving Set Relations with Truth Tables**

It is possible to prove identities of sets using truth tables. For example one of DeMorgan's laws

$$\overline{(A \cup B)} = \bar{A} \cap \bar{B}$$

is verified from Table 2.5, replacing the union by "or," the intersection by "and," and set complementation by "not."

Prove the following identities using truth tables.

a) $A \cap \bar{A} = \emptyset$
b) $A \cup \bar{A} = U$
c) $A \cup (B \cup C) = (A \cup B) \cup C$
d) $A \cup B = (A \cap B) \cup (A - B) \cup (B - A)$

Table 2.5 Verifying set relations with truth tables.

		(1)	(2)	(3)	(4)	(5)
$x \in A$	$x \in B$	$A \cup B$	$\overline{(A \cup B)}$	$\bar{A}$	$\bar{B}$	$\bar{A} \cap \bar{B}$
T	T	T	F	F	F	F
T	F	T	F	F	T	F
F	T	T	F	T	F	F
F	F	F	T	T	T	T

Same truth values

19. **Sets and Their Power Sets**

Prove

$$A \subseteq B \Leftrightarrow P(A) \subseteq P(B)$$

20. **Computer Representation of Sets**

Finite sets can be represented by vectors of 0s and 1s. For the set $U = \{a_1, a_2, \dots, a_n\}$, we can represent a subset S of this set by a bit string, where the ith bit is 1 if $a_i \in S$ and 0 if $a_i \notin S$. The following problems relate to this representation of sets. Take as the universe, the set $U = \{0, 1, 2, 3, 4, 5, 6, 7\}$.

a) If $U = \{3, 9, 2, 5, 6\}$, what is $S \subseteq U$ for the bit string 11001?
b) If $U = \{1, 2, 3, 4, 5, 6\}$, what is the bit string for $A = \{2, 6\}$?

c) If $U = \{1, 2, 3, 4, 5, 6\}$ and $S = \{1, 4, 5\}$, $T = \{1, 2, 4, 6\}$, what is the bit string for $S \cap T$?

d) If $U = \{1, 2, 3, 4, 5, 6\}$ and $S = \{1, 4, 5\}$, $T = \{1, 2, 4, 6\}$, what is the bit string for $S \cup T$?

e) How would you represent the complement of a subset of $U = \{1, 2, 3, 4, 5, 6\}$?

21. De Morgan's Law

Prove the De Morgan law $\overline{A \cap B} = \bar{A} \cup \bar{B}$.

22. Internet Research

There is a wealth of information related to topics introduced in this section just waiting for curious minds. Try aiming your favorite search engine toward *mathematical rigor, naive set theory, George Boole, power set of a set.*

2.2

Families of Sets

Purpose of Section   We now extend the operations of union and intersection of sets from two sets to several sets, even an infinite number. In many areas of mathematics, such as topology, analysis, measure theory, probability, and others, finding the intersection and union of large families of sets is common.

2.2.1   Introduction

The union and intersection of sets can be extended to the union and intersection of many sets, even an infinite number. When dealing with a collection of several sets, it is usual practice to refer to them as **families** or **classes** of sets. To denote a family of sets, one often uses indices such as $\{A_1, A_2, A_3, \ldots, A_{10}\}$ and for an infinite family, we might write $\{A_1, A_2, A_3, \ldots\}$ or

$$\{A_k\}_{k=1}^{\infty}.$$

Other common ways to denote families of sets are

$$\{A_i : i \in \Lambda\} \text{ or } \{A_k\}_{k\in\Lambda}$$

where the set Λ is called an **index set**.
   For example

$$I_n = \left\{ \left[0, \frac{1}{n}\right) : n \in \mathbb{N} \right\} = \left\{ [0,1), \left[0, \frac{1}{2}\right), \left[0, \frac{1}{3}\right), \ldots \right\}.$$

The reader might recall the notation for infinite sums and products as

$$\sum_{k=1}^{\infty} a_k = a_1 + a_2 + \cdots, \quad \prod_{k=1}^{\infty} a_k = a_1 a_2 \cdots$$

which motivates the following notation and definition for sets.

Definition   Unions and Intersections of Families of Sets
The union of the family of n subsets $A_1, A_2, \ldots, A_n$ in a universe U is defined as

$$A_1 \cup A_2 \cup \cdots \cup A_n = \bigcup_{k=1}^{n} A_k = \{x \in U : x \in A_k \text{ for some } k = 1, 2, \ldots, n\}$$

The **intersection** of n sets $A_1, A_2, \ldots, A_n$ is

$$A_1 \cap A_2 \cap A_3 \cdots \cap A_n = \bigcap_{k=1}^{n} A_k = \{x \in U : x \in A_k \text{ for all } k = 1, 2, \ldots, n\}$$

Sometimes one has a family of sets $\{A_\alpha\}_{\alpha \in \Lambda}$, where the index α ranges over some **index set** Λ. The index set Λ might be a finite or infinite set. For these cases, the **union of a family** $\{A_\alpha\}_{\alpha \in \Lambda}$ is defined as

$$\bigcup_{\alpha \in \Lambda} A_\alpha = \{x \in U : x \in A_\alpha \text{ for at least one } \alpha \in \Lambda\}$$

and the **intersection of a family** $\{A_\alpha\}$ is

$$\bigcap_{\alpha \in \Lambda} A_\alpha = \{x \in U : x \in A_\alpha \text{ for all } \alpha \in \Lambda\}$$

The following examples illustrate these ideas.

Example 1   Infinite Intersections and Unions
Define the family of closed intervals

$$A_k = \left[0, \frac{k-1}{k}\right], \quad k = 2, 3, \ldots$$

where a few are drawn in Figure 2.8. Find the following unions and intersections.

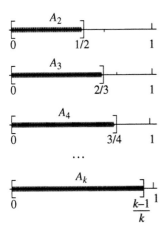

Figure 2.8 Increasing family of closed intervals.

a) $\displaystyle\bigcup_{k=2}^{4} A_k$

b) $\displaystyle\bigcup_{k=2}^{\infty} A_k$

c) $\displaystyle\bigcap_{k=2}^{4} A_k$

d) $\displaystyle\bigcap_{k=2}^{\infty} A_k$

Solution
The sets $A_1 \subseteq A_2 \subseteq A_3 \subseteq \cdots$ in Figure 2.8 form an increasing family of closed intervals, each member in the family is a subset of the next.

a) $\displaystyle\bigcup_{k=2}^{4} A_k = \left[0, \frac{3}{4}\right]$

b) $\displaystyle\bigcup_{k=2}^{\infty} A_k = [0, 1)$

c) $\displaystyle\bigcap_{k=2}^{4} A_k = \left[0, \frac{1}{2}\right]$

d) $\displaystyle\bigcap_{k=1}^{\infty} A_k = \left[0, \frac{1}{2}\right]$

Important Note   We have introduced five number systems $\mathbb{N}, \mathbb{Z}, \mathbb{Q}, \mathbb{R}$, and $\mathbb{C}$. We know that $\mathbb{N}$ stands for *natural* numbers, $\mathbb{Q}$ for *quotients*, $\mathbb{R}$ for *real* numbers, and $\mathbb{C}$ for *complex* numbers, but where does the letter $\mathbb{Z}$ which represents the *integers* come from? The answer is that $\mathbb{Z}$ refers to the first letter of the word "Zahlen," the German word for number.

Example 2   Infinite Intersections of Unions
Define the sequence of open intervals

$$A_k = \left(0, \frac{k+1}{k}\right), \quad k = 1, 2, \ldots$$

Find the following infinite union and intersection.

a) $\displaystyle\bigcup_{k=1}^{\infty} A_k = A_1 \cup A_2 \cup \ldots$ b) $\displaystyle\bigcap_{k=1}^{\infty} A_k = A_1 \cap A_2 \cap \ldots$

Solution

A few sets in the family are drawn in Figure 2.9.

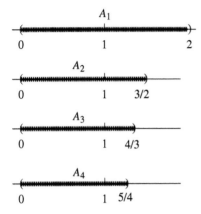

Figure 2.9 Decreasing family of open intervals.

We find

$$\bigcup_{k=1}^{\infty} A_k = \bigcup_{k=1}^{\infty} \left(0, \frac{k+1}{k}\right) = \left(0, \frac{2}{1}\right) \cup \left(0, \frac{3}{2}\right) \cup \left(0, \frac{4}{3}\right) \cup \cdots = (0,2)$$

$$\bigcap_{k=1}^{\infty} A_k = \bigcap_{k=1}^{\infty} \left(0, \frac{k+1}{k}\right) = \left(0, \frac{2}{1}\right) \cap \left(0, \frac{3}{2}\right) \cap \left(0, \frac{4}{3}\right) \cap \cdots = (0,1]$$

Example 3   Indexed Family

Define the sequence of sets

$$A_k = \{k+1, k+2, ..., 2k\}, \quad k = 1, 2, ...$$

Here, we have

$$\bigcup_{k=1}^{\infty} A_k = \{n \in N : n \geq 2\} \qquad \bigcap_{k=1}^{\infty} A_k = \emptyset$$

Example 4   Set Projection

The set $S \subseteq \{(x, y) : x, y \in \mathbb{R}\}$ denotes the large oval-shaped set of points in the plane drawn in Figure 2.10. For each real number x, we define the set

$$A_x = \{y \in \mathbb{R} : (x, y) \in S\}$$

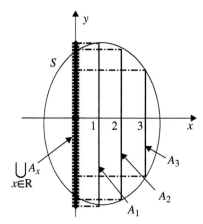

Figure 2.10 Projection of a set.

which defines the values of y such that $(x, y) \in S$. The union

$$\bigcup_{x \in \mathbb{R}} A_x$$

is the projection of S on the y-axis.

2.2.2 Extended Laws for Sets

Many rules for the intersection and union of sets introduced in Section 2.1 can easily be extended to families of sets. We leave the proofs to many of these laws to the reader.

Laws for Families of Sets

a) $A \cap \left(\bigcup_{\alpha \in \Lambda} B_\alpha \right) = \bigcup_{\alpha \in \Lambda} (A \cap B_\alpha)$

b) $A \cup \left(\bigcap_{\alpha \in \Lambda} B_\alpha \right) = \bigcap_{\alpha \in \Lambda} (A \cup B_\alpha)$

c) $\overline{\bigcap_{\alpha \in \Lambda} A_\alpha} = \bigcup_{\alpha \in \Lambda} \overline{A_\alpha}$ (De Morgan's Law)

d) $\overline{\bigcup_{\alpha \in \Lambda} A_\alpha} = \bigcap_{\alpha \in \Lambda} \overline{A_\alpha}$ (De Morgan's Law)

Proof of d)

($\subseteq$): To show $\overline{\bigcup_{\alpha\in\Lambda} A_\alpha} \subseteq \bigcap_{\alpha\in\Lambda} \overline{A_\alpha}$ we let

$$x\in\overline{\bigcup_{\alpha\in\Lambda} A_\alpha} \Rightarrow x\notin \bigcup_{\alpha\in\Lambda} A_\alpha$$

$$\Rightarrow (\forall \alpha\in\Lambda)(x\notin A_\alpha)$$

$$\Rightarrow (\forall \alpha\in\Lambda)(x\in \overline{A_\alpha})$$

$$\Rightarrow x\in \bigcap_{\alpha\in\Lambda} \overline{A_\alpha}$$

Hence,

$$\overline{\bigcup_{\alpha\in\Lambda} A_\alpha} \subseteq \bigcap_{\alpha\in\Lambda} \overline{A_\alpha}.$$

($\supseteq$): The proof of the set containment $\supseteq$ as well as the proofs of a), b), and c) follows along similar lines and is left to the reader.

2.2.3 Topologies on a Set

A topology on a set is a family of subsets of the set that places a "structure" on the set that allows for the study of convergence and limits of points in the set. The study of point-set topology, which we introduce in Section 5.4, forms the "metrical" structure for several areas of mathematics, such as real and complex analysis.

The idea is to introduce a family J of subsets of a given set U, such as a family J of subsets of the real numbers $U = \mathbb{R}$. The sets in the family J are called **open sets,** and these sets act as "neighborhoods" of points in U, allowing for the discussion of convergence sequences in U. The family of open sets J is called a **topology** on U. But not any collection of subsets of U is a topology. There are three restrictions on a family J in order that it be a topology on U, They are as follows:

Definition A **topology** J on a set U is any collection of subsets of U that satisfies the following three conditions:

1) The empty set $\emptyset$ and U itself belong to the family J.
2) The *union* of *any* collection of sets in J also belongs to J.
3) The *intersection* of any *finite* number sets in J also to J.

The sets in the topology J are called the open **sets in the topology**
(or just **open sets**). Properties 2) and 3) say that the family J is **closed** under **all unions** (finite or infinite) and **finite intersections**.

Example 5   Tiny Topologies

As examples of topologies on a set, consider the set U of three elements

$$U = \{a,b,c\}$$

where we have listed below five families J_1, J_2, J_3, J_4, J_5 of subsets of U. We leave it to the reader to verify that each of these families is a topology on U. See Problem 8.

a) $J_1 = \{\emptyset, U\}$ indiscrete topology
b) $J_2 = \{\emptyset, \{a\}, U\}$
c) $J_3 = \{\emptyset, \{a\}, \{b, c\}, U\}$
d) $J_4 = \{\emptyset, \{a, b\}, U\}$
e) $J_5 = \{\emptyset, \{a\}, \{b\}, \{c\}, \{a, b\}, \{a, c\}, \{b, c\}, U\}$ discrete topology

Theorem 1   Show That the Family of Subsets

$$J = \{\emptyset, \{a\}, \{b,c\}, \{a,b,c\}\} \subseteq P(U)$$

of $U = \{a, b, c\}$ is a topology on U.

Proof

We must verify the three conditions required for a family J of subset to be a topology. The first condition is verified since the topology J contains both the empty set $\emptyset$ and U. To verify that J is closed under unions, we simply take all possible unions of sets in J and verify that the union also belongs to J For example

$$\{a\} \cup \{b,c\} = \{a,b,c\} \in J$$
$$\emptyset \cup \{a\} = \{a\} \in J$$
$$\{a\} \cup \{a,b,c\} = \{a,b,c\} \in J$$

To show the family J is closed under intersections, we take intersections of open sets of J and observe their intersections also belong to J. For example

$$\{a\} \cap \{b,c\} = \emptyset \in J$$
$$\{b\} \cap \{b,c\} = \{b\} \in J$$
$$\{b,c\} \cap \{a,b,c\} = \{b,c\} \in J$$

Hence, we say J is a topology on U, and that the pair (U, J) is a topological space.

To understand how a topological structure on a set allows for the study of convergence, limits, etc., wait until Section 5.4 and the introduction of point-set topology on the real line.   ∎

Problems

1. **Unions and Intersections**
   Let $A_1 = \{1, 2\}$, $A_2 = \{2, 3\}$, $A_3 = \{3, 4\}$ and in general $A_k = \{k, k+1\}$. Write explicitly the following sets:

   a) $\bigcup_{k=1}^{5} A_k$

   b) $\bigcup_{k \in \mathbb{N}} A_k$

   c) $\bigcup_{k \geq 5} A_k$

   d) $\bigcup_{1 \leq k \leq 4} A_k$

   e) $\bigcap_{k=1}^{5} A_k$

   f) $\bigcap_{k \in \mathbb{N}} A_k$

2. **More Unions and Intersections**
   Find the infinite union and intersections

   $$\bigcup_{k=1}^{\infty} A_k \text{ and } \bigcap_{k=1}^{\infty} A_k$$

   of the following sets:

   a) $A_k = \left[0, \dfrac{k-1}{k}\right]$

   b) $A_k = \left[-\dfrac{1}{k}, \dfrac{1}{k}\right]$

   c) $A_k = \left(0, \dfrac{1}{k}\right)$

   d) $A_k = \{k\} \cup \left[\dfrac{1}{k}, 2k\right]$

   e) $A_k = [k, k+1]$

   f) $A_k = \left[0, 1 + \dfrac{1}{k}\right]$

3. **Families of Sets in the Plane**
   Define a family of subsets

   $$A_{m.n} = \left\{(x,y) \in \mathbb{R}^2 : x \geq m, \ y \geq n\right\}$$

   of the plane $\mathbb{R}^2$ by where $m, n \in \mathbb{N}$. Find the following sets. Hint: Proceed like one does with double series.

a) $\displaystyle\bigcup_{m=1}^{3}\left(\bigcup_{n=2}^{3} A_{m,n}\right)$

b) $\displaystyle\bigcup_{m=1}^{3}\left(\bigcap_{n=2}^{3} A_{m,n}\right)$

4. Identity of an Indexed Family

Prove the following distributive property for indexed families of sets:

$$B\cap\left(\bigcup_{\alpha\in\Lambda} A_\alpha\right) = \bigcup_{\alpha\in\Lambda}(B\cap A_\alpha)$$

5. Algebra of Sets

Let A be a set and $\mathfrak{I}$ a collection of subsets of A. The collection $\mathfrak{I}$ is called an **algebra**[1] of sets if

c) $C\cup D$ is in $\mathfrak{I}$ whenever C and D are in $\mathfrak{I}$

d) $\bar{C}$ is in $\mathfrak{I}$ whenever C is in $\mathfrak{I}$

When this happens, we say the family $\mathfrak{I}$ is **closed** under **unions** and **complementation**. Which of the following collections of subsets of $A = \{a, b, c\}$ constitute an algebra of subsets of A?

a) The power set $\mathfrak{I} = P(A)$
b) $\mathfrak{I} = \{\emptyset, A\}$
c) $\mathfrak{I} = \{\emptyset, \{a\}, A\}$
d) $\mathfrak{I} = \{\emptyset, \{a\}, \{b, c\}, A\}$

6. Sets of Length Zero

In measure theory, a subset A of the real numbers is said to have **length** (or **measure**) zero if $\forall \varepsilon > 0$ and there exists a sequence of intervals $A_k = (a_k, b_k)$ that satisfy

$$A \subseteq \bigcup_{k=1}^{\infty} (a_k, b_k)$$

where their *total length* is less than ε; that is

$$\sum_{k=1}^{\infty} |b_k - a_k| < \varepsilon.$$

Show that any sequence of real numbers $\{c_k, k = 1, 2, ...\}$ has measure zero. Hint: Cover $\{c_k\}$ by a union of intervals (a_k, b_k), where the length of each

---

1 Algebras of sets and sigma algebras (families of sets closed under *countable* unions) are fundamental in the study of measure theory. Note the difference between an algebra of subsets and a topology of subsets on a universe; just a minor difference makes for vastly different structures on the universe.

interval satisfies $|b_k - a_k| = \varepsilon/2^k$. Use this result to show that the rational numbers have a total length of 0.

7. **Compact Sets**

A subset A of the real numbers is said to be **compact** if for every collection $\mathfrak{I} = \{(a_\alpha, b_\alpha) : \alpha \in \Lambda\}$ of open intervals whose union contains (or covers) A; i.e.

$$A \subseteq \bigcup_{\alpha \in \Lambda} (a_\alpha, b_\alpha)$$

there exists a *finite* subcollection of intervals of $\mathfrak{I}$ whose union also contains (or covers) A. Show the set $A = (0, 1)$ is not compact by showing the following.

a) A is covered by $\quad \mathfrak{I} = \left\{ \left(0, 1 - \dfrac{1}{k} \right) : k = 1, 2, \dots \right\}$

b) There does not exist a finite subcollection of $\mathfrak{I}$ whose union contains A.

8. **Topologies I**

Verify that the following families of subsets of A form topologies on $A = \{a, b, c\}$.

$J_1 = \{ \varnothing, A \}$

$J_2 = \{ \varnothing, \{a\}, A \}$

$J_3 = \{ \varnothing, \{a\}, \{b,c\}, A \}$

$J_4 = \{ \varnothing, \{a,b\}, A \}$

$J_5 = \{ \varnothing, \{a\}, \{b\}, \{c\}, \{a,b\}, \{a,c\}, \{b,c\}, A \}$

9. **Topologies II**

Which of the families of subsets of $\{a, b, c\}$ are topologies on $\{a, b, c\}$?

10. **Finding Intersections**

Find an infinite family of sets whose intersection is

a) $\{1\}$

b) $[0, \infty)$

11. **Finding Unions**

Find an infinite family of sets whose union is

a) $(0, \infty)$

b) $\mathbb{R}$

12. **Internet Research**

There is a wealth of information related to topics introduced in this section just waiting for curious minds. Try aiming your favorite search engine toward *unions and intersections of families of sets, topology on a finite set.*

2.3

Counting

The Art of Enumeration

> **Purpose of Section** To introduce some basic tools of counting the members of a set, and to do this, we introduce a variety of tools of the trade, such as the multiplication principle, permutations, and combinations. We show how the multiplication rule gives rise to counting permutations and combinations. We close with a brief introduction to the pigeonhole principle, and show how such a ridiculously simple idea can solve some ridiculous difficult problems.

2.3.1   Introduction

Although counting is one of the first things we learn at a tender age, be assured there are counting problems that test the mental agility of the brightest minds among us. In this section, we will learn to count, although we refer to it as *combinatorics*. We will answer such counting questions as

how many ways can 10 people choose sides and play a game 5 against 5?

Counting problems often require almost no technical background and are often characterized by being easy to understand and hard to solve. Finding the number of ways to cover an 8 x 8 checkerboard with dominoes is a good example. The problem is easy to understand, but the number of coverings was not determined until 1961 by a M. E. Fischer, who found the number to be $2^4 \times (901)^2 = 12\,988\,816$.

To assist you in perfecting counting skills, we introduce one of the basic tools of the trade, the multiplication principle.

Advanced Mathematics: A Transitional Reference, First Edition. Stanley J. Farlow.
© 2020 John Wiley & Sons, Inc. Published 2020 by John Wiley & Sons, Inc.
Companion website: www.wiley.com/go/farlow/advanced-mathematics

2.3.2 Multiplication Principle

One of the basic principles of counting is the **multiplication principle,** although the principle is simple, it has far-reaching consequences. For example, how many dots are there in Figure 2.11?

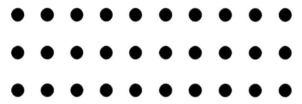

Figure 2.11 Counting exercise.

We suspect you said 30 after about three seconds, and you did not even bother to count all the dots. You simply counted the number of columns in the first row and then multiplied by 3. If so, you used the multiplication principle.

As a small step up the counting ladder, count the number of paths in Figure 2.12 going from A to C. No doubt this problem did not stump you either, getting $4 \cdot 5 = 20$ paths. Again, you used the multiplication principle.

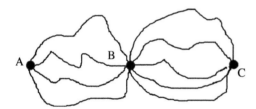

Figure 2.12 How many paths from A to C?

This leads us to the formal statement of the multiplication principle.

Multiplication Rule   If a procedure can be broken into successive stages, and if there are s_1 outcomes for the first stage, s_2 outcomes for the second stage, ..., and s_n for the nth stage, then the entire procedure has $s_1 \, s_2 \ldots s_n$ outcomes.

We now let you test your counting skills using the multiplication rule with something a little harder.

Example 1   Counting Subsets
Show that a set $A = \{a_1, a_2, \ldots, a_n\}$ containing n elements has 2^n subsets.

Proof

Any subset of A can be formed in n successive steps. On the first step, pick or not pick a_1, on the second step pick or not pick a_2, and so on. On each step, there are two options, to pick or not to pick. Using the multiplication principle, the number of subsets that can be picked is

$$\underbrace{2 \cdot 2 \cdots \cdots 2}_{n \text{ twos}} = 2^n$$

Hence, the set $A = \{a, b, c\}$ containing three members has $2^3 = 8$ the eight subsets, and since they are so few we can enumerate them:

$$2^A = \{\emptyset, \{a\}, \{b\}, \{c\}, \{a,b\}, \{a,c\}, \{b,c\}, \{a,b,c\}\}$$

Example 2   Counting Functions

How many functions are there from the set $A = \{0, 1\}$ to itself?[1]

Solution

For each of the two members in A, the function can take on two values and so by the multiplication rule, the number of functions is $2 \cdot 2 = 2^2 = 4$. We graph these functions in Figure 2.13.

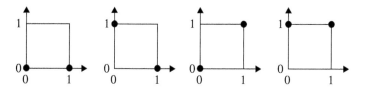

Figure 2.13 Four functions from {0, 1} to {0, 1}.

In general, the number of functions from a set with n members to a set with m members is m^n. For example, the number of functions from $A = \{0, 1\}$ to $B = \{0, 1, 2\}$ is $3^2 = 9$. Can you draw their graphs?

2.3.3   Permutations

One of the important uses of the multiplication principle is counting permutations. A **permutation** is simply an **arrangement** of things. For example, there are six permutations of the three letters *abc*, which are

   abc, acb, bac, bca, cab, cba

---

1 We will talk more about functions in Chapter 4.

The number of permutations increases dramatically as the size of the set increases. The number of permutations of the letters of the alphabet is

$$26! = 403\ 291\ 461\ 126\ 605\ 635\ 584\ 000\ 000$$

To determine the number of permutations of these letters, we begin by selecting one member of the alphabet as the left-most member in the permutation (or arrangement), which can be done in 26 ways. Once this is done, there are 25 remaining letters for the second member, the third member 24 ways, and so on, down to the last member. Using the principle of multiplication, the number of permutations of 26 objects is 26!.

2.3.4 Permutations of Racers

Suppose four individuals *a*, *b*, *c*, *d* are in a foot race, and we wish to find the number of ways four runners can finish first and second. Since each of the four runners can finish first, and for each winner, there are three possible second-place finishers, there are $4 \cdot 3 = 12$ possible ways the runners can go 1–2. The tree diagram in Figure 2.14 illustrates these finishes.

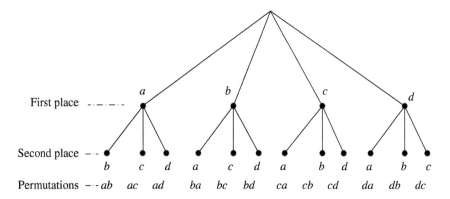

Figure 2.14 Permutations of two elements from a set of four elements.

This leads to the general definition of permutations of different sizes.

Definition A **permutation** of *r* elements taken from a set of *n* elements is an arrangement of *r* elements chosen from a set of *n* elements. The number of such arrangements (or permutations) is denoted by $P(n, r)$.

Using the multiplication principle, we can find the number of permutations of *r* elements taken from a set of *n* elements.

Theorem 1   Number of Permutations

The number of permutations of r elements taken from a set of n elements is

$$P(n,r) = n \cdot (n-1) \cdot (n-2) \cdots (n-r+1) = \frac{n!}{(n-r)!}$$

Proof

Choosing r elements from a set of size n, we have

- the first element can be selected n ways.
- the second element can be selected $n-1$ ways.
- the third element can be selected $n-2$ ways.
- the rth element can be selected $n-r+1$ ways.

Hence, by the principle of sequential counting (or the multiplication rule), we have

$$P(n,r) = \underbrace{n \cdot (n-1) \cdot (n-2) \cdots (n-r+1)}_{r \text{ factors}}.$$

When $r = n$, we have the number of permutations of n objects is n factorial, or

$$P(n,n) = n(n-1)(n-2)\cdots(2)(1) = n! \qquad \blacksquare$$

Important Note   To evaluate $P(n, r)$ start at n and multiply r factors.

$$P(4,2) = 4 \cdot 3 = 12$$
$$P(7,3) = 7 \cdot 6 \cdot 5 = 210$$
$$P(4,1) = 4 = 4$$
$$P(10,3) = 10 \cdot 9 \cdot 8 = 720$$

Big Number   Factorials grow very fast. The number 50! is *30 vigintillion, 404 novemdecillion, 93 octodecillion, 201 septendecillion, 713 sexdecillion, 378 quindecillion, 43 quatuordecillion, 612 tredecillion, 608 dodecillion, 166 undecillion, 64 decillion, 768 nonillion, 844 octillion, 377 septillion, 641 sextillion, 568 quintillion, 960 quadrillion, 512 trillion.* That is

$$50! = 30\,404\,093\,201\,713\,378\,043\,612\,608\,166\,064\,768\,844\,377\,641\,568$$
$$960\,512\,000\,000\,000\,000$$

Example 3 Permutations of a Set with Three Elements
Find the permutations of size $r = 1, 2, 3$ selected from $\{a, b, c\}$.

Solution
The permutations of size $r = 1, 2, 3$ taken from the set $\{a, b, c\}$ with $n = 3$ elements are listed in Table 2.6.

Table 2.6 Permutations $P(n, r)$.

$r = 1$	$r = 2$	$r = 3$
a	ab	abc
b	ac	acb
c	ba	bac
	bc	bca
	ca	cab
	cb	cba

We do not use set notation for writing permutations[2] since order is important. The permutation ab is not the same as ba.

2.3.5 Distinguishable Permutations

The number of permutations of elements in a set of the three letters is $3! = 6$, however, for the three letters in the word "too" we cannot distinguish one "o" from the other, hence, we only have three **distinguishable permutations** that we can distinguish with our eyes, which are *too, oto, oot*.

Example 4 Distinguishable Permutations
How many distinguishable permutations are there of the letters of the word SYSTEM?

Solution
Since the two S's in SYSTEM are indistinguishable,[3] we must modify our counting strategy. If we momentarily denote the two S's as S_1 and S_2, we have

---

2 Sometimes permutations are written with round parenthesis, such as $ab = (ab)$.
3 If we interchanged the two S's, we would get two different permutations, SYSTEM and SYSTEM, but they are not *distinguishable* permutations since the S's look alike.

S_1YS_2TEM, which has 6! permutations. But the two letters S_1 and S_2 have two permutations, so we must divide 6! by 2!, getting the number of distinguishable

$$\frac{6!}{2!} = 6 \cdot 5 \cdot 4 \cdot 3 = 360$$

A few more distinguishable permutations are given in Table 2.7.

Table 2.7 Distinguishable permutations.

Word	Indistinguishable permutations
Too	$\dfrac{3!}{2!} = 3$
Error	$\dfrac{5!}{3!} = 20$
Toot	$\dfrac{4!}{2!2!} = 6$
Mississippi	$\dfrac{11!}{4!4!2!} = 34\,650$

2.3.6 Combinations

Now that we have mastered permutations, we turn to combinations. A combination of things is simply a subset of things selected from larger collection of things. For the set $\{a, b, c\}$ of three elements, a typical combination of two elements is $\{a, c\}$. We write combinations in set notation $\{\ \}$ since the order of the members in the combination does not matter. The combination $\{a, c\}$ is the same as the combination $\{c, a\}$. Just remember, order counts in permutations but not in combinations.

Example 5   Combinations
Find the combinations of size $r = 1, 2, 3$ selected from the set $\{a, b, c\}$.

Solution
The combinations are listed in Table 2.8, which are the eight subsets of $\{a, b, c\}$ with the exception of the empty set Ø.

Table 2.8 Nonempty subsets of $\{a, b, c\}$.

$r = 1$	$r = 2$	$r = 3$
$\{a\}$	$\{a, b\}$	$\{a, b, c\}$
$\{b\}$	$\{a, c\}$	
$\{c\}$	$\{b, c\}$	

Theorem 2   Number of Combinations
The number of **combinations** of size r taken from a set of size n is

$$C(n,r) = \binom{n}{r} = \frac{n!}{r!(n-r)!}.$$

Proof
Since the number of permutations of size r taken from a set of size n is

$$P(n,r) = \frac{n!}{(n-r)!}.$$

and since the r elements can be permuted $r!$ ways, the number of combinations is the number of permutations divided by $r!$ or

$$C(n,r) = \frac{P(n,r)}{r!} = \frac{n!}{r!(n-r)!}. \qquad \blacksquare$$

We denote combinations by either notation

$$C(n,r) \quad \text{or} \quad \binom{n}{r}$$

The second notation is possibly familiar to the reader since it is often the notation for the coefficients in binomial expansion formulas:

$$(a+b)^2 = \binom{2}{0}a^2 + \binom{2}{1}ab + \binom{2}{2}b^2 = a^2 + 2ab + b^2$$

$$(a+b)^3 = \binom{3}{0}a^3 + \binom{3}{1}a^2b + \binom{3}{2}ab^2 + \binom{3}{3}b^3 = a^3 + 3a^2b + 3ab^2 + b^3$$

$$\cdots \qquad \cdots \qquad \cdots$$

It helps in thinking about combinations to read $C(n, r)$ as "*n choose r*" since it represents the number of ways one can choose r things from a set of n things. For example $C(4, 2) = 6$ is read "4 choose 2" meaning there are six ways you can choose two things from four things. Can you visualize in your mind the six subsets of size two from the set $\{a, b, c, d\}$?

Example 6   Game Time

How many ways can 10 players choose sides to play five-against-five in a game of basketball?

Solution

As in many counting problems, there is more than one way to do the counting. Perhaps the simplest way here is to determine the number of ways one player can choose his or her four teammates from nine players. In other words, determine the number of subsets of size four taken from a set of size nine, or "nine choose four," which is

$$\binom{9}{4} = \frac{9!}{4!5!} = \frac{9 \cdot 8 \cdot 7 \cdot 6}{4 \cdot 3 \cdot 2 \cdot 1} = 126 \text{ ways}.$$

Example 7   Number of Seven-Game-Series

How many ways can two teams play a best-of-seven game series, like the World Series[4] in the United States? By "ways" we do not distinguish the two teams. For example, Team A or Team B can sweep the series in four games, but we call this the single outcome WWWW, not *two* outcomes as AAAA when Team A sweeps or BBBB when Team B sweeps.

Solution

Since the number of games played will be four, five, six, or seven, we begin by finding the number of each of those series. Since the winner of the series wins a total of four games and always wins the last game, we ask how many ways can the series be played *before* the last game. For example, for a six-game series, we find the number of ways the winner can win three games in the first five games, which is "5 choose 3" or $C(5, 3) = 10$. For all four, five, six, and seven game series, the number of each of those series is

Number of four-game series	$= C(3,3) = 1$
Number of five-game series	$= C(4,3) = 4$
Number of six-game series	$= C(5,3) = 10\dot{}$
Number of seven-game series	$= C(6,3) = 20$

---

4 The World Series is a best-of-seven game series.

Adding these series gives 35 possible outcomes for a best-of-seven game series. If we distinguish between the teams A and B, the total is 70.

Example 8   Going to the Movies

Three boys and two girls are going to the movies.

a) How many ways can they sit next to each other if no boy sits next to another boy.
b) How many ways can they sit next to each other if the two girls sit next to each other.

Solution

a) Since the only arrangement is boy-girl-boy-girl-boy, and since the boys can be permuted 3! = 6 ways, and the girls 2! = 2 ways, the total number of ways is 3! 2! = 12.
b) Momentarily, think of the two girls as "one girl," so there are now four individuals, and so we have 4! = 24 permutations of the "four" individuals. However, there are 2! = 2 permutations of the two girls, hence, the total number of permutations is 4! 2! = 48.

Example 9   How Many Ways Home?

How many ways can Mary go home from school in the road system in Figure 2.15 when she always moves to the right or down?

School

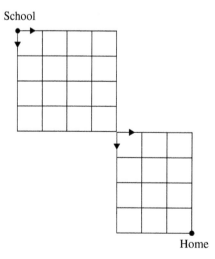

Home

Figure 2.15 Counting paths.

Solution

Since all paths pass through the one-point gap, we find the number of paths before the gap and the number of paths after the gap, then multiply the results. From start to the gap, we travel a total of eight blocks, four blocks to the right and four blocks down. Labeling each block as R or D, depending whether the move is to the right or down, all paths can be written $\{x, x, x, x, x, x, x, x\}$, where four xs are R and four x's are D. Hence, the total number of paths is the number of ways one can select four D's (or four R's) from a set of size 8, which is "8 choose 4" or $C(8, 4) = 70$. Similarly, the number of paths from the "gap" to end is $C(7, 3) = 35$ Hence, the total number is

$$C(8,4)C(7,3) = 70 \cdot 35 = 2450 \text{ paths.}$$

2.3.7 The Pigeonhole Principle

The pigeonhole principle (or Dirichlet Principle) illustrated in Figure 2.16 is based on the observation that if n items are placed in m containers, where $n > m$, then at least one container will contain more than one item. Although the principle seems almost too trivial to yield useful ideas, nothing is further from the truth. Its applications are far-reaching and deep and widely used in many fields, including computer science, mathematical analysis, probability, number theory, geometry, and statistics.

Figure 2.16 Pigeonhole principle.

> **Important Note** People often misinterpret the difference between axioms and definitions. An axiom is an assumed truth, whereas a definition assigns names and symbols to given concepts to make it easier to talk about things.

Example 9   Pigeonhole Principle at Work

Given *any* set A of n natural numbers, there are always two numbers in the set whose difference is divisible by $n - 1$.

Solution

When any number in the set, regardless of its size, is divided by $n - 1$, its remainder will be one of the $n - 1$ values $0, 1, 2, ..., n - 2$. But if the set A has n members, then by the pigeonhole principle, *at least* two members of A have the same remainder, say r, when divided by $n - 1$. Letting N_1, N_2 be two numbers with the same remainder, we can write

$$\frac{N_1}{n-1} = Q_1 + \frac{r}{n-1}$$
$$\frac{N_2}{n-1} = Q_2 + \frac{r}{n-1}$$

and subtracting gives

$$\frac{N_1 - N_2}{n-1} = Q_1 - Q_2$$

which means their difference $N_1 - N_2$ is divisible by $n - 1$. The reader may try a few examples to verify this result. ∎

Should the pigeonhole principle be taken as an axiom or should it be proven from more fundamental principles? The word "obvious" is a loaded word in mathematics, since many famous "obvious" claims of the past have turned out not only nonobvious but nontrue.[5] Although the pigeonhole principle seems obvious, it can be proven from more "basic principles" of set theory, although one wonders what could be more basic than the pigeonhole principle. Interested readers can find references on the Internet.

---

5 Many geometric ideas that were taken as facts at the beginning of the nineteenth century were overturned by the "arithmetization" of mathematics prevalent in the nineteenth century.

Example 10 Pigeonhole Principle Goes to a Party

There are 20 people at a dinner party, where each person shakes hands with someone at least once. Show there are at least two people who shook hands the same number of times.[6]

Solution

Label each person with the number of handshakes they make. Each person then has a label of 1 through 19. But there are 20 people at the party, so by the pigeonhole principle, at least two persons must have the same label.

Example 11 Points in a Square

Show that for any five points in a square whose sides have length 1, there will always be two points whose distance is less than or equal to $\sqrt{2}/2$.

Solution

Figure 2.17 shows five points, pigeons if you like, randomly placed in the unit square. If the square is subdivided into four equal subsquares, the pigeonhole principle says one of the subsquares (pigeonholes) contains at least two numbers. But the diameter of each subsquare has length $\sqrt{2}/2$, hence, the distance between any two points in the subsquare is less than or equal to $\sqrt{2}/2$.

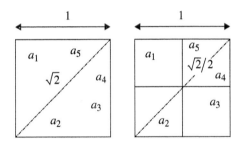

Figure 2.17 Pigeonhole principle at work.

Important Note The popular example of the pigeonhole is the claim that there are at least two nonbald persons in New York City with exactly the same number of hairs on their head. The argument being no one person has more than a million hairs on their head, and since there are over a million people living in New York, the pigeonhole principle states that at least two people have the same number of hairs.

---

6 We assume that people do not shake hands with themselves.

> **Historical Note**   The first statement of the pigeonhole principle was made in 1834 by the German mathematician Peter Gustav Lejeune Dirichlet (1805–1859) who called it *Schubfachprinzip* ("the drawer principle").

Problems

1. **Compute the following**
   a) $P(5, 3)$
   b) $P(30, 2)$
   c) $C(4, 1)$
   d) $C(10, 8)$
   e) $\binom{7}{2}$
   f) $\binom{9}{2}$
   g) $(a + b)^6$

2. **Distinguishable Permutations**
   Find the number of distinguishable permutations in the following words.
   a) SNOOT
   b) DALLAS
   c) TENNESSEE
   d) ILLINOIS

3. **Going to the Movies**
   Find the numbers of ways in which four boys and four girls can be seated in a row of eight seats if they sit alternately as boy and girl.

4. **Movies Again**
   Three couples go to the movies and sit together in six seats. How many ways can they be arranged so that each couple sits together?

5. **Counting Softball Teams**
   A college softball team is taking 25 players on a road trip. The traveling squad consists of three catchers, six pitchers, eight infielders, and eight outfielders. Assuming each player can only play her own position, how many different teams can the coach put on the field?

6. **Counting Functions**
   How many functions are there from $A = \{a, b, c\}$ to $B = \{0, 1, 2\}$?

7. **Interesting Problem**
   How many three-digit numbers $d_1d_2d_3$ are there whose digits add up to eight? That is, $d_1 + d_2 + d_3 = 8$, Note: 063 is not a three-digit number, it is the two-digit number 63.

8. **World Series Time**
   Two teams compete in a best-of-five game series. In other words, the team that wins three games wins the series. How many possible ways are there to play a five-game series when we do not distinguish between the two teams?

9. **Bell Numbers**
   A partition of a set A is defined as a collection of nonempty subsets of A that are pairwise disjoint and whose union is A. The number of such partitions of a set A of size n is called the Bell number B_n of the set. For example when $n = 3$, the Bell number $B_3 = 5$ since there are five partitions of a set of size three. For example, if $A = \{a, b, c\}$, then the five partitions are

   $\{\{a\},\{b\},\{c\}\}$

   $\{\{a\},\{b,c\}\}$

   $\{\{b\},\{a,c\}\}$

   $\{\{c\},\{a,b\}\}$

   $\{\{a,b,c\}\}$

   Enumerate the partitions of the set $\{a, b, c, d\}$ to find the Bell number B_4.

10. **Two Committees**
    What is the total number of ways the Snail Darter Society, which consists of 25 members, can elect an executive committee of 2 members and an entertainment committee of 4 members, if no member can serve on both committees?

11. **Serving on More than One Committee**
    How many different ways can the Snail Darter Society, which has 25 members, elect an executive committee of 2 members, an entertainment committee of 3 members, and a welcoming committee of 2 members, if members can serve on more than one committee?

12. **Single-Elimination Tournaments**
    In a 64-team single elimination tournament, what is the total number of possible outcomes of the tournament?

13. Counting Handshakes

There are 20 people in a room and each person shakes hands with everyone else. How many handshakes are there?

14. Not Serving on More than One Committee

How many ways can the Snail Darter Society, who has 25 members, elect an executive committee of 2 members, an entertainment committee of 3 members, and a welcoming committee of 2 members if members can not serve on more than one committee?

15. Counting Softball Teams

A college softball team is taking 25 players on a road trip. The traveling squad consists of three catchers, six pitchers, eight infielders, and eight outfielders. Assuming each player can only play her own position, how many different teams can the coach put on the field?

16. Catalan Numbers

The Catalan numbers C_n, $n = 1, 2, 3, \ldots$ are the number of ways a convex polygon with $n + 2$ sides can be subdivided into triangles. The following Figure 2.18 illustrates the Catalan numbers C_1, C_2, and C_3. Can you find the Catalan number C_4 which is the number of ways to triangulate a convex hexagon which has six sides?

$C_1 = 1$

$C_2 = 2$

$C_3 = 5$

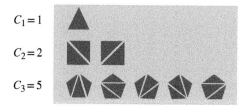

Figure 2.18 First three Catalan numbers.

17. Famous Apple Problem

We wish to distribute eight identical apples to four children. How many ways is this possible if each child gets at least one apple?

18. Basic Pizza Cutting

You are given a circular pizza and ask to cut it into as many pieces as possible, where a cut means any line that passes all the way through the pizza although not necessarily through the center.

a) How many different pieces can you obtain with three slices?

b) How many different pieces can you obtain with four slices?

19. **Pizza Cutter's Formula**
    A pizza cutter wants to cut a pizza in such a way that it has the maximum number of pieces, not necessarily the same size or shape. The only restriction on how the pizza is cut is that each cut much pass all the way through the pizza, not necessary through the center. See Figure 2.19.
    a) If $P(n)$ is the maximum number of pieces from n cuts, then

    $$P(n) = P(n-1) + n.$$

    b) Show the pizza cutter's formula is given by

    $$P(n) = \frac{n^2 + n + 1}{2}.$$

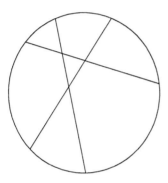

Figure 2.19 Typical pizza after three cuts with seven pieces.

20. **Round Robin Tournaments**
    In a round robin tournament, every team plays every other team exactly once. Normally, an even number n teams compete and the tournament goes $n - 1$ rounds, where each team competes on each round. Find the total number of games played in a round robin tournament of n teams, assuming n is an even number.

21. **Lottery Problem**
    A lottery works as follows. A container is filled with 36 marbles numbered 1, 2, ..., 36 whereupon six marbles are selected at random by the lottery organizers. You buy a ticket and fill in six blanks with numbers between 1 and 36. You win if you choose the same six numbers selected by the organizers, regardless of order. How many different choices of six numbers can you select and since only one choice is a winner, the probability you win is 1 over that number. How many ways can you pick your six numbers and what are your chances of winning?

22. **Relatively Prime Light**

    How many positive integers ≤32 are relatively prime to 32? Two numbers are relatively prime if their greatest common divisor is 1. For example, the positive integers relatively prime to 10 are 1, 3, 7, and 9.

23. **Relatively Prime Medium**

    How many positive integers ≤35 are relatively prime to 35? Two numbers are relatively prime if their greatest common divisor is 1. For example the positive integers relatively prime to 10 are 1, 3, 7, and 9.

24. **Relatively Prime Hard**

    How many positive integers ≤70 are relatively prime to 70? Two numbers are relatively prime if their greatest common divisor is 1. For example the positive integers relatively prime to 10 are 1, 3, 7, and 9.

25. **Internet Research**

    There is a wealth of information related to topics introduced in this section just waiting for curious minds. Try aiming your favorite search engine toward *problems in combinatorics, applications of combinatorics,* and *pigeonhole principle.*

2.4

Cardinality of Sets

> **Purpose of Section** To introduce the concept of **cardinality** of a set and the **equivalence** of two sets. We also define what it means for a set to be finite and infinite and how to compare sizes of sets, both finite and infinite. We end our discussion by defining the cardinality of the natural numbers.

2.4.1   Introduction

No one knows exactly when people first started counting, but a good guess might be when people started accumulating things. However, long before number systems were invented, two people might have determined whether they had the same number of goats and sheep as illustrated in Figure 2.20 by simply placing them in a one-to-one correspondence with each other.

Another person might have designated a stone for each goat, thus obtaining a one-to-one correspondence between the goats and a pile of stones. Today, we no longer use stones to enumerate things since we have symbolic stones in the form of 1, 2, ... To determine the number of goats, we simply "count," 1, 2, ... and envision the rocks $R = \{1, 2, 3, 4, 5\}$ in our mind as illustrated in Figure 2.21.

Throughout the history of mathematics, the subject of infinity has been mostly taboo, more apt to be included in discussions on religion or philosophy. The Greek philosopher Aristotle (c. 384–322 BCE), was one of the first mathematicians to think seriously about the subject, said there were two kinds of infinity, the *potential* and *actual*. He said the natural numbers 1, 2, 3, ... are *potentially* infinite since they can never be completed. The philosopher and theologian Thomas Aquinas (1225–1275) argued that with the exception of God, nothing is actual infinite.

In the seventeenth century, the Italian astronomer Galileo made an observation concerning the perfect squares 1, 4, 9, 16, 25, ... He said since they

Advanced Mathematics: A Transitional Reference, First Edition. Stanley J. Farlow.
© 2020 John Wiley & Sons, Inc. Published 2020 by John Wiley & Sons, Inc.
Companion website: www.wiley.com/go/farlow/advanced-mathematics

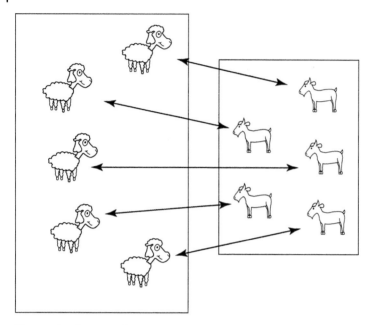

Figure 2.20 Counting sheep.

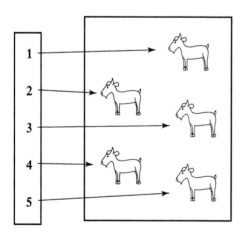

Figure 2.21 Modern way to count.

constitute a subset of the natural numbers, there should be "fewer" of them than the natural numbers, and Table 2.9 would seem to bear this out.

However, he also observed that when one *lines up* the perfect squares as in Table 2.10, it appears that both sets contain the *same* number of members.

Table 2.9 More natural numbers than perfect squares.

1	2	3	4	5	6	7	8	9	10	11	12	13	14	15	16
1			4					9							16

Table 2.10 Equal number of perfect squares as natural numbers.

1	2	3	4	5	6	7	8	9	10	11	$\cdots$	n
$\updownarrow$	$\updownarrow$	$\updownarrow$	$\updownarrow$	$\updownarrow$	$\updownarrow$	$\updownarrow$	$\updownarrow$	$\updownarrow$	$\updownarrow$	$\updownarrow$		$\updownarrow$
1	4	9	16	25	36	49	64	81	100	121	$\cdots$	n^2

His argument was that for every perfect square n^2, there is exactly one natural number n, and conversely for every natural number n, there is exactly one square n^2. Using this reasoning, he concluded terms like *less than, equal,* and *greater than* apply to the size of finite sets but not infinite sets.

2.4.1.1 Early Bouts with Infinity

To get an idea how difficult it is to think about infinity, the ancient Greek philosopher Zeno of Elea (fifth century BCE) argued that motion is an illusion since a person who wishes to travel a fixed distance must first travel half the distance, then half the remaining distance, and then half that remaining distance and so on. So how could a person travel an infinite number of distances in a finite amount of time? Surprisingly, the paradox was not satisfactorily resolved until the tools of calculus were rigorously developed in the nineteenth century. Today, we understand that an infinite number of objects can add ad infinitum and still yield a finite result, much like the one-foot square in Figure 2.22, whose

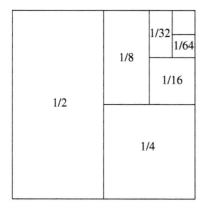

Figure 2.22 Visual response to Zeno's paradox.

area is the union of an unending series of areas of smaller and smaller and smaller squares and rectangles, but yielding a total area of

$$\frac{1}{2} + \frac{1}{4} + \frac{1}{8} + \frac{1}{16} + \cdots = 1$$

Cantor's Seminal Contribution to Infinity   The ground-breaking work of German mathematician, Georg Cantor (1845–1918), whose seminal insights into infinity transformed our thinking about "potential" versus "actual" infinities. Many mathematicians at the time resisted Cantor's ideas, but by the time of Cantor's death in 1918, his ideas were accepted for their importance.

Cantor's big discovery was the realization that although it is not possible to "count" the members of an infinite set, it *is* possible to determine if one infinite set contains the same number of members as another infinite set by simply matching members of one set with members of another set.

This is not any different from the way you determine that you have the same number of fingers on one hand as you have on your other hand. You place your thumb of one hand against the your thumb on the other hand, then place your index finger of one hand against your index finger of your other hand, and so on. When you are finished your fingers are matched with each other in a *one-to-one correspondence;* every finger on one hand having a "matched finger" on the other hand. You may not know how many fingers you have, but Figure 2.23 shows both hands have the same number.

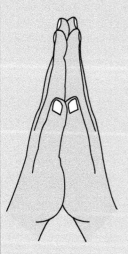

Figure 2.23 One-to-one correspondence.

The above discussion motivates the formal definitions related to Cantor's fundamental ideas.

2.4.2 Cardinality, Equivalence, Finite, and Infinite

We begin with several important definitions.

Definition

- The **cardinality** of a set A is the number of members of A and denoted by $|A|$. Two sets A and B have the **same cardinality** or **cardinality number** when their members can be placed in a one-to-one correspondence with each other. Such sets are called **equivalent sets**, which we denote by $A \approx B$.

  The sets $A = \{a, b, c\}$ and $B = \{1, 2, 3\}$ are equivalent sets since we have a one-to-one correspondence as[1] $a \leftrightarrow 1, b \leftrightarrow 2, c \leftrightarrow 3$ describes and one-to-one correspondence between the members of A and B. Thus, we write $A \approx B$.

- A nonempty set A is **finite** if and only if it is equivalent to a set of the form $\mathbb{N}_n = \{1, 2, \dots, n\}$ for some natural number n. In this case, the set has **cardinality** n and is denoted by $|A| = n$.

  The set $A = \{a, b, c\}$ has cardinality $|A| = 3$ since its members can be put in a one-to-one correspondence with members of $\mathbb{N}_3 = \{1, 2, 3\}$.

- A set is **infinite** if it is not finite. The natural, rational, and real numbers are all examples of infinite sets. The members of any of these sets cannot be placed in a one-to-one correspondence with members of a set of the form $\{1, 2, 3, \dots, n\}$ for any natural number n.

We now know there are two kinds of sets when it comes to size, finite and infinite. We know that some finite sets are larger, smaller, or equal in size to other sets. The cardinality of $A = \{a, b, c\}$ is $|A| = 3$ and the cardinality of $B = \{a, b, c, d\}$ is $|B| = 4$, and so we have $|A| < |B|$. But what about the cardinality of infinite sets, such as $\mathbb{N}$, $\mathbb{Z}$, and $\mathbb{Q}$? We know these sets have infinite cardinalities, but are the cardinalities the same, or is one cardinality larger than the others? And if that is so, exactly how many different cardinalities are there? Is there more than one infinity?

---

1 There are, of course, many one-to-one correspondences between these sets. In fact the reader may list a few, and even ask, how *many* one-to-one correspondences there are between these sets?

When it comes to comparing sizes of sets, finite or infinite, we have a good friend, and a friend the reader is no doubt familiar. It is the concept of a function, and although we do not formally introduce functions until the next chapter, the reader no doubt will have an adequate understanding of functions to understand the ideas presented here. We begin by introducing three important types of functions $f: A \rightarrow B$ that map a domain A into a set B; the one-to-one function, the onto function, and the one-to-one correspondence function.

One-to-One Function   If a function $f: A \rightarrow B$ sends *different* members $x \in A$ to *different* values $f(x) \in B$, the function f is called **one-to-one** (1–1) or an **injective function**. In the language of predicate logic, we write this as

$$(\forall x_1, x_2 \in A)[x_1 \neq x_2 \Rightarrow f(x_1) \neq (x_2)]$$

or equivalently, its contrapositive:

$$(\forall x_1, x_2 \in A)[f(x_1) = f(x_2) \Rightarrow x_1 = x_2]$$

We might think of 1–1 functions as those functions that "spread out" points of A, resulting in images not doubling up in B. The following Figure 2.24 illustrates a 1–1 function and illustrates how 1–1 functions are related to the sizes of A and B.

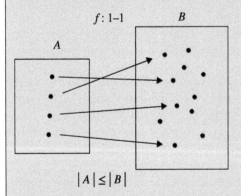

$$|A| \leq |B|$$

Figure 2.24 One-to-one correspondence.

The second type of function is onto function.

Onto Function   A function $f: A \rightarrow B$ from A to B is said to be from A **onto** B (or a **surjection**) if every member $y \in B$ is the image of *at least* one preimage $x \in A$. In other words

$$(\forall y \in B)(\exists x \in A)(y = f(x))$$

We might think of onto functions as those that "cover up" members of *B*. The following Figure 2.25 illustrates an onto function and how onto functions are related to the sizes of *A* and *B*.

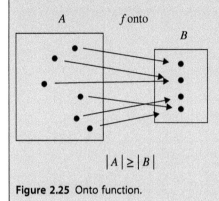

$$|A| \geq |B|$$

Figure 2.25 Onto function.

The third type of function we introduce is the one-to-one correspondence.

One-to-One Correspondence   A function $f: A \rightarrow B$ is called a **one-to-one-correspondence**[2] (or **bijection**) between *A* and *B* if it is both 1–1 and onto. We can think of a one-to-one correspondence between sets as a relation connecting each member of one set to exactly one member of the other set. The following Figure 2.26 illustrates a one-to-one correspondence between *A* and *B* and how a one-to-one correspondence between sets is related to their cardinality.

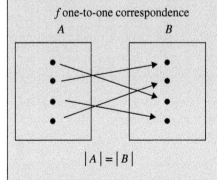

$$|A| = |B|$$

Figure 2.26 One-to-one correspondence.

---

2 Be careful not to confuse a 1–1 function with a function that is a one-to-one-correspondence. The language is a bit ambiguous. You might want to use the alternate names, injection and bijection.

2.4.3 Major Result Comparing Sizes of Finite Sets

Theorem 1 (Comparing Finite Sets)
Let f be a function that assigns each member of a set A a member of a set B, denoted by $f: A \rightarrow B$. We can compare the sizes of A and B using the following rules[3]:

a) If $f: A \rightarrow B$ is 1–1, then $|A| \leq |B|$.
b) If $f: A \rightarrow B$ is 1–1 but not onto, then $|A| < |B|$.
c) If $f: A \rightarrow B$ is onto, then $|A| \geq |B|$.
d) If $f: A \rightarrow B$ is onto but not 1–1, then $|A| > |B|$.
e) If $f: A \rightarrow B$ is a one-to-one correspondence, then $|A| = |B|$.

Proof
We prove (a) and leave the proofs of others to the reader.

a) The proof that

$$f : A \rightarrow B \text{ is } 1-1 \Rightarrow |A| \leq |B|$$

is by a contradiction of the form

$$[(C \wedge \sim D) \Rightarrow \sim C] \equiv (C \Rightarrow D)$$

In other words, if we assume that f is 1–1 and $|A| > |B|$, this leads to the contradiction that f is *not* 1–1, hence, we conclude $|A| \leq |B|$. We begin by using the assumption $|A| > |B|$ where

$$A = \{a_1, a_2, ..., a_m\}, B = \{b_1, b_2, ..., b_n\}$$

and $m > n$. Since f is 1–1, we can write[4]

$$f(a_1) = b_1, f(a_2) = b_2, ..., f(a_n) = b_n, f(a_{n+1}) = \text{no where to go}$$

where all the b_js are distinct. But, there are more members in B than in A, so we have run out of places to map members of A, thus f must map a_{n+1} to a previous values of b_j, thus proving the contradiction that that f is *not* 1–1. Hence, the result is proven. ∎

---

3 These "if–then" results can be stated as "if and only if," but we are only interested in the given direction of the theorems.
4 There is no reason the specific a_js have to map to the b_js with the same index, but there is no reason we cannot write the function in this way.

For finite sets A, B we know what is means when we say $|A| < |B|$, but this meaning breaks down for infinite sets when sets cannot be counted. However, the language of sets helps us overcome this difficulty. For finite sets, it is intuitively clear that $|A| \leq |B|$ when there is an injection (1–1) from A to B. We now use this motivation do define what it means by $|A| \leq |B|$ for infinite sets.

Comparing Infinite Sets   The previous discussion for finite sets motivates how functions can aid in comparing the cardinalities of infinite sets. If A and B are two infinite arbitrary sets, we define:

a) $|A| = |B|$ means there is a bijection $A \rightarrow B$
b) $|A| \leq |B|$ means there is a 1–1 function $A \rightarrow B$
c) $|A| < |B|$ means there is a 1–1 function but no onto function $A \rightarrow B$
d) $|A| \geq |B|$ means there is an onto function $A \rightarrow B$
e) $|A| > |B|$ means there is an onto function but no 1–1 function $A \rightarrow B$

Useful Note   Sometimes words in mathematics do not suggest the type of information that is most useful. It might be useful to think of 1–1 functions as "expanding" functions since 1–1 maps $f : A \rightarrow B$ result in $|A| \leq |B|$. By the same token, it might be useful to think of onto functions as "contracting" functions since onto maps $f : A \rightarrow B$ result in $|A| \geq |B|$.

2.4.4   Countably Infinite Sets

We defined an infinite set as a set that is not finite. We now define a countably infinite (or denumerable) set.

Definition
An infinite set A is **countably infinite** (or **denumerable**) if and only if its members can be arranged in an infinite list $a_1, a_2, a_3, \ldots$, and the cardinality (i.e. number of elements) is given the name **aleph null** and written $\aleph_0$. The most obvious example of a countably infinite set is the natural numbers $\mathbb{N} = \{1, 2, 3, \ldots\}$, thus we write $|\mathbb{N}| = \aleph_0$.

Infinity can be tricky. If we add a new member to a finite set, we increase its cardinality by one, but things behave much differently for infinite sets. If we add "0" to the natural numbers $\mathbb{N} = \{1, 2, 3, \ldots\}$ getting $\mathbb{N} \cup \{0\}$, do we increase the

Table 2.11 $\mathbb{N} \cup \{0\} \approx \mathbb{N}$.

$\mathbb{N} \cup \{0\}$	0	1	2	3	4	5	$\cdots$	n
$\updownarrow$	$\updownarrow$	$\updownarrow$	$\updownarrow$	$\updownarrow$	$\updownarrow$	$\updownarrow$	$\updownarrow$	$\updownarrow$
$\mathbb{N}$	1	2	3	4	5	6	$\cdots$	$n+1$

"size" of the set? The answer is no since we can relate the two sets with the one-to-one correspondence shown in Table 2.11.

Example 1
$\mathbb{Z} \approx \mathbb{N}$ Show that there is a one-to-one correspondence between the natural numbers $\mathbb{N}$ and the integers $\mathbb{Z}$, which means the sets have the same cardinality.

Solution
There are many one-to-one correspondences between $\mathbb{N}$ and $\mathbb{Z}$, but we require just one. A single one-to-one correspondence is illustrated graphically[5] in Figure 2.27, which implies the sets have the same cardinality, namely $|\mathbb{Z}| = |\mathbb{N}| = \aleph_0$.

Figure 2.27 One-to-one correspondence between $\mathbb{N}$ and $\mathbb{Z}$.

The rows in Table 2.12 illustrate the same correspondence in tabular form as the visual correspondence in Figure 2.27.

Table 2.12 Bijection between $\mathbb{N}$ and $\mathbb{Z}$.

$\mathbb{N}$	1	2	3	4	5	6	$\cdots$
$\updownarrow$	$\updownarrow$	$\updownarrow$	$\updownarrow$	$\updownarrow$	$\updownarrow$	$\updownarrow$	$\updownarrow$
$\mathbb{Z}$	0	1	-1	2	-2	3	$\cdots$

---

5 A one-to-one correspondence between two sets can be an equation or some type of visual diagram.

Cardinal and Ordinal Numbers   Numbers are used in two different ways. The number 3 is called a **cardinal number** when we say the "three little pigs," but when we say "the third little pig built his house out of bricks," the number three (or third) is called an **ordinal number**.

Important Note   The symbol "∞," from calculus is not intended to stand for an infinite number. The phrase $x \to \infty$ simply refers to the fact that we allow the variable x to grow without bound.

Theorem 2   Rational Numbers Are Countably Infinite
The cardinality of the rational numbers $\mathbb{Q}$ is $\aleph_0$.

Proof
Again, we resort to a visual one-to-one correspondence illustrated in Figure 2.28. Here, every rational number, reduced to lowest form, lies somewhere in the array, and the given path passes through every rational number one and only one time. If we assign "1" to the first member in the path, "2" to the second member, and so on, we have found a one-to-one correspondence between the natural numbers $\mathbb{N}$ and the rational numbers $\mathbb{Q}$. Hence, $\mathbb{Q} \approx \mathbb{N}$ or $|\mathbb{Q}| = |\mathbb{N}| = \aleph_0$.   ∎

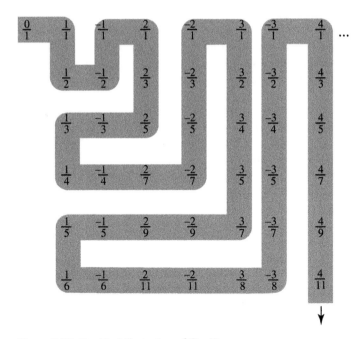

Figure 2.28 Graphical illustration of $\mathbb{Q} \approx \mathbb{N}$.

Interesting Note   We sometimes call 1–1 function an **injective function** and an onto function a **surjective functions**. The words "injection" and "surjection" come from the French language, where the French word *injectif* means injecting something into another. The word "sur" in French means "on," hence, the word surjection for an onto mapping. The terminology was originally coined by the Bourbaki group, a society of French mathematicians named after a French general. Interested readers can learn about this group of mathematicians online.

Cantor was initially convinced that all infinite sets were countably infinite as suggested from $|\mathbb{N}| = |\mathbb{Z}| = |\mathbb{Q}|$. The next obvious question was what about the cardinality of the real numbers? What is $|\mathbb{R}|$? Cantor tried and tried to prove the real numbers are countably infinite by proving they could be put in a one-to-one correspondence with the natural numbers, but in the process discovered something amazing. What did he discover? Wait until the next section.

Problems

1. **Countable Sets**
   Show that the union of two countable sets is countable.

2. **Equivalent Sets**
   For the following sets, find an explicit one-to-one correspondence that shows the intervals are equivalent.
   a) $\{a, b, c\} \approx \{1, 2, 3\}$
   b) $[0, 1) \approx [0, \infty)$
   c) $(0, 1) \approx \mathbb{R}$
   d) $[0, 1] \approx [3, 5]$

3. **Even and Odd Natural Numbers**
   Let E be the set of even positive integers and O be the set of odd positive integers. Given is an explicit function to show the following equivalences:
   a) $E \approx O$
   b) $\mathbb{N} \approx O$
   c) $\mathbb{N} \approx E$
   d) $\mathbb{N} \times \mathbb{N} \approx \mathbb{N}$

4. **Infinite Sets**
   A set is infinite if and only if it is equivalent to a proper subset of itself. Use this definition of an infinite set to show the following sets are infinite:
   a) $\mathbb{N}$
   b) $\mathbb{R}$

5. **Bijection from the Prime Numbers**
   The following sets are equivalent. Find a bijection $f : A \to B$.
   a) $A = \mathbb{N}$, $B =$ prime numbers
   b) $A = \mathbb{N}$, $B = \{10, 12, 14, ...\}$
   c) $A = \mathbb{R}$, $B = (0, \infty)$

6. **Cardinality Test**
   Use the results of Theorem 1 to show the following.
   a) Prove the cardinality of $A = \{a, b\}$ is less than or equal to the cardinality of $B = \{1, 2, 3\}$ by finding a 1–1 function $f : A \to B$.
   b) Prove the cardinality of $A = \{a, b\}$ is strictly less than the cardinality of $B = \{1, 2, 3\}$ by finding a 1–1 function $f : A \to B$ that is not onto.
   c) Prove that the cardinality of $A = \{a, b, c\}$ is greater than or equal to the cardinality of $B = \{1, 2\}$ by finding an onto function $f : A \to B$.
   d) Prove the cardinality of $A = \{a, b, c\}$ is strictly greater than that of $B = \{1, 2\}$ by finding an onto function $f : A \to B$ that is not 1–1.
   e) Prove that the cardinality of $A = \{a, b, c\}$ is equal to the cardinality of $B = \{1, 2, 3\}$ by finding a one-to-one correspondence between the sets.

7. **Internet Research**
   There is a wealth of information related to topics introduced in this section just waiting for curious minds. Try aiming your favorite search engine toward *countably infinite*, *Georg Cantor*, *Zeno of Elea*, and *cardinality of a set*.

2.5

Uncountable Sets

> **Purpose of Section**   We now present Cantor's seminal result that the cardinality of real numbers is uncountable. We then continue with the equally exciting result that the cardinality of the two-dimensional plane is the same as the cardinality of the one-dimensional real line.

2.5.1   Introduction

Thus far, the only infinity considered has been countable infinity, which is the cardinality of the natural numbers or any set of many sets that can be written as a sequence. This begs the question are all infinite sets equivalent to the natural numbers? In 1874, the German mathematician Georg Cantor proved that there are in fact sets with even larger cardinalities than of the natural numbers, which led to the development of modern set theory.

At the time Cantor believed, as did all mathematicians, that infinity was infinity. Thinking along those lines, Cantor tried to prove that the real numbers have the same cardinality as the natural numbers, but failed and his failure was one of the greatest failures in the history of mathematics since in the process he proved that the real numbers have a larger cardinality than that of the natural numbers.

Definition   An infinite set with cardinality larger than that of the natural numbers is said to have **uncountable cardinality**, or simply an **uncountable set**.

The above discussion leads us to Cantor's famous diagonalization theorem.

Advanced Mathematics: A Transitional Reference, First Edition. Stanley J. Farlow.
© 2020 John Wiley & Sons, Inc. Published 2020 by John Wiley & Sons, Inc.
Companion website: www.wiley.com/go/farlow/advanced-mathematics

Theorem 1   Cantor's Diagonalization Theorem
The set of real numbers in the open interval (0, 1) has a cardinality larger than the natural numbers called an uncountable set.[1]

Proof
First, note that mapping $f: \mathbb{N} \to (0, 1)$, defined by

$$f(1) = 2/3$$

$$f(n) = 1/n, \quad n = 2, 3, \ldots$$

maps different natural numbers into different numbers in the interval (0, 1) points and is thus 1–1 (an "expanding" function), which implies $|\mathbb{N}| \leq |(0, 1)|$. To prove strict inequality $|\mathbb{N}| < |(0, 1)|$, Cantor assumed the contrary that $|\mathbb{N}| = |(0, 1)|$, meaning there exists a one-to-one correspondence between $\mathbb{N}$ and (0, 1). We illustrate Cantor's assumed correspondence in Table 2.13, where natural numbers are listed on the left and numbers in the interval (0, 1), expressed in decimal form,[2] on the right.

Table 2.13 Hypothesized one-to-one correspondence $\mathbb{N} \leftrightarrow (0, 1)$.

$\mathbb{N}$															
↓					↓										
1	↔	0	.	1	4	3	2	0	2	8	1	4	...
2	↔	0	.	3	5	5	4	4	4	6	2	6	...
3	↔	0	.	6	3	0	3	5	3	4	1	5	...
4	↔	0	.	8	7	8	7	3	5	5	3	3	...
5	↔	0	.	0	5	8	6	5	8	8	7	5	...
6	↔	0	.	4	9	6	5	8	7	7	5	4	...
...		...	.	...	...	...	...	...	...	...	...		⋱

Cantor showed that regardless how this correspondence is formed there is always a real number *not* on the list, which implies there is no one-to-one correspondence between $\mathbb{N}$ and $\mathbb{R}$.

---

1 Cantor proved this result known as his *diagonalization* proof in 1877. He had another proof which he published earlier in 1874.
2 Every real number can be expressed uniquely in decimal form $a_0 a_1 a_2 a_3 \ldots$, where a_0 is an integer and the numbers $a_1, a_2, \ldots$ after the decimal are integers between $0 \leq a_i \leq 9$, provided the convention is made that if the decimal expansion ends with an infinite string of 9's, such as 0.499 999... this is taken as the same as 0.5. Making this convention provides a one-to-one correspondence between the real numbers and their decimal expansion.

To show how Cantor creates a "rogue" real number 0. $a_1a_2a_3...$ not on the list in Table 2.13, Cantor picks the first digit a_1 *different* from the first digit of the real number corresponding to 1 (i.e. the number 0.143 202 814...). In other words, pick a_1 anything other than 1, say $a_1 = 3$. See Table 2.14. Now select the second digit a_2 anything other than the second digit of the real number corresponding to the 2. So we select a_2 different from 5, which we (randomly) pick $a_2 = 2$.

Continuing this process, working down the diagonal of the table, the first six digits of our rogue number might be 0.327 245... and continuing this process indefinitely, we arrive at a real number that does *not* correspond to any natural number. Hence, the assumed one-to-one correspondence $f: \mathbb{N} \to (0, 1)$ described in Table 2.14 is not a one-to-one correspondence between $\mathbb{N}$ and $(0, 1)$ since we have found an element of $(0, 1)$ that does not correspond to any natural number. Thus we conclude that the open interval $(0, 1)$ has a larger cardinality than the natural numbers, or $|\mathbb{N}| < |(0, 1)|$. ∎

Table 2.14 Cantor's diagonalization process.

$\mathbb{N}$				(0, 1)									
↓				↓									
1	↔	0	.	[1]	4	3	2	0	2	8	1	4	...
2	↔	0	.	3	[5]	5	4	4	4	6	2	6	...
3	↔	0	.	6	3	[0]	3	5	3	4	1	5	...
4	↔	0	.	8	7	8	[7]	3	5	5	3	3	...
5	↔	0	.	0	5	8	6	[5]	8	8	7	5	...
6	↔	0	.	4	9	6	5	8	[7]	7	5	4	...
...	...	...		...	...	...	...	...	...	...	...	...	...
...	...	↓		↓	↓	↓	↓	↓	↓				
...	...	0	.	3	2	7	2	4	5	...	...	...	...

Example 1   More Real Numbers than Natural Numbers
Show the open interval $(0, 1)$ has the same cardinality of the real numbers, i.e. $|(0, 1)| = |\mathbb{R}|$, thus proving $|\mathbb{N}| < |(0, 1)| = |\mathbb{R}|$.

Solution
We can visualize in Figure 2.29 a one-to-one correspondence between members of the interval $(0, 1)$ and real numbers by wrapping the interval $(0, 1)$ around the bottom half of a circle and whose center lies on the upper y axis of the Cartesian plane. Hence, we have equal cardinalities $|(0, 1)| = |\mathbb{R}|$ and thus $|\mathbb{N}| < |(0, 1)| = |\mathbb{R}|$.

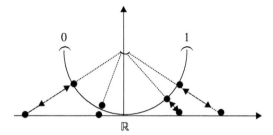

Figure 2.29 Visual proof of $|(0, 1)| = |\mathbb{R}|$.

Definition
The cardinality of the real numbers is called the **cardinality of the continuum** and denoted by the letter c. Sets with cardinality c are called **uncountable sets** (or **uncountably infinite**).

> **Important Note** Roughly speaking, an uncountable set has so many points that its members cannot be arranged in a sequence.

You may now ask what other sets have the cardinality of the continuum. The answer will surprise you.

Example 2 Equivalent Intervals
If (a, b) and (c, d) are any two intervals of finite length on the real line, then $(a, b) \approx (c, d)$.

Proof
The function

$$y = \left(\frac{d-c}{b-a}\right)(x-a) + c$$

provides a one-to-one correspondence between members of (a, b) and members of (c, d). Figure 2.30 gives the visual representation of this one-to-one correspondence.

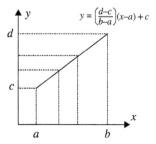

Figure 2.30 Equivalence of two intervals of real numbers.

> **Important Note**   Cantor's proof that there are different "sizes" of infinity is one of the cornerstones of all mathematics and plays an important role in many areas of pure mathematics such as topology and analysis.

Example 3   All Lines are Equal

Show that all line segments have the same cardinality, namely the cardinality of the continuum.

Solution

The drawing in Figure 2.31 shows one-to-one correspondence between line segments of different lengths. In other words, any two line segments $A'B'$ and AB of different lengths can be placed in one-to-one correspondence by the one-to-one $x \leftrightarrow x'$. This fact, combined with the fact that $(0, 1) \approx \mathbb{R}$, we have that any line segment, regardless of its length, has cardinality c, the cardinality of the continuum.

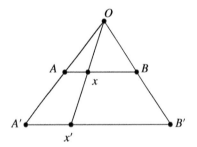

Figure 2.31 AB and $A'B'$ have the same cardinality.

> **Important Note**   The intervals
> $$(a,b), [a,b], [a,b), (a,b], (-\infty,b], [a, \infty),$$
> all have cardinality c.

> **Important Note**   The lazy figure eight symbol "∞" does not represent infinity, but is simply a symbol used to denote that a set of real numbers is unbounded, such as (a, ∞), $(-\infty, b)$, $(-\infty, \infty)$, and so on.

2.5.2 Cantor's Surprise

Cantor had not finished surprising the mathematics world with his discovery of more than one kind of infinity. It completely destroyed previous thinking that "infinity was infinity." Cantor's next project was to prove there are more points in the plane than there are real numbers, and again Cantor failed, and again his failure made another monumental discovery. Cantor worked tirelessly from 1871 to 1874 to prove the cardinality of the plane is greater than the cardinality of the real line, but to his amazement he proved they are the same. In a letter to his good friend Richard Dedekind, he said, "I see it, but I don't believe it." Here is a summary of Cantor's proof.

Theorem 2 Cantor's Surprise
The cardinality of the open interval (0, 1) drawn in Figure 2.32 is the same as the cardinality of the open unit square

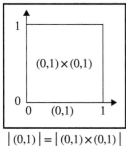

$$|(0,1)| = |(0,1) \times (0,1)|$$

Figure 2.32 $|(0, 1)| = |(0, 1) \times (0, 1)|$.

Proof
The function $f: (0, 1) \rightarrow (0, 1) \times (0, 1)$, defined by

$$f(x) = \left(x, \frac{1}{2}\right), \quad 0 < x < 1$$

maps different numbers in (0, 1) to different points in the unit square $(0, 1) \times (0, 1)$, hence, the function is 1–1 (remember an "expanding" function) which implies $|(0, 1) \leq |(0, 1) \times (0, 1)||$. Going the other way, the function $g: (0, 1) \times (0, 1) \rightarrow (0, 1)$ that maps the unit square to the unit interval, defined by[3]

$$g : (0.a_1 a_2 a_3 ..., 0.b_1 b_2 b_3 ...) \rightarrow 0.a_1 b_1 a_2 b_2 a_3 b_3 ... \in (0,1)$$

---

3 To avoid ambiguity, choose 0.5 instead of 0.499 99... for the number 1/2. In this way, each decimal form represents exactly one number.

is also 1–1 ("expanding" function), which implies $|(0, 1) \times (0, 1)| \leq |(0, 1)|$. Hence, we are left with $|(0, 1) \times (0, 1)| = |(0, 1)|$, which showed that both sets have the cardinality of the continuum c. Figure 2.33 illustrates the mapping g from the unit square to the unit interval.

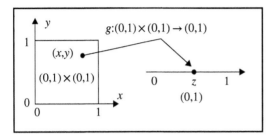

Figure 2.33 One-to-one map showing $|(0, 1) \times (0, 1)| \leq |(0, 1)|$.

Final Step: The mapping $(x, y) \rightarrow (X, Y)$ from the open unit square $(0, 1) \times (0, 1)$ to the Cartesian plane $\mathbb{R}^2$ defined by

$$X = \tan\left(\pi(x - 1/2)\right), \quad 0 < x < 1$$
$$Y = \tan\left(\pi(y - 1/2)\right), \quad 0 < y < 1$$

with inverse

$$x = \frac{1}{\pi}\arctan(X) + \frac{1}{2}, \quad -\infty < X < \infty$$
$$y = \frac{1}{\pi}\arctan(Y) + \frac{1}{2}, \quad -\infty < Y < \infty$$

is a one-to-one correspondence. You can envision the nature of this transformation. First, realize that the point $(x, y) = (0.5, 0.5)$ maps to the origin $(0, 0)$. Now envision tiny squares (horizontal and vertical sides) centered at $(0.5, 0.5)$ and visually expand them so they approach the unit square. Doing this, by the nature of the tangent function, the images of these tiny expanding squares will be large expanding squares centered at $(0, 0)$ that will approach the Cartesian plane.

Hence, step by step we have shown

$$\left|\mathbb{R}^2\right| = |(0, 1) \times (0, 1)| = |(0, 1)| = |\mathbb{R}| = c$$

In fact, this line of reasoning can continue, the net result being that the cardinality of n-dimensional space $\mathbb{R}^n$ for any $n = 2, 3, \ldots$ is the same as the cardinality c of the real numbers. ∎

> **Historical Note**   Between the years 1871–1884, Cantor created a new and special mathematical discipline, the theory of infinite sets.

2.5.2.1   Cardinality of the Irrational Numbers

The interval $(0, 1)$ is the disjoint union of rational and irrational numbers. Since the interval $(0, 1)$ has cardinality c and the rational numbers have cardinality $\aleph_0$ and the union of two countably infinite sets is countably infinite, this implies that the irrational numbers have cardinality c. In other words, there are more irrational numbers than rational numbers.

> **Important Note**   One may think that the study of infinite sets is simply an academic curiosity for set theorists, but nothing could be further from the truth. Set theory is intimately related to many branches of pure mathematics, notably topology, real analysis, probability, and measure theory. A major concept in topology is connectedness and connectedness is related to cardinality by the property that only uncountable sets (like the real numbers) are connected and countable sets (like the rational numbers) are disconnected. In measure theory, countably infinite sets always have measure zero, a concept important in the study of integration theory.

2.5.2.2   Algebraic and Transcendental Numbers

A number x is called a real **algebraic** number if it is a real root of a polynomial equation[4]

$$x^k + a_{k-1}x^{k-1} + \cdots + a_1 x + a_0 = 0$$

where the coefficients a_k's are integers. The irrational number $\sqrt{2}$ is an example of an algebraic number being a root of $x^2 - 2 = 0$. Numbers that are not algebraic are called **transcendental**. One would guess there are more algebraic numbers than transcendental numbers since the first transcendental number, called the Liouville constant after who had discovered it, was only discovered in 1851. It was later discovered that π and Euler's constant e are also transcendental. Later, Cantor shocked the mathematical world when he proved there are more transcendental numbers than algebraic ones.

---

4 We restrict ourselves here to real algebraic numbers.

Theorem 3   The Set of Algebraic Numbers Has Cardinality $\aleph_0$

Proof
The idea is to express the algebraic numbers as a countable union of finite sets and thus countable. To do this, we let A_n be the collection of polynomial equations[5]

$$P_k(x) = x^k + a_{k-1}x^{k-1} + \cdots + a_1 x + a_0 = 0. \quad k = 1,2,...,n$$

where the coefficients $a_0, a_1, \ldots, a_{k-1}$ of each polynomial $P_k(x)$ are allowed to take on any of the $2n+1$ integers $\{-n, -n+1, \ldots, -1, 0, 1, \ldots, n\}$. For example A_{10} is the collection of all polynomial equations of degrees 1 through 10, where the coefficients of each polynomial are allowed any of the 21 integers ranging from -10 to $+10$. Three such equations in A_{10} are

$$9x + 4 = 0$$

$$x^4 + 8x^3 - 2x + 3 = 0$$

$$x^8 + 10x^3 - 4 = 0$$

To show each A_n, $n = 1, 2, \ldots$ is finite, we count the polynomial equations $P_k(x) = 0$ for $k = 1, 2, \ldots, n$ in the special case when $n = 10$. These equations are listed in Table 2.15, along with the number of their roots, real and complex.

Table 2.15 Polynomial equations of A_{10} and their roots.

k	Polynomial equation $P_k(x) = 0$	# Equations	# Roots
1	$P_1(x) = x + a_0 = 0$	$2n+1 = 21$	$1 \cdot (2n+1) = 21$
2	$P_2(x) = x^2 + a_1 x + a_0 = 0$	$(2n+1)^2 = 21^2$	$2 \cdot (2n+1)^2 = 2 \cdot 21^2$
3	$P_3(x) = x^3 + a_2 x^2 + a_1 x + a_0 = 0$	$(2n+1)^3 = 21^3$	$3 \cdot (2n+1)^3 = 3 \cdot 21^3$
...	...	...	...
10	$P_{10}(x) = x^{10} + a_9 x^9 + \cdots + a_1 x + a_0 = 0$	$(2n+1)^{10} = 21^{10}$	$n \cdot (2n+1)^{10} = 10 \cdot 21^{10}$

The sum in the middle column of Table 2.14 yields the total number of polynomial equations in A_{10} and the rightmost column yields the total number of roots, real and complex, of the polynomial equations in A_n. We now observe (and this is the important part) that all polynomial equations belong to A_n for some $n = 1, 2, \ldots$ For example take a random the polynomial equation like

$$x^{55} + 3304x^3 + 3 = 0$$

---
5 There is no loss of generality of taking the coefficient of x^k as 1 since one can always divide the equation by the coefficient of the highest power and not change the roots of the equation.

This polynomial equation belongs to A_{3304} since A_{3304} consists of all polynomial equations of degrees from 1 to 3304 whose coefficients take on values ranging from -3304 to $+3304$.

Hence, *every* algebraic number is a solution of some polynomial equation A_n for some $n = 1, 2, \ldots$ and since there are a countably infinite number of these equations, and since each polynomial equation has a finite number of roots, real and complex, we conclude that the algebraic numbers can be written as a countably infinite number of finite sets, and thus countable. ∎

The fact that the algebraic numbers are countable implies the transcendental numbers are uncountable since if the transcendental numbers were countable that would imply the real numbers would also be countable being the union of countable sets. But the real numbers are uncountable and so the transcendental numbers must be uncountable.

Problems

1. Which of the following sets are finite, countable, or uncountable?
    a) $[0, 1] \cap \mathbb{N}$
    b) $\{\mathbb{N}, \mathbb{Z}, \mathbb{Q}, \mathbb{R}, \mathbb{C}\}$
    c) $\{1/n : n \in \mathbb{N} - \{0\}\}$
    d) $\{x \in \mathbb{R} : x^2 + 1 < 0\}$
    e) The set of all 2 by 2 matrices with natural numbers for elements.

2. **Cardinality of Functions**
    Show that the cardinality of the set of functions $F = \{f : \mathbb{N} \to \mathbb{N}\}$ is uncountable.

3. **Irrational Numbers**
    Show that the irrational numbers in the interval $[0, 1]$ is uncountable.

4. **Algebraic Numbers**
    Show that the numbers

    $$\frac{1 + \sqrt{3}}{2} \text{ and } \frac{1 - \sqrt{3}}{2}$$

    are algebraic numbers.

5. **Countable Plus Singleton**
    Prove that if we add a member to a countable set A, we still have a countable set.

6. **Proving Cardinality ($\mathbb{R} \approx \mathbb{R}^3$)**

    Show that the unit cube

    $$C = \{(x,y,z) : x,y,z \in \mathbb{R}, 0 < x < 1, 0 < y < 1, 0 < z < 1\} \subseteq \mathbb{R}^3$$

    is equivalent to the unit interval $(0, 1) \subseteq \mathbb{R}$. Hint: Use the technique Cantor used to prove $(0, 1)$ is equivalent to the unit square.

7. **Liouville Constant**

    Some famous transcendental numbers are π, e, e^π, π^e, $\ln 2$, ...The first provable transcendental number was found 1851 by the French mathematician Joseph Louiville (1809–1882), who constructed and proved the constant

    $$L = \sum_{n=1}^{\infty} 10^{-n!}$$

    is transcendental. Write out the first few decimal digits of this number. Google "Liouville's constant" and read more about it.

8. **Irrational Numbers**

    Prove or disprove. There exists a countably infinite subset of the irrational numbers.

9. **Internet Research**

    There is a wealth of information related to topics introduced in this section just waiting for curious minds. Try aiming your favorite search engine toward *uncountable set, Cantor's diagonization proof, and examples of uncountable sets.*

2.6

Larger Infinities and the ZFC Axioms

> **Purpose of Section** We state and prove Cantor's power set theorem, which guarantees the existence of infinities larger than the cardinality of the continuum. We also state and discuss the Zermelo–Fraenkel axioms, which are the most commonly used system of axioms for set theory, as well as the Axiom of Choice (AC) and the Continuum Hypothesis (CH), topics that lie at the center of the foundation of set theory and mathematics.

2.6.1   Cantor's Discovery of Larger Sets

The nineteenth-century German mathematician Georg Cantor must have felt that he was on a great adventure. He had just discovered that there were two kinds of infinity, the counting infinity $\aleph_0$ of the natural numbers and the larger continuum infinity c of the real numbers. He then wondered if there were still more infinities; infinities even larger than that of c. He also wondered if there was an infinity between $\aleph_0$ and c, or in other words, is there a set whose cardinality is greater than the cardinality $\aleph_0$ of the natural numbers and less than the cardinality c of the real numbers. He would spend the remainder of his life trying to answer that question and the result will amaze you.

Little ideas often lead to big ideas. We have seen that for finite sets, the power set of a set is larger than the set itself. For example, the set $A = \{a, b, c\}$ contains three elements, whereas its power set

$$P(A) = \{\phi, \{a\}, \{b\}, \{c\}, \{a,b\}, \{a,c\}, \{b,c\}, \{a,b,c\}\}$$

has $2^3 = 8$ elements. This prompted Cantor to ask if the same property held for infinite sets. This question was answered in the affirmative by the following theorem that allows one to create *an infinity of infinites*.

Advanced Mathematics: A Transitional Reference, First Edition. Stanley J. Farlow.
© 2020 John Wiley & Sons, Inc. Published 2020 by John Wiley & Sons, Inc.
Companion website: www.wiley.com/go/farlow/advanced-mathematics

Theorem 1   Cantor's Power Set Theorem
The power set $P(A)$ of any set A, finite or infinite, has a cardinality strictly larger than the cardinality of A.

Proof
For any point $a \in A$, we define the 1–1 mapping $f: a \to \{a\}$ that sends a point $a \in A$ into the set that contains only a, that is $\{a\}$. Hence, we have $|A| \le |P(A)|$. To show the inequality is strict, that is $|A| < |P(A)|$ Cantor used his favorite method of proof, contradiction, and assumed the contrary $A \approx P(A)$. To make the proof more visual, we let A be the countable set[1] $A = \mathbb{N} = \{1, 2, 3, ...\}$ and assume there is a one-to-one correspondence between the natural numbers $\mathbb{N}$ and its power set $P(\mathbb{N})$. A typical one-to-one correspondence is shown in Table 2.16.

Here is where the proof gets tricky. Note that some numbers, like 1, 3, 4 are not members of the subset of which they are paired, while 2 and 5 belong to the sets they are paired. We call the numbers 1, 3, 4 *unmatched* and the numbers 2 and 5 *matched*. We now form the *sets* of matched and unmatched numbers:

$$U = \text{unmatched set} = \{n \in \mathbb{N} : n \text{ unmatched}\} \in P(\mathbb{N})$$

$$M = \text{matched set} = \{n \in \mathbb{N} : n \text{ matched}\} \in P(\mathbb{N})$$

But the unmatched set $U \subseteq \mathbb{N}$ *itself* is a subset of natural numbers itself so by hypothesis, it must be possible[2] to pair with *some* natural number u, which we indicate $u \leftrightarrow U$ and illustrate in Figure 2.34.

Table 2.16 Pairing of the natural numbers with its power set.

$\mathbb{N}$	$\leftrightarrow$	$P(\mathbb{N})$	Matched or unmatched
1	$\leftrightarrow$	$\{3, 5\}$	**Unmatched**
2	$\leftrightarrow$	$\{1, 2, 7\}$	Matched
3	$\leftrightarrow$	$\{9, 13, 20\}$	**Unmatched**
4	$\leftrightarrow$	$\{1, 5, 6\}$	**Unmatched**
5	$\leftrightarrow$	$\{2, 5, 11, 23\}$	Matched
$\vdots$	$\vdots$	$\vdots$	$\vdots$

---

1 We present the proof for countable infinite sets. The proof for uncountable sets follows along similar lines.
2 We will see this assumption leads to a contradiction.

Cantor now asks the fascinating (and tongue-twisting) question, does the number u that is matched with the *unmatched set U belong* to the unmatched set? That is, does $u \in U$ or does $u \notin U$? Take your best guess.

Yes: If you say u belongs to the *unmatched set U* you are in trouble since U consists only of those numbers that are *not* members of the set which they are matched. So your answer must be no. Right?

No: If you say u does *not* belong to the *unmatched set U*, you are again in trouble since again U consists of numbers that are not members of their matched set, which implies that $u \in U$. So regardless of whether you say $u \in U$ or $u \notin U$ you arrive at a contradiction. Hence, Cantor concludes there does *not* exist a one-to-one correspondence from A to $P(A)$, hence $|\mathbb{N}| < |P(\mathbb{N})|$. ∎

We illustrated these ideas in Figure 2.34

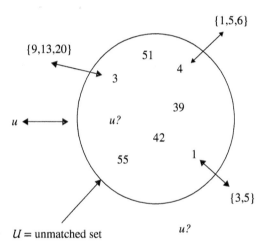

Figure 2.34 $u \in U$ or $u \notin U$?.

2.6.1.1 Summary

Since the power set of a set has more members than the set itself, regardless of whether the set is finite or infinite, it is possible to construct a sequence of larger and larger infinities by simply taking the power set of a power set as illustrated

$$\aleph_0 = |\mathbb{N}| < |P(N)| < |P(P(\mathbb{N}))| < |P(P(P(\mathbb{N})))| < \cdots$$

which are all uncountably infinite except for smallest infinity $\aleph_0$. Cantor called all numbers that are not finite **transfinite numbers,** which means numbers larger than finite numbers.

Since Cantor's theorem shows that for any set, there is always a larger set, it follows there does *not* exist a set of all sets. That is, *there is no set of everything*[3] *since there is always more.*

Now that we know the power set of a set has larger cardinality than the set itself, how does the cardinality of the power set of natural numbers compare with the cardinality of the real numbers? The answer is that they are equal, that is $|P(\mathbb{N})| = |\mathbb{R}|$, but to prove this, we need the help of the Cantor–Bernstein–Schroeder theorem.

2.6.2 The Cantor–Bernstein Theorem

We know from real numbers that $a \leq b$ and $b \leq a$ imply $a = b$. We now ask if the same result holds for cardinalities, does $|A| \leq |B|$ and $|B| \leq |A|$ imply $|A| = |B|$? The answer is yes, and you are probably saying the result if obvious since the cardinalities $|A|$ and $|B|$ are just real numbers. However, keep in mind how cardinalities are defined. The statement $|A| \leq |B|$ between infinite sets means there is a 1–1 map $f : A \rightarrow B$ and $|B| \leq |A|$ means there is a 1–1 map $g : B \rightarrow A$, but the 1–1 maps f and g do not have to be the same. The Cantor–Bernstein–Schroeder theorem says that if there exists a 1–1 map $f : A \rightarrow B$ and a 1–1 map $g : B \rightarrow A$, then there exists a one-to-one correspondence $h : A \rightarrow B$, thus proving $|A| = |B|$.

Theorem 2   Cantor-Bernstein
Given sets A and B if there exists a 1–1 mapping $f : A \rightarrow B$ and a 1 – 1 mapping $g : B \rightarrow A$, then $|A| = |B|$.

Proof
For finite sets $A = \{a_1, a_2, \ldots, a_m\}$ and $B = \{b_1, b_2, \ldots, b_n\}$, the proof is straightforward since $f : A \rightarrow B$ being 1–1 implies $m \leq n$, and if $g : B \rightarrow A$ being 1–1 implies $n \leq m$, hence, we are left with $m = n$. For infinite sets, however, this argument breaks down, and the proof is quite deep and is not given here. Interested readers can find discussions online.

We now get to the main event.   ∎

Theorem 3   The cardinality of the power set of the natural numbers is equal to the cardinality of the real numbers. In other words $|P(\mathbb{N})| = |\mathbb{R}|$.

Proof
The goal is to find a one-to-one correspondence between the subsets of $\mathbb{N}$ and members of the interval $[0, 1]$, and since $[0, 1] \approx \mathbb{R}$, we have $P(\mathbb{N}) \approx \mathbb{R}$.

---

3 The German mathematician David Hilbert said in 1910 that "No one shall drive us from the paradise which Cantor created." Later in 1926, Hilbert said, "It appears to me the most admirable flower of the mathematical intellect and one of the highest achievements of purely rational human activity."

We begin by observing that any real number $x \in [0, 1]$ can be expressed in binary decimal form $x = 0. b_1 b_2 b_3 ...$ where each b_j is 0 or 1. For example

$$x = 0.110\,1000... = \frac{1}{2} + \frac{1}{2^2} + \frac{0}{2^3} + \frac{1}{2^4} + \cdots = \frac{13}{16}$$

$$x = 0.111\,1111... = \frac{1}{2} + \frac{1}{2^2} + \frac{1}{2^3} + \frac{1}{2^4} + \cdots = 1$$

We now associate to each real number $x = 0.b_1 b_2 ...$ in $[0, 1]$ the subset of $\mathbb{N}$ consisting of those natural numbers j for which $b_j = 1$. For example,

$$x = 0.000\,0...0 \in [0, 1] \leftrightarrow \{\} = \varnothing \in P(\mathbb{N})$$
$$x = 0.110\,1... \in [0, 1] \leftrightarrow \{1, 2, 4, ...\} \in P(\mathbb{N})$$
$$x = 0.011\,011... \in [0, 1] \leftrightarrow \{2, 3, 5, 6, ...\} \in P(\mathbb{N})$$
$$x = 0.111\,11... \in [0, 1] \leftrightarrow \{1, 2, 3, 4, ...\} = \mathbb{N} \in P(\mathbb{N})$$

This relationship defines a one-to-one correspondence between the decimal expansions of numbers in $[0, 1]$ and subsets of $\mathbb{N}$, or $P(\mathbb{N}) \approx [0,1]$. But we have seen $[0,1] \approx \mathbb{R}$ and so $P(\mathbb{N}) \approx \mathbb{R}$ or $|P(\mathbb{N})| = |\mathbb{R}|$. ∎

Important Note We must be careful since some numbers in $[0, 1]$ have two decimal representations for the same number, so we must omit one of them to get a one-to-one correspondence. For example, the number 1 can be represented both by 1 and 0.999... and 0.5 can be expressed both by 0.5 and by 0.049 99.... In order that each real number in $[0, 1]$ has exactly one decimal expansion, we use the convention that an infinite string of 9s is never used, but replaced by its numeric equivalent. In other words, 0.009 99... is replaced by 0.01. Also, infinite strings of 0s are omitted, hence, 0.101 000... is replaced by 0.101. Making these conventions results in a one-to-one correspondence between $[0, 1]$ its binary representations.

2.6.3   The Continuum Hypothesis

We have seen that the cardinality of the real numbers is larger than the cardinality of the natural numbers, which begs the question is there an infinity between the two? It was Cantor's belief that c was the next largest infinity after $\aleph_0$, but was never able to prove it. The hypothesis or belief that there is no infinity between c and $\aleph_0$ is known as the **continuum hypothesis.**

The Continuum Hypothesis (CH)   There is no set S whose cardinality satisfies $|\mathbb{N}| < |S| < |\mathbb{R}|$.

A proof of the continuum hypothesis (CH) would confirm the belief that the cardinality of the real numbers c is the smallest uncountable set and bridges the gap between countable and uncountable sets.

After Cantor's death, due to the logical paradoxes of Bertrand Russell and other logicians like Ernst Zermelo and Abraham Fraenkel, set theory was placed on an axiomatic foundation based on the Zermelo–Fraenkel axioms. In 1938, under the framework of these axioms, the Austrian logician Kurt Gödel (1906–1978) proved that the *negation* of CH cannot be proven, which means the CH is either false, cannot be proven false, or is true. Later in 1963, the American logician Paul Cohen proved that the CH cannot be proved true under the Zermelo–Fraenkel axioms. In other words, the CH cannot be proven true or be proven false, which means it is independent of the Zermelo–Fraenkel axioms.

It is analogous to Euclid's fifth axiom of geometry in the sense that you treat it an axiom and accept it as true or accept it as false. If the CH is accepted as true and included with the other Zermelo–Fraenkel axioms, the resulting theory is called **Cantorian set theory**, whereas if the CH is assumed false, the theory is called **non-Cantorian set theory**. In either case, one has a valid set of axioms, albeit much different. Very strange indeed![4]

Gödel's Incompleteness Theorem In 1931, the Austrian logician Kurt Gödel (1906–1978) stated and proved, what arguably is, the most famous and far-reaching theorem in the foundation of mathematics. The theorem states that

Any axiom system containing sufficient axioms to deduce elementary arithmetic cannot be both consistent and complete.

A set of axioms is **consistent** if one cannot prove contradictions from the axioms, and a set of axioms is **complete** if there are statements in the language of the axioms that are not provable using the given axioms. The implication of Gödel's theorem is immense since it says in any consistent axiom system, there will always be statements that cannot be proved and hence, their truth value has no meaning in the framework of the axioms. And which statements are those "undecidable" statements? We will never know.

2.6.4 Need for Axioms in Set Theory

The reader should not entertain the belief that Cantor was the first mathematician to think about sets and their operations, such as union, intersection, and compliment. The concept of a set has been known for centuries. Cantor's

---

4 Most logicians accept Cantorian set theory.

contribution was to introduce the formal study of *infinite sets* and a deep analysis of the infinite and its many unexpected properties. Although Cantor produced many deep ideas, his interpretation of a set was intuitive. That is, to him a set was simply a collection of objects. It was Bertrand Russell who upset that view of sets with his 1901 discovery, called **Russell's Paradox**, which showed that Cantor's the *naive* view sets as "a collection of objects" leads to contradictions. Thus, it became clear in order to have a consistent theory of sets one must formalize its study with "rules-of-the-game." That is, axioms.

Russell's Paradox Russell's paradox is the most famous of all set-theoretic paradoxes that arises in the naive (nonaxiomatic) set theory and motivates the need for an axiomatic formulation of set theory. The paradox was constructed by English logician Bertrand Russell (1872–1970) in 1901, who proposed the rather strange set R, consisting of all sets that do *not* contain themselves. He then asked whether the family R itself was a member of R. The claim that $R \in R$ leads to a contradiction since R consists of sets that do *not* contain themselves. Also, the claim that $R \notin R$ also leads to a contradiction since R *contains* sets that do not contain themselves, which means $R \in R$. Hence, we are left with the contradictory statement

$$R \in R \Leftrightarrow R \notin R$$

Hence, we conclude that we must "restrict" the meaning of a set from the naive point of view of simply being a "collection of things" to being defined by well-thought-out axioms.

2.6.5 The Zermelo–Fraenkel Axioms

The most commonly accepted axioms of set theory are the Zermelo–Fraenkel[5] (ZFC) axioms, developed by German logicians Ernst Zermelo (1871–1953) and Abraham Fraenkel (1891–1965). The original axioms, proposed by Zermelo,[6] restricts the wide latitude of Cantor's interpretation of a set, thus avoiding "bad" things like Russell's paradox, but allowing enough "objects" to be taken as sets for ordinary use in mathematics. The "C" in "ZFC" refers to the Axiom of Choice (AC), the most controversial and debated of the 10 ZFC axioms.

---

5 There are other axioms of set theory than the ZFC axioms, such as the Von Neumann–Bernays–Gödel axioms (which are logically equivalent to ZFC) and the Morse-Kelly axioms, which are "stronger" than ZF.

6 Zermelo published his original axioms in 1908 and were modified in 1922 by Fraenkel and Skolem, and today are called the Zermelo–Fraenkel (ZF) axioms.

Although there is no agreement on the names for each of the 10 axioms and are often written in varying notation, we have settled on the following list.

Zermelo–Fraenkel Axioms

1) **Axiom of the Empty Set:** There is a set Ø with no elements.[7]
   Comment: This set is called the **empty set** and denoted by Ø. This axiom says that the whole theory defined by these 10 axioms is not vacuous by stating *at least one* set exists, albeit nothing in it.
2) **Axiom of Equality:** Two sets are **equal** if and only if they have the same members.
   Comment: The Axiom of Equality (often called the Axiom of Extension) defines what it means for two sets to be equal. The first two axioms say that the empty set (guaranteed by Axiom 1) is unique.
3) **Axiom of Pairs:** Sets $\{a, b\}$ exist.

$$(\forall a)(\forall b)(\exists A)(\forall z)[z \in \{a,b\} \Leftrightarrow (z = x) \vee (z = y)]$$

   Comment: If we pick $a = b$, then the axiom implies there exists sets with one element $\{a\}$. Note that the existential quantifier $\exists A$ quantifies a set, which is not allowed in the first-order predicate logic we studied in Chapter 1.
4) **Axiom of Unions (Unions of Sets Exist):** For any *collection* of sets $F = \{A_\alpha\}_{\alpha \in \Lambda}$, there exists a set

$$Y = \bigcup_{\alpha \in \Lambda} A_\alpha,$$

   called the union of all sets in F. That is

$$(\forall F)(\exists A)(\forall x)[(x \in A) \Leftrightarrow (\exists C \in F)(x \in C)]$$

   Comment: From Axioms 3 and 4, we can construct finite sets. (We are making progress.)
5) **Power Set Axiom (Power Sets Exist):** For every set A, there is a set B (think power set) such that the members of B are subsets of A.
6) **Axiom of Infinity (Infinite Sets Exists)** There exists a set A (think infinite) such that

$$[(\emptyset \in A) \wedge (\forall x)(x \in A)] \Rightarrow [(x \cup \{x\}) \in A]$$

---

7 Zermelo's original axioms were stated in the language of second-order logic, i.e. logical system where *sets* are quantified (like $(\forall A)$, $(\exists A)$) in addition to variables like we studied in Chapter 1. There are versions of the ZFC axioms that use only first-order logic notation, but for convenience, we have quantified *sets* in a few instances.

Comment: This axiom "creates" the numbers $A = \{0, 1, 2, 3, \ldots\}$ according to the correspondence: $0 = \varnothing$, $1 = \{0\}$, $2 = \{1\}$, $3 = \{2\}$, and so on. We conclude the set of natural numbers exist.

7) **Correct Sets Axiom (Axiom that Avoids Russell's Paradox):** If $P(x)$ is a statement and A any set, then

$$\{x \in A : P(x)\}$$

defines a set.

Comment: This axiom rules out sets of the form $\{x : x \notin x\}$ which is the basis of the Russell paradox. The axiom is also an example of what logicians call an *axiom schema*, meaning it is really an infinite number of axioms, one for each statement $P(x)$.

8) **Images of Sets Are Sets:** The image of a set under a function is again a set. In other words if A, B are sets and if $f : A \to B$ is a function with domain A and codomain B, then the image

$$f(A) = \{y \in B : y = f(x)\}$$

is a set.

9) **Axiom of Regularity (No Set Can Be an Element of Itself):** Every nonempty set contains an element disjoint from the set. In other words

$$(\forall A \neq \varnothing) \Rightarrow (\exists x)[(x \in A) \wedge (x \cap A = \varnothing)]$$

Comment: As an example, if $A = \{a, b, c\}$, we can pick "a" and observe $a \cap \{a, b, c\} = \varnothing$, since $\{a\} \cap \{a. b. c\} = \{a\} \neq \varnothing$ but $a \cap \{a, b, c\} = \varnothing$. What this axiom accomplishes is to "keep out" sets that contain *themselves*, like $A = \{1, 2, 3, A\}$.

10) **Axiom of Choice (AC):** Given *any* collection of nonoverlapping, nonempty sets, it is possible to choose one element from each set.

Comments: The AC is a pure *existence* axiom that claims the existence of a set formed by selecting one element from each set in a collection of nonempty sets. The objection to this axiom lies in the fact that the axiom provides no rule for how the items are selected from the sets, simply that it can be done. Some mathematicians argue that mathematics should not allow such a vague rule for declaring the existence of a set.

2.6.6 Comments on the AC

A group of mathematicians are attending a buffet dinner, and as they pass the dessert line containing plates of cookies, each individual plate containing identical cookies, one mathematician says she will appeal to the AC and selects one cookie from each plate, that is the AC. The AC says, given a collection of nonempty disjoint sets, it is possible to define a new set by selecting one member from each set in the collection, regardless of the fact the objects are identical and there is no logical way to decide on what object to select. So what is the fuss over AC? It seems so trivial.

The AC is a pure *existence* axiom, and for that reason, it is a bone of contention for some mathematicians who prefer a constructive approach to sets. Should the AC be an accepted axiom of a set theory. The majority of set theorists say, yes.

Bertrand Russell's Shoe Model for AC   Bertrand Russell once gave an intuitive reason why the AC is sometimes required and why sometimes it is not. Suppose you are supplied with an infinite number of pairs of shoes and are told to pick one shoe from each pair. How would you do it? It is easy, simply choose the left shoe (or right) from each pair.

However, if you are given an infinite pair of socks and told to pick a sock from each pair, then that is a different matter. If you are the kind of person who demands a constructive reason for everything, then you are in trouble. There is no way to pick a sock from each pair if you require a reason for your selection. However, the AC comes to your rescue, which says there *is* such a rule, although not stated, that allows you to pick one sock from each pair. So you simple say, I am appealing to the AC, and randomly pick one of the shoes. It is the non*constructive* aspect of the AC that causes angst to some mathematicians and logicians. To some intuitionists, the word "exists" belongs more to religion than mathematics.[8]

2.6.7   Axiom of Choice ⇔ Well-Ordering Principle

When Zermelo introduced the AC to set theory, most of the mathematicians accepted it and never gave it much thought.[9] However, in 1905, Zermelo proved a theorem that did give people some thoughts. We are getting ahead of ourselves, but consider the less than relation "<" which you have known since middle school, comparing the size of two real numbers $\mathbb{R}$. Now, unlike the interval $[0, \infty)$, the real numbers do *not* have a smallest element. What Zermelo proved was that for *any* nonempty set, the real numbers being an example, it is possible to *order them* in such a way that the ordered set and any subset of it *always has a smallest element*. This theorem is called Zermelo's **Well-Ordering Theorem**. Unfortunately, the theorem *does not say* how to make this ordering; only that

---

8 My wife is notorious about not making up her mind about what dress, coat, socks, ... to buy. When she says, "How can I decide, they look exactly all alike, I just tell her to use the Axiom of Choice."
9 However, many of the foremost mathematicians of the day objected to the axiom of choice, including measure theory pioneers Henri Lebesgue and Emile Borel.

one exists. To this day, no one has ever discovered a way to order the real numbers so they and any subset have a smallest number under this new ordering.

What makes Zermelo's Well-Ordering Theorem so perplexing is that on an intuitive level, it seems very hard to accept, but what is even more perplexing is that Zermelo proved that it is equivalent to the AC, which most people feel is obvious.

Section Summary   So what do we know about infinite sets?

Question: Is there more than one type of infinity (or cardinality)?

Answer: Yes, $\aleph_0$ is the smallest infinity, but there are an infinite number of larger infinities.
Question: Is there a largest infinity (or cardinality)?

Answer: There is no largest infinity, Cantor's theorem shows that power sets of sets always yield larger infinities.
Question: Is there a set of all sets?

Answer: No. If S is the set of all sets, Cantor's theorem states that $|S| < |2^S|$. But $2^S \subseteq S$ since we are assuming S is everything, and so $|2^S| \leq |S|$ which is a contradiction. Hence, the statement there is a set of all sets is meaningless. There is no everything.

Problems

1. **Cardinality of Sets of Functions**
   Show that the set of all functions defined on the natural numbers with values 0 and 1 has cardinality c. Hint: Relate each sequence of 0s and 1s to a subset of natural numbers and then use Cantor's theorem.

2. **Well-Ordered Integers**
   A set is said to be well-ordered if every nonempty subset of the set has a least element. The usual ordering $\leq$ of the integers is not a well-ordering of the integers since the set itself has no smallest element. However, the following relation $\prec$ is a well-ordering of the integers

   $$x \prec y \Leftrightarrow [(|x| < |y|) \vee (|x| = |y| \wedge (x \leq y))]$$

   Order the following subsets of $\mathbb{Z}$ with the above ordering.
   a) $\mathbb{Z}$
   b) $\{n \in \mathbb{Z}: n \leq -1\}$
   c) $\{2,-3,5,-9,-2\}$
   d) $\{3,2,1\}$

3. **Internet Research**

   There is a wealth of information related to topics introduced in this section just waiting for curious minds. Try aiming your favorite search engine toward *Zermelo Fraenkel axioms, Cantor's theorem, Zermelo, well-ordering principle, and Cantor's paradox.*

Chapter 3

Relations

3.1

Relations

> **Purpose of Section** To introduce the concept of a (binary) **relation** between two objects: the objects being almost anything you can imagine, although in mathematics they normally are integers, real numbers, functions, and so on. This section acts as the background for the relations studied in the following sections: the *order* relations, the *equivalence* relation, and the *function* relation.

3.1.1 Introduction and the Cartesian Product

The idea of things being related enters our consciousness a dozen times a day. We talk about people being related in many way, such as gender, race, height, age, and so on. In mathematics, the word relation is used to show the relations between pairs of objects, as when we say

"is less than"	"is a subset of"
"is perpendicular to"	"divides"
"is greater than"	"is congruent to"
"is parallel to"	"is equivalent to"
"is homeomorphic to"	"is isomorphic to?"

However, before defining a relation, it is necessary that we introduce the **Cartesian product** of two sets.

Definition of Cartesian Product Let A and B be arbitrary sets. The Cartesian product of A and B, denoted $A \times B$, and read "A cross B," is the set of ordered pairs

$$A \times B = \{(a,b) : a \in A, b \in B\}.$$

You are already familiar with one Cartesian product, namely the Cartesian plane, which is $\mathbb{R} \times \mathbb{R}$, often denoted by $\mathbb{R}^2$.

Advanced Mathematics: A Transitional Reference, First Edition. Stanley J. Farlow.
© 2020 John Wiley & Sons, Inc. Published 2020 by John Wiley & Sons, Inc.
Companion website: www.wiley.com/go/farlow/advanced-mathematics

Example 1   Cartesian Product and Its Graph
If $A = \{1, 2\}$ and $B = \{1, 3, 4\}$, then their Cartesian product is

$$A \times B = \{(1,1),(1,3),(1,4),(2,1),(2,3),(2,4)\}$$

The dots in Figure 3.1 are the **graph** of the Cartesian product, where we have labeled the horizontal and vertical axes by x and y.

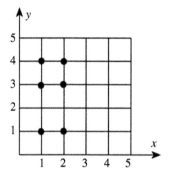

Figure 3.1 Graph of the $\{1, 2\} \times \{1, 3, 4\}$.

Order is important for Cartesian products. For the sets A, B in Example 1, we have

$$A \times B = \{(1,1),(1,3),(1,4),(2,1),(2,3),(2,4)\}$$
$$B \times A = \{(1,1),(3,1),(4,1),(1,2),(3,2),(4,2)\}$$

which illustrates that in general $A \times B \neq B \times A$. Note that the graph of the Cartesian product $B \times A$ consists of points reflected through the 45° line $y = x$ from the graph of $A \times B$ as illustrated in Figure 3.2.

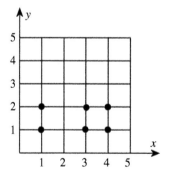

Figure 3.2 Graph of $\{1, 3, 4\} \times \{1, 2\}$.

The Cartesian product interacts with the union and intersection of sets in the following way as one might suspect.

Example 2   Identities of Relations
If A, B, and C are sets, then

a) $A \times (B \cup C) = (A \times B) \cup (A \times C)$
b) $A \times (B \cap C) = (A \times B) \cap (A \times C)$
c) $(A \times B) \cap (C \times D) = (A \cap C) \times (B \cap D)$
d) $(A \times B) \cup (C \times D) = (A \cup C) \times (B \cup D)$

Proof
a) Note how the proof employs the distributive property from sentential logic involving "and" and "or."

$$(x,y) \in A \times (B \cup C) \Leftrightarrow x \in A \text{ and } y \in B \cup C$$
$$\Leftrightarrow x \in A \text{ and } (y \in B \text{ or } y \in C)$$
$$\Leftrightarrow (x \in A \text{ and } y \in B) \text{ or } (x \in A \text{ and } y \in C)$$
$$\Leftrightarrow (x,y) \in A \times B \text{ or } (x,y) \in A \times C$$
$$\Leftrightarrow (x,y) \in (A \times B) \cup (A \times C).$$

Figure 3.3 illustrates a typical example of this identity, although in general A, B, C need not be intervals of real numbers.

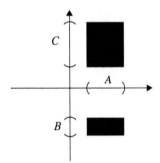

Figure 3.3 Visualization of $A \times (B \cup C) = (A \times B) \cup (A \times C)$.

A few typical Cartesian products are drawn in dark in Figure 3.4.

3.1.2   Relations

A (binary) **relation** is a rule that assigns truth values (true or false) to two things, which normally are numbers, sets, functions, and so on. When the value assigned by the relation is true, we say the things are related. For example, the pair of

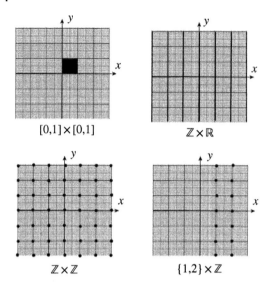

Figure 3.4 Typical Cartesian products.

numbers (2, 3) would be assigned "true" for the relation "is less than" since 2 < 3. The reader is well familiar with many relations, such as the *equal* relation "=," which would assign a truth value to the pair (3, 3) but not to the pair (3, 4). What this chapter does is it provides a general theory behind many of the relations you already known, and some you do not know.

Definition   Relation
Let A and B be sets and $A \times B$ their Cartesian product. A **binary relation** R from A to B is simply a subset $R \subseteq A \times B$. The elements in a relation R form the **graph** of the relation. If $(x, y) \in R$, we say that "x and y are related," and we denote this by writing xRy. If $(x, y) \notin R$, we say "x and y are *not* related" and denote this by $x\cancel{R}y$. When the two sets A and B are the same set $A = B$, we say that the relation $R \subseteq A \times A$ is a relation *on* R.

> **Important Note**   You might think of a relation from A to B as similar to a function from A to B, like $y = \sin x$, which assigns to each $x \in A$ the value $y = \sin x$, except that for a relation R from A to B, an element x in A does not necessarily map to a *single* element y in B, but possibly to several. Functions are a special type of relation that we will study in Section 3.4.

The following examples will familiarize you with some common relations.

Example 3   Typical Relation

Let A be a set of five students and B a set of four university classes, where

$$A = \{Mary, John, Ann, Sally, Jim\}$$

$$B = \{Math, Literature, Chemistry, Psychology\}$$

and consider the relation $R \subseteq A \times B$ defined by

$$xRy = x \text{ likes class } y.$$

This relation is represented by the set of x_s in Table 3.1, showing which student likes which classes. Since (John, Math) $\in R$, this means John likes the math class, and since (Ann, Literature) $\notin R$, this means Ann does not like the literature class.

Table 3.1 Relation "likes class" on {Students} × {Classes}.

	Mary	John	Ann	Sally	Jim
Math		x			x
Literature	x			x	x
Chemistry			x		x
Psychology	x		x		x

Important Note   The definition of a relation can be a bit confusing at first, thinking of it as both a *set* and a *relation* between things. Just realize when $(x, y) \in R$, we denote this by writing xRy just like we do when we write $x \le y$, $x = y$, and so on.

Important Note   Binary relations are standard fare in many areas of mathematics and computer science. Common binary relations are "is greater than," "is equal to," "divides" in arithmetic, "is congruent to" in geometry, "is adjacent to" in graph theory; "is orthogonal to" in linear algebra; "is linked to" in computer science, and so on.

Example 4   Identify the Relation

The Cartesian product of the sets $A = \{1, 2\}$, $B = \{1, 2, 3\}$ consists of the $2 \cdot 3 = 6$ pairs of points

$$A \times B = \{(1,1), (1,2), (1,3), (2,1), (2,2), (2,3)\}$$

A popular relation from A to B is given by

$$R = \{(1,2), (1,3), (2,3)\} \subseteq A \times B$$

Have you seen this relation before? Name the relation.

Solution

We write the relation in the more suggestive form:

a) $(1, 2) \in R \Leftrightarrow 1R2$ more commonly written $1 < 2$
b) $(1, 3) \in R \Leftrightarrow 1R3$ more commonly written $1 < 3$
c) $(2, 3) \in R \Leftrightarrow 2R3$ more commonly written $2 < 3$

Hence, the relation is "is less than," which we normally express by writing "<." The graph of this relation is illustrated by the dots in Figure 3.5.

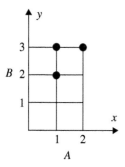

Figure 3.5 Relation "<" on $A \times B$.

3.1.3 Visualization of Relations with Directed Graphs

Another way to visualize a relation is with a **directed graph**, which consists of a collection of dots representing members of A, and arrows connecting the dots if members are related. For example, the relation

$$R = \{(1,2),(2,5),(2,3),(3,1),(4,5),(5,5)\}$$

on $A = \{1, 2, 3, 4, 5\}$ can be visualized by the directed graph drawn in Figure 3.6.

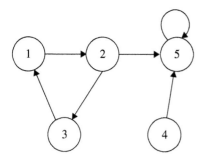

Figure 3.6 Directed graph of a relation on $\{1, 2, 3, 4, 5\}$.

Common Relations in Mathematics

1.	$=$	Equals
2.	$\leq, \geq$	Less than or equal, greater than or equal
3.	$<, >$	Less than, greater than
4.	$\parallel$	Lines parallel
5.	$\perp$	Lines perpendicular
6.	$\cong$	Congruent figures
7.	$\approx$	Has the same cardinality
8.	$\equiv \bmod(n)$	Numbers equivalent modulo n
9.	$\mid$	Divides
10.	$\equiv$	Similar figures
11.	f	Is a function of
12.	$\sim$	Is homeomorphic to
13.	$\sim$	Is isomorphic to

Example 5   Directed Graph of the Division Relation
Let $A = \{1, 2, 3, 4, 5, 6\}$ and define the relation R on A by

$$xRy \Leftrightarrow x \text{ divides } y$$

Draw the directed graph that represents this relation.

Solution
The directed graph in Figure 3.7 illustrates the division relation on the set $\{1, 2, 3, 4, 5, 6\}$.

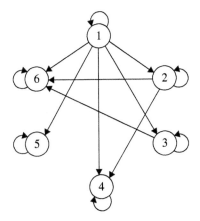

Figure 3.7 Division relation on $\{1, 2, 3, 4, 5, 6\}$.

Higher-Order Relations   In set theory and logic, a *n-ary* is a relation that assigns a truth value to *n*members of a given set, and a **ternary** relation is a relation *R(a, b, c)* that assigns a truth value to three members. For example, for natural numbers *a, b, c* the relation $a^3 + b^3 = c^3$ is a ternary relation on $\mathbb{N}$, which according to Andrew Wiles,[1] is false.

3.1.4   Domain and Range of a Relation

A relation R is a generalization of a function f in the sense that every function is also a relation, but not vice versa. When the relation is a function, the relation maps a single element $x \in A$ into a single value $f(x) \in B$, whereas for a general relation, a single value of $x \in A$ can be related to none or many values of $y \in B$. But like functions, relations also have domains and ranges.

Definition   Domain and Range of a Relation
Let R be a relation from A to B. The **domain** of the relation is the set

- $\mathrm{Dom}(R) = \{x \in A : \exists\, y \in B \text{ such that } xRy\} \subseteq A$

The **range** of the relation R is

- $\mathrm{Range}(R) = \{y \in B : \exists\, x \in A \text{ such that } xRy\} \subseteq B$

In Plain English: The domain of a relation R from A to B is the set of first members of the ordered pairs in R and the range of R is the set of second members of the ordered pairs. By definition, we have

$$\mathrm{Dom}(R) \subseteq A, \mathrm{Range}\,(R) \subseteq B.$$

For example, if $A = \{1, 2, 3, 4\}$ and $B = \{2, 3, 4\}$ with relation R from A to B, given by

$$R = \{(1,2),(1,3),(2,3),(2,4)\}.$$

we have

$$\mathrm{Dom}(R) = \{1,2\} \subseteq A$$
$$\mathrm{Range}(R) = \{2,3,4\} \subseteq B$$

---

1 English mathematician Andrew Wiles verified Format's Last Theorem in the affirmative by proving that if a, b, c are positive integers, the equation $a^n + b^n = c^n$ has no integer solutions when $n \geq 3$.

3.1.5 Inverses and Compositions

Two common ways of constructing new relations from old ones are **inverse relations** and **compositions** of relations.

Definition Inverse Relation

If R is a relation from A to B, the **inverse** of R is defined as

$$R^{-1} = \{(y,x) : (x,y) \in R\}.$$

If $R = \{(1, 2), (3, 5), (4, 1)\}$ then $R^{-1} = \{(2, 1), (5, 3), (1, 4)\}$.

Example 6 Relation and Its Inverse

Define a relation on the set $A = \{1, 2, 3, 4\}$ by

$$R = \{(1,2),(1,3),(2,4),(3,4)\}.$$

The inverse of R is

$$R^{-1} = \{(2,1),(3,1),(4,2),(4,3)\}.$$

Both R and R^{-1} are drawn in Figure 3.8, where R is denoted by round dots and R^{-1} by square dots. Note that the graphs of R and R^{-1} are reflections of each other through the 45° line $y = x$.

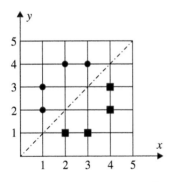

Figure 3.8 Graphs of R and R^{-1}.

3.1.6 Composition of Relations

The composition of two (or more) relations is similar to the composition of functions that the reader might be familiar. Recall that the composition of functions f and g, written $f \circ g$, is defined by

$$(f \circ g)(x) = f(g(x))$$

for all x in the domain of g, where $g(x)$ in the domain of f. The generalization of this definition to relations goes as follows.

Definition   Composition of Relations
If R is a relation from A to B, and S a relation from B to C, then the **composition** (or **composite**) of the relations R and S is the relation

$$S \circ R = \{(a,c) \in A \times C : \exists b \in B \text{ such that } (a,b) \in R \text{ and } (b,c) \in S\}.$$

In plain language, the composition is the collection of all "paths" from A to C as illustrated in Example 7.

Example 7   Composition of Relations
Given sets

$$A = \{1,2,3,4\}$$
$$B = \{a,b,c\}$$
$$C = \{\text{cat}, \text{dog}, \text{horse}\}$$

and relation R from A to B and relation S from B to C defined by

$$R = \{(1,a),(1,c),(2,a),(3,b),(4,b)\} \subseteq A \times B$$
$$S = \{(a,\text{dog}),(b,\text{horse}),(b,\text{cat}),(c,\text{dog}),(c,\text{horse})\} \subseteq B \times C$$

the composition $S \circ R$ is illustrated in Figure 3.9 and displays all paths starting at A and ending at C. They are

$$S \circ R = \{(1,\text{dog}),(1,\text{horse}),(2,\text{dog}),(3,\text{cat}),(3,\text{horse}),(4,\text{cat}),(4,\text{horse})\}$$

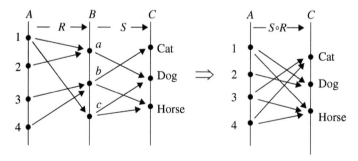

Figure 3.9 Composition $S \circ R$.

Problems

1. **True or False**
    Tell whether the following statements are true or false.
    a) $\mathbb{Q} \times \mathbb{Q} \subseteq \mathbb{R} \times \mathbb{R}$
    b) $aRb \Rightarrow (a, b) \in R$
    c) $aRb \Rightarrow bRa$
    d) aRa
    e) For any two sets A, B, $A \times B = B \times A$
    f) For some sets A, B, $A \times B = B \times A$

2. **Relations**
    For the set $A = \{1, 2, 3, 4\}$, write out the ordered pairs in the relation R on A if
    a) $xRy \Leftrightarrow x < y$
    b) $xRy \Leftrightarrow x = y$
    c) $xRy \Leftrightarrow x$ divides y
    d) $xRy \Leftrightarrow x$ is a multiple of y

3. **Four Basic Cartesian Products**
    Given $A = \{1, 2, 3\}$, $B = \{a, b\}$. Find the following Cartesian products.
    a) $A \times B$
    b) $B \times A$
    c) $A \times A$
    d) $B \times B$

4. **Cartesian Products**
    For each of the following pair of sets A and B, find the Cartesian products $A \times B$ and $B \times A$.
    a) $A = \{0, 2\}$, $B = \{-1, 0\}$
    b) $A = \{a, b\}$, $B = \{b, c\}$
    c) $A = \mathbb{R}$, $B = \mathbb{N}$
    d) $A = \mathbb{Z}$, $B = \mathbb{N}$
    e) $A = \mathbb{R}$, $B = \{-1, 0, 1\}$

5. **Graphing a Relation**
    Draw a sketch of the following relations.
    a) $R = \{(x, y) \in \mathbb{R} \times \mathbb{R} : x^2 + y^2 = 1\}$
    b) $R = \{(x, y) \in \mathbb{R} \times \mathbb{R} : y = \sin x\}$
    c) $R = \{(x, y) \in \mathbb{R} \times \mathbb{R} : x = y^2\}$
    d) $R = \{(x, y) \in \mathbb{R} \times \mathbb{R} : |x| \le 1, |y| \ge 1\}$
    e) $R = \{(x, y) \in \mathbb{N} \times \mathbb{N} : x \equiv 0(\mathrm{mod}3), y \equiv 1(\mathrm{mod}3)\}$
    f) $R = \{(x, y) \in \mathbb{N} \times \mathbb{N} : x$ divides $y\}$

6. **Algebra of Relations**

Given are the closed intervals

$$A = [0,3], B = [2,5], C = [1,4]$$

on the real line $\mathbb{R}$. Sketch the following relations in the plane.

a) $R = (A \cup B) \times C \subseteq \mathbb{R} \times \mathbb{R}$

b) $R = (A \cap B) \times C \subseteq \mathbb{R} \times \mathbb{R}$

c) $R = (A \times B) \cup (A \times C) \subseteq \mathbb{R} \times \mathbb{R}$

d) $R = A \times (A \cup C) \subseteq \mathbb{R} \times \mathbb{R}$

7. **Naming a Relation**

Give common names that describe the following relations on $A = \{1,2,3\}$. Then find the inverse relation. What is a name for the relation and inverse relation?

a) $R = \{(1, 1), (2, 2), (3, 3)\}$

b) $R = \{(1, 2), (1, 3), (2, 3)\}$

c) $R = \{(1, 1), (1, 2), (1, 3), (2, 2), (2, 3), (3, 3)\}$

8. **Important Types of Relations**

Two important types of relations are injections (1–1) and surjections (onto), whose definitions we have seen in Section 2.4 in conjunction with the cardinality of sets. The definition of an injection and surjection defined especially for relations are as follows:

Surjective Relation: A relation $R \subseteq X \times Y$ is surjective if

$$(\forall y \in Y)(\exists x \in X)(xRy)$$

Injective Relation: A relation $R \subseteq X \times Y$ is injective if

$$(\forall x_1, x_2 \in X)(\forall y \in X)[(x_1 R y) \wedge (x_2 R y) \Rightarrow (x_1 = x_2)]$$

For $X = Y = \{1, 2, 3\}$, give an example of an injective and surjective relation from X to Y.

9. **Meaning of Relations**

For each of the following, describe the members of the relation R.

a) $R \subseteq \mathbb{Q} \times \mathbb{Q}, R = \left\{ \left(\dfrac{p}{q}, \dfrac{r}{s} \right) : ps = rq \right\}$

b) $R \subseteq \mathbb{Q} \times \mathbb{Q}, R = \emptyset$

c) $R \subseteq \mathbb{R} \times \mathbb{R}, R = \mathbb{R} \times \mathbb{R}$

10. **Inverse Relation of Compositions**

Given $A = \{1, 2, 3\}$ verify the following identity for the inverse of a composition:

$$(S \circ R)^{-1} = R^{-1} \circ S^{-1}$$

for relations

$$R = \{(1,2),(2,3)\} \subseteq A \times A$$
$$S = \{(2,2),(3,1)\} \subseteq A \times A$$

11. **Composition of Relations**
Find the composition of the relations $S \circ R$ given the sets

$$A = \{a,b,c,d\}, B = \{1,2,3\}, C = \{x,y,z\}$$

and the relations

$$R = \{(a,2),(a,3),(b,2),(c,1),(d,3)\} \subseteq A \times B$$
$$S = \{(2,x),(1,y),(1,z),(3,y)\} \subseteq B \times C$$

12. **Cartesian Product Identities**
Prove the identity

$$(A - B) \times B = (A \times B) - (B \times B).$$

13. **Number of Relations**
If a set A has m elements and B has n elements, show that the number of relations from A to B is 2^{mn}.

14. **Graphing Relations and Their Inverses**
Graph the following relations and their inverses.
a) $R \subseteq \mathbb{R} \times \mathbb{R}$, $xRy \Leftrightarrow y = 1/x$
b) $R \subseteq \mathbb{R} \times \mathbb{R}$, $xRy \Leftrightarrow y = e^x$
c) $R \subseteq \mathbb{R} \times \mathbb{R}$, $R = \{(x, y) : |x| + |y| = 1\}$

15. **Counting Relations I**
What is the total number of relations that can be defined on the set $A = \{1, 2\}$?

16. **Counting Relations II**
What is the total number of relations that can be defined on the set $A = \{1, 2, 3\}$?

17. **Converse of a Binary Relation**
If R is a binary relation on a set A, then the converse relation $\tilde{R}$ on A is defined by the relation $x\tilde{R}y \Leftrightarrow yRx$. State in English or write out in ordered pairs the converse of the following relations R.

a) The relation "is the mother of."
b) The relation "is a uncle of."
c) The relation "<" on the real numbers.
d) The relation "=" on the real numbers.

18. **Blood Typing**
There are four blood, types A, B, AB, and O.
- Type A can receive blood from type A and O.
- Type B can receive blood from type B and O.
- Type AB can receive blood from all types.
- Type O can receive blood from only type O.

Given the set $S = \{A, B, AB, O\}$, define a binary relation R on S as follows:

$xRy \Leftrightarrow$ type x person can receive blood from type y persons

a) Write out the ordered pairs of R.
b) Draw a directed graph of the relation R.
c) Define the converse $\tilde{R}$ of R by $x\tilde{R}y \Leftrightarrow yRx$. What is the interpretation of the converse?

19. **Internet Research**
There is a wealth of information related to topics introduced in this section just waiting for curious minds. Try aiming your favorite search engine toward *important relations in mathematics, visualizing mathematical relations, and composition of relations.*

3.2

Order Relations

Purpose of Section   To introduce some important types of **orderings**, such as a **partial order** and **strict order** as well as what it means for a set to be **totally ordered**. Using the order relation, we introduce the concept of upper and lower bounds of a set, the least upper bound, and greatest lower bound of a set. We also show how order relations can be illustrated graphically by means of **Hasse diagrams**.

3.2.1   Let There Be Order

The British philosopher Edmund Burke once said, "Order is the foundation of all that is good," and although he probably was not referring to inequalities of numbers, nevertheless, order is as important in mathematics as it is anywhere else. The reader is familiar with the inequality relations ≤ and < that impose an ordering of real numbers, and the relation ⊆, which imposes an "order" on sets. Other objects can be "ordered" as well, such as functions, matrices, and points in the plane. Ordering objects according to given rules brings structure to an area that might otherwise be difficult to analyze. In computer science, order not only brings understanding, but efficiency. Imagine trying to find information on the Internet if search engines did not have clever "ordering" strategies for searching for information.

Definition   Simple, Partial, and Strict Order

1) A **partial order** on a set A, denoted by ≤, is a binary relation on A such that for all x, y, z in A, the following RAT conditions hold.

Advanced Mathematics: A Transitional Reference, First Edition. Stanley J. Farlow.
© 2020 John Wiley & Sons, Inc. Published 2020 by John Wiley & Sons, Inc.
Companion website: www.wiley.com/go/farlow/advanced-mathematics

Reflexive : $x \le x$

Antisymmetric property : $(x \le y \text{ and } y \le x) \Rightarrow x = y$

Transitive property : $(x \le y \text{ and } y \le z) \Rightarrow x \le z$

When a partial order is defined on a set, the set is said to be **partially ordered** or a **partially ordered** set. When $x \le y$, we say x **precedes** y. The usual "less than or equals to" is a prime example of a partial order on the real numbers.

2) A **strict order**, denoted by <, is a *binary relation* on a set A such that for all x, y, z of A, the following IAT conditions hold.

Irreflexive property : It is not true that $x < x$.

Asymmetric property : If $x < y$, then $y < x$ is not true.

Transitive property : $(x < y \text{ and } y < z) \Rightarrow x < z$.

When a strict order is defined on a set, the set is said to be **strictly ordered** or a **strictly ordered set**. The usual "less than" inequality (<) is a prime example of a strict order on the real numbers, and we would say the real numbers are strictly ordered by the less than order.

Both partially ordered and strictly ordered sets are said to be **totally ordered** if every two members of the set are comparable. The usual "less than or equal to" order ($\le$) is a total order on the real numbers since any two real numbers x, y satisfy $x \le y$ or $y \le x$. On the other hand, the set inclusion relation ($\subseteq$) is a partial order on the power set $P(A)$ of $A = \{a, b, c\}$, but does not totally order the power set since not all sets in the power set are comparable, an example being the sets $\{a, c\}$ and $\{b\}$.

Example 1   Relations
Given the set $A = \{1, 2, 3\}$ and the relation on A

$$R = \{(1,1),(1,2),(2,3),(1,3)\} \subseteq A \times A$$

a) Is R a partial order on A?
b) Is R a strict order on A?
c) Is R a total order on A?

Solution
a) $(2, 2) \notin R$ so R is not reflexive, hence not a partial order.
b) $(1, 1) \in R$ so R is not irreflexive, hence not a strict order.
c) $(3, 3) \notin R$ so R is not a total order, 3 is not comparable with itself.

Example 2   A Famous Relation You Might Know
Given the set $A = \{1, 2, 3\}$ whose members are related by the following relation:

$$R = \{(1,1),(2,2),(3,3),(2,1),(3,1),(3,2)\} \subseteq A \times A$$

Verify that R is a partial order on A. Do you recognize the ordering? It is the *greater than or equal to* relation "$\geq$." The relation becomes clear when written in a more common form as

$$1 \geq 1, 2 \geq 2, 3 \geq 3, 2 \geq 1, 3 \geq 1, 3 \geq 2.$$

Three ways to represent this partial order are illustrated in Figure 3.10.

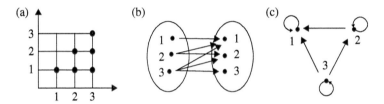

Figure 3.10 Picturing partial orders (a) graphs, (b) arrow illustration, and (c) directed graph.

Example 3   Ordering by Division Property
Let "|" denote the relation "divides" on the set of natural numbers $\mathbb{N}$. For example, $1 \mid 7$, $2 \mid 8$, $3 \mid 9$, $7 \mid 21$. If m does not divide n, we denote this by $m \nmid n$ as in $3 \nmid 7$. Show that "|" defines a partial order on the natural numbers.

Solution
We show that division is reflexive, antisymmetric, and transitive.

- **Reflexive** $[n \mid n]$: The relation is reflexive since natural numbers divide themselves.
- **Antisymmtric:** We must show

$$(\forall m, n \in \mathbb{N})[(m \mid n) \wedge (n \mid m)] \Rightarrow m = n.$$

Using what it means for one number to divide another, we write

$$\begin{cases} m \mid n \\ n \mid m \end{cases} \Rightarrow \begin{cases} \exists k_1 \in \mathbb{N} \text{ such that } n = k_1 m \\ \exists k_2 \in \mathbb{N} \text{ such that } m = k_2 n \end{cases}$$

so we have

$$m = k_2 n = k_2 (k_1 m) = (k_2 k_1) m$$

which implies $k_2 k_1 = 1$, and since k_1 and k_1 are positive integers, we have the relation $k_1 = k_2 = 1$, which implies $m = n$.

- **Transitive:** We must prove

$$(\forall m, n \in \mathbb{N})[(m \mid n) \wedge (n \mid p)] \Rightarrow m \mid p$$

Using what it means for one number to divide another, we write

$$\begin{cases} m \mid n \\ n \mid p \end{cases} \Rightarrow \begin{cases} \exists k_1 \in \mathbb{N} \text{ such that } n = k_1 m \\ \exists k_2 \in \mathbb{N} \text{ such that } p = k_2 n \end{cases}$$

Hence,

$$p = k_2 n = k_2 (k_1 m) = (k_2 k_1) m$$

which implies $m \mid p$.

Example 4   Partially Ordered Sets
The power set of $A = \{a, b, c\}$ is

$$P(A) = \{\varnothing, \{a\}, \{b\}, \{c\}, \{a,b\}, \{a,c\}, \{b,c\}, \{a,b,c\}\}.$$

Show that inclusion relation "$\subseteq$" is defined by $xRy \Leftrightarrow x \subseteq y$, where $x, y \in P(A)$ defines a partial order on $P(A)$.

Proof
We show "$\subseteq$" satisfies the following properties:

- **Reflexive:** Any set is a subset of itself; hence, $\subseteq$ is reflexive.
- **Antisymmetric:** The relation is antisymmetric since

$$(\forall B, C \in P(A))[(B \subseteq C) \wedge (C \subseteq B)] \Rightarrow (B = C)$$

- **Transitive:** The relation is transitive since

$$(\forall B, C, D \in P(A))[(B \subseteq C) \wedge (C \subseteq D)] \Rightarrow (B \subseteq D),$$

Hence, $\subseteq$ is a partial order on $P(A)$.

3.2.2   Total Order and Symmetric Relations

Although, we have seen that "$\leq$" and "$\subseteq$" are partial orders on $\mathbb{R}$ and $P(A)$, respectively, there is an important difference. The partial order "$\leq$" on the real numbers is also a total order, inasmuch as any two real numbers x and y are *comparable* as in $x \leq y$ or $y \leq x$. On the other hand, "$\subseteq$" is not a total order on the power set of $\{a, b, c\}$ since it is not true that

$$\{a, c\} \subseteq \{a, b\} \quad \text{or} \quad \{a, c\} \supseteq \{a, b\}.$$

3.2.3 Symmetric Relation

We say a relation R on a set A is symmetric if $xRy \Leftrightarrow yRx$ for all $x, y \in R$. One might think offhand that antisymmetry is the negation of being symmetric, but this is not true. The equal relation "=" is both symmetric and antisymmetric. However, the relation "is married to" is a symmetric relation but is not antisymmetric, whereas the relation $\leq$ on the real line is antisymmetric but not symmetric.

> **Historical Note** The person who defined the order relation was a German mathematician Felix Hausdorff (1868–1942) who did so in 1914. Hausdorff was one of the founders of modern topology.

3.2.4 Hasse Diagrams and Directed Graphs

Although a partially ordered set can contain an infinite number of elements, many important examples are finite. A useful way to represent finite partially ordered sets is a **Hasse Diagram**,[1] where each element of the ordered set is denoted by a dot (node) and a line segment that goes *upward* from node x to node y means that $x \leq y$.

A Hasse diagram for a partial order on $A = \{a, b, c, d, e, f, g\}$ is drawn in Figure 3.11, where a few orderings are $e \leq d$, $f \leq d$, $g \leq d$. Also $e \leq c$ by transitivity since one can move upwards from e to c moving through the nodes d and b. On the other hand, $e \nleq f$ and $a \nleq c$, so the order is not a total order since some elements are not comparable.

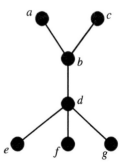

Figure 3.11 Hasse diagram.

---

1 A Hasse (pronounced Ha suh) diagram is named after the German mathematician Helmut Hasse (1898–1979).

Example 5 Ordering the Power Set
Draw the Hasse diagram for the power set

$$P(A) = \{\,\emptyset,\{a\},\{b\},\{c\},\{a,b\},\{a,c\},$$
$$\{b,c\},\{a,b,c\}\}$$

of $\{a, b, c\}$ ordered by set inclusion $\subseteq$.

Solution
The Hasse diagram is shown in Figure 3.12.

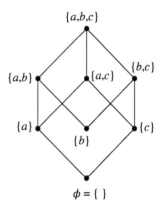

Figure 3.12 Hasse diagram for set inclusion on P(A).

Example 6 Hasse Diagram
The Hasse diagram for the power set of four elements $\{a, b, c, d\}$ is shown in Figure 3.13.

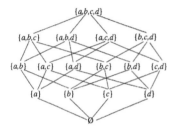

Figure 3.13 Hasse diagram for the power set of four elements.

The concept of ordering objects introduces a whole collection of new ideas and concepts.

3.2.5 Upper Bounds, Lower Bounds, glb, and lub

Definition Let $\leq$ be a partial order on a set U and $S \subseteq U$ a subset of U.

- **Upper Bound of S:** An element $u \in U$ is an **upper bound** of S if and only if $(\forall s \in S)(s \leq u)$.
- **Least Upper Bound (lub) of S:** An element $lub(S) \in U$ is the **least upper bound** (or **supremum**) of S if $lub(S)$ it is an upper bound for S and lub $(S) \leq u$ for every other upper bound u of S.
- **Lower Bound:** An element $l \in U$ is a **lower bound** of S if and only if $(\forall s \in S)$ $(l \leq s)$.
- **Greatest Lower Bound (glb):** An element glb $(S) \in U$ is the **greatest lower bound** (or **infimum**) of S if glb (S) is a lower bound for S and $l \leq glb (S)$ for every other lower bound l of S.

- **Maximum and Minimum:** An element $M \in S$ is the **maximum** element of S if and only if

$$(\forall s \in S)(s \leq M)$$

An element $m \in S$ is the **minimum** element of S if and only if

$$(\forall s \in S)(m \leq s)$$

- **Maximal Element:** An element $M \in S$ is a **maximal element** of S if and only if

$$\sim (\exists s \in S)(M \leq s).$$

In other words, a maximal element is an element of the set that is not "smaller" than any other member of the set.

- **Minimal Element:** An element $m \in S$ is a **minimal element** of S if and only if

$$\sim (\exists s \in S)(s \leq m).$$

In other words, a minimal element is an element of the set such than no other member is "less" than the minimal member. See Figure 3.14.

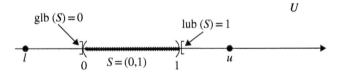

Figure 3.14 Ordering properties.

Example 7   Bounds on an Open Interval

The open interval, $S = (0, 1) \subseteq \mathbb{R}$, is a partially ordered subset of the real numbers $U = \mathbb{R}$ ordered by $\leq$. Do you understand that

a) S has many upper bounds, 1, 3, 5.3, π, ...
b) S has many lower bounds; -1, -5, -10.3, ...
c) S has the least upper bound of 1
d) S has the greatest lower of 0
e) S has no maximum, no maximal, no minimum, and no minimal element.

> **Important Note**   Order theory is an area of mathematics that provides a framework for statements like "is greater than," or "A precedes B," and so on. Although the history of ordering objects in mathematics is vague, an early explicit mention of "order" is found in the nineteenth-century works of English

mathematician George Boole. Other early researches who explicitly studied the concept of "order" were Charles Saunders Peirce and Richard Dedekind. The word "poset," which was first used as an abbreviation for a "partially ordered set" was coined by the American mathematician Garrett Birkhoff in his book *Lattice Theory*.

Example 8 Partially Ordered Set
The set

$$S = \{A, B, C, D, E, F, G, H, I, J, L, M, N, O\}$$

is partially ordered according to the Hasse diagram in Figure 3.15.

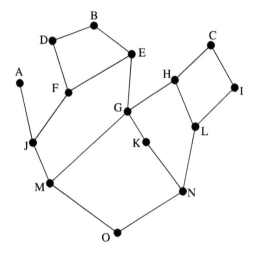

Figure 3.15 Hasse diagram partial order.

Do you understand why the following properties are valid?

a) The set has no upper bound.
b) O is a lower bound of the set.
c) The set has no least upper bound since it has no upper bound.
d) The greatest lower bound of the set is O.
e) The maximal elements of the set are A, B, C.
f) The minimal element of the set is O.
g) The set has no maximum.
h) The minimum of the set is O.

Example 9   Multiples and Divisors of 24

The following set A consists of the divisors of 24.

$$A = \{1,2,3,4,6,8,12,24\}$$

Two partial orders on A are

$aMb \Leftrightarrow a$ is a multiple of b

$aDb \Leftrightarrow a$ divides b

Note that $24M8$ denotes 24 is a multiple of 8 and $4D12$ since 4 divides 12. Draw Hasse diagrams for the two partial orders.

Solution

The Hasse diagram is drawn in Figure 3.16.

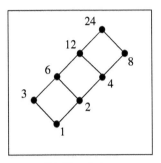
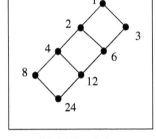

Divide relation   Multiple relation

Figure 3.16 Hasse diagrams for division and multiplication.

The 24 at the bottom of the Hasse diagram for the multiple relation denotes the fact that 24 is a multiple of the other divisors. The 24 at the top of the divide relation means all other divisors divide 24. The reader can verify that both relations are partial orders, but not total orders.

> **Important Note** The inverse relation R^{-1} of a partial order relation R is also a partial order. The inverse relation of $\leq$ on the real numbers is $\geq$.

Table 3.2 lists some common relations and their properties. The set over which the relation is defined is given in parenthesis next to the relation.

Table 3.2 Properties of common relations.

Relations	Reflexive xRx	Antisymmetric $xRy \land yRx \Rightarrow x = y$	Transitive $xRy \land yRz \Rightarrow xRz$	Symmetric $xRy \Rightarrow xRx$
$\leq$ ($\mathbb{R}$)	Yes	Yes	Yes	No
$<$ ($\mathbb{R}$)	No	No	Yes	No
$\equiv$ (mod n)	Yes	No	Yes	Yes
$\approx$ (sets)	Yes	No	Yes	Yes
$\subseteq$ (sets)	Yes	Yes	Yes	No
$\perp$ (lines)	No	No	No	Yes
$\parallel$ (lines)	Yes	No	Yes	Yes
$\mid$ on $\mathbb{Z}$	Yes	No	Yes	No
$\mid$ on $\mathbb{N}$	Yes	Yes	Yes	No
$=$ ($\mathbb{R}$)	Yes	Yes	Yes	Yes

Problems

1. **Testing for an Order Relation**

   Tell whether the following relations on $A = \{1, 2, 3\}$ are reflexive, antisymmetric, and transitive. Plot the points of the relation in the Cartesian product $A \times A$ and denote the members of $R \subseteq A \times A$. If the relation is an order relation, draw a Hasse diagram and directed graph for the relation.

   a) $R = \{(1, 1), (2, 2), (3, 3)\}$
   b) $R = \{(1, 1), (1, 2), (2, 1)\}$
   c) $R = \{(1, 1), (2, 2), (3, 3), (1, 2), (1, 3), (2, 3)\}$
   d) $R = \{(1, 2), (2, 3), (1, 3)\}$

2. **Finding Relations**

   Find a relation on $A = \{1, 2, 3, 4\}$ with the following properties.

   a) Reflexive, but not antisymmetric
   b) antisymmetric and reflexive
   c) not reflexive, but transitive
   d) not reflexive, not antisymmetric, not transitive

3. **Ordering of Functions**

   Let $C[0, 1]$ be the set of continuous functions defined on $[0, 1]$. For $f, g \in C[0, 1]$, define the ordering

   $$f \leq g \Leftrightarrow (\forall x \in [0,1])[f(x) \leq g(x)].$$

   Show that "$\leq$" defines a partial order on $C[0, 1]$.

4. **Upper and Lower Bounds**

The Hasse diagram for the power set $P(A)$ of $A = \{a, b, c, d\}$ is drawn in Example 6 with order relation $\subseteq$. Use the Hasse diagram to find an upper bound, a least upper bound, a lower bound, and the greatest lower bound of the following subsets of $P(A)$

a) $B = \{\{a\}, \{a, b\}\}$
b) $B = \{\{a\}, \{b\}\}$
c) $B = \{\{a\}, \{a, b\}\{a, b, c\}\}$
d) $B = \{\{a\}, \{c\}, \{a, c\}\}$
e) $B = \{\varnothing, \{a, b, c\}\}$
f) $B = \{\{a\}, \{b\}, \{c\}\}$

5. **Sups and Infs**

If they exist, find the sup, inf, max, and min of the following sets.

a) $(-\infty, 2)$
b) $(-\infty, 2]$
c) $\left\{1 + \dfrac{1}{n} : n \in \mathbb{N}\right\}$
d) $\left\{\dfrac{1}{m} + \dfrac{1}{n} : m, n \in \mathbb{N}\right\}$
e) $\left\{\dfrac{1}{m} - \dfrac{1}{n} : m, n \in \mathbb{N}\right\}$
f) $\left\{\dfrac{1}{2}, \dfrac{1}{3}, \dfrac{2}{3}, \dfrac{1}{4}, \dfrac{3}{4}, \dfrac{1}{5}, \dfrac{2}{5}, \dfrac{3}{5}, \dfrac{4}{5}, \dots\right\}$ (omit fractions not in reduced form)
g) $\left\{\dfrac{n}{n^2 + 1} : n \in \mathbb{N}\right\}$
h) $\{y : y = x^2 + x - 2, x \in \mathbb{R}\}$

6. **Hasse Diagram**

Jane is getting a degree in mathematics and has several courses to take. Some of the courses have prerequisites as shown below.

Course needed	Prerequisites
• Calculus I	
• Calculus II	Calculus I
• Calculus III	Calculus II
• Linear Algebra	Calculus III
• Differential Equations	Calculus III
• Intro to Pure Math	Linear Algebra, Calculus II
• Abstract Algebra	Linear Algebra, pure math
• Advanced Calculus	Calculus III, pure math

Draw a Hasse diagram that illustrates the order which Jane must take courses.

7. **Hasse Diagram for Multiples**
   Let M be the order relation "a is a multiple of b" defined on the positive divisors of 15. Draw a Hasse diagram for M.

8. **A Partial Order for Points in the Plane**
   There are various ways to construct new orders from existing orders. A partial order can be constructed on the Cartesian product of two partially ordered sets by defining

   $$(a,x) \le (b,y) \Leftrightarrow [(a \le b) \wedge (x \le y)]$$

   a) Construct a Hasse diagram that represents a partial order for

   $$A = \{(-1,3),(0,3),(1,7),(0,6),(0,5)\}.$$

   b) Draw a directed graph for the partial order.

9. **Equivalent form of Antisymmetry**
   State the contrapositive form of the antisymmetry condition

   $$(\forall x, y \in \mathbb{R})[(x \le y) \wedge (y \le x)] \Rightarrow x = y.$$

10. **Ordering the Complex Numbers**
    Suppose we try to order the complex numbers $z = a + bi$ according to magnitudes

    $$z_1 \le z_2 \Leftrightarrow |z_1| \le |z_2|$$

    where $|z| = \sqrt{a^2 + b^2}$ is the magnitude of the complex number. Is this a partial order on the complex numbers?

11. **Total Order of the Complex Numbers**
    The complex numbers can be totally ordered as follows. Given two complex numbers in polar form

    $$z_1 = r_1 e^{i\theta_1}, \quad z_2 = r_2 e^{i\theta_2}, \quad r_1, r_2 \ge 0, \quad 0 \le \theta < 2\pi$$

    order the complex numbers by

    $$z_1 \le z_2 \Leftrightarrow \begin{cases} r_1 < r_2 \\ r_1 = r_2, \theta_1 < \theta_2 \end{cases}$$

    Compare the following complex numbers.
    a) $z_1 = i$, $z_2 = 1 + i$
    b) $z_1 = i$, $z_2 = -1$

c) $z_1 = 6i$, $z_2 = 2 + 3i$
d) $z_1 = 0$, $z_2 = 1$

12. **Counting Partial Orders**
    There are a total of three partial orders on $A = \{1, 2\}$. Can you find them?

13. **Hasse Diagram**
    Given the subset $S \subseteq U$ represented by "stars" in the Hasse diagram in Figure 3.17 which describes a partial ordering of the set:

    $$U = \{A, B, C, D, E, F, G, H, I, J, K, L\}$$

    find the following quantities of S:
    a) upper bound(s)
    b) lower bound(s)
    c) the least upper bound
    d) greatest lower bound
    e) maximal element(s)
    f) minimal element(s)
    g) maximum
    h) minimum

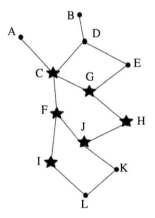

Figure 3.17 Hasse diagram.

14. **Test Your Knowledge of Sups and Infs**
    A Hasse diagram representation of a partial order is shown in Figure 3.18. If they exist, find the sup and inf of the following sets. A partially ordered set is called a **lattice** if every pair of elements in the set has a sup and inf. Is the set under this partial order a lattice?

a) {2, 3}
b) {8, 9}
c) {1, 2}

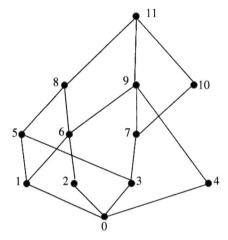

Figure 3.18 Sups and Infs.

15. **SUP or MAX?**
    If it exists, find the maximum value of the set

    $$S = \left\{ \frac{n}{n+1} : n = 1, 2, 3, \ldots \right\}$$

    If it does not exist, find the least upper bound of S.

16. **Lattices**
    A lattice L is a partially ordered set in which every two members $a, b \in L$
    has a supremum and infimum in L. The supremum is called the **join** of a

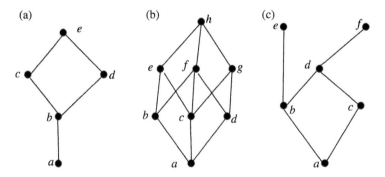

Figure 3.19 Lattice or non lattice.

and b and denoted by $a \vee b$, and the infimum is called the **meet** of a and b, and denoted by $a \wedge b$. Determine which of the partially ordered sets in Figure 3.19 represented by Hasse diagrams, are lattices.

17. **Lattice of Partitions**
A **lattice** is a partially ordered set in which every two elements has a unique least upper bound (called their **join**) and a unique greatest lower bound (called their **meet**.) Figure 3.20 shows a Hasse diagram for the set of all partitions of $\{1.2.3.4\}$ into disjoint subsets, partially ordered by "increasing merging of sets." The slashes between numbers represent different partitions. For instance 1/2/3/4 means the partition $\{1\}, \{2\}, \{3\}, \{4\}$ and 14/23 denotes the partition $\{1, 4\}, \{2, 3\}$. Draw the lattice for the set of partitions of the set $\{a, b, c\}$.

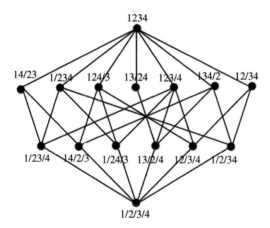

Figure 3.20 Lattice of partitions.

18. **True or False**
Assuming the usual partial ordering $\leq$ for the rational and real numbers, tell which of the following are true or false.
a) Every partially ordered set has a least upper bound.
b) Every set that is bounded above has a least upper bound.
c) Every set of rational numbers bounded above has a least upper bound.
d) Every set of real numbers that is bounded above has a least upper bound.

19. **Dense Orders**
A partial order R on a set A is said to be dense in A if

$$(\forall x, y \in A)[xRy \Rightarrow (\exists z \in A)(xRz \wedge zRy)]$$

Which of the following partial orders are dense on the given set?
a) "less than or equal to" $\leq$ on the rational numbers.
b) "is a subset of" $\subseteq$ on a power set $P(A)$.
c) "less than" $<$ on the real numbers $\mathbb{R}$.
d) "is younger than" on a collection of people.

20. **Inverse of a Partial Order**
If R is a partial order on a set A, then show the inverse R^{-1} relation is a partial order on A.

21. **An Upper Bounded Set with No Sup**
Show that the set

$$S = \left\{ x \in \mathbb{Q} : 0 \leq x \leq \sqrt{2} \right\}$$

is bounded above but has no least upper bound.

22. **Composition of Partial Orders**
Let R be the usual "less than or equal to" ordering and S be the usual "greater than or equal to" ordering on the set $A = \{1, 2, 3\}$. Find the composition $R \circ S = \leq \circ \geq$ of the two orders.

23. **Partitions of a Natural Number**[2]
A partition of a positive integer is a way of writing the number as a sum of positive integers where the order is not important and numbers can be repeated. The five partitions of the number 4 are shown in Figure 3.21.

$4 = 1 + 1 + 1 + 1$
$\quad = 2 + 1 + 1$
$\quad = 2 + 2$
$\quad = 3 + 1$
$\quad = 4$

Figure 3.21 Partition of 4.

The function that gives the number of partitions of a natural number n is called the **partition function** $p(n)$, and in this case $p(4) = 5$. A convenient way to organize the partitions on a number is a **Ferrers diagram**, where the number of rows in the diagram represent the number of terms in the partition and the number of squares in the row is the size of the term. The

---

2 The study of integer partitions arises in combinatorial problems in unexpected ways. The subject got its big start with Euler in the eighteenth century and today is an active area of research among additive number theorists. There are many unsolved problems in additive number theory, including whether (asymptotically) half the values of $p(n)$ are even and half are odd.

Ferrers diagram for the partitions of numbers 1–5 is shown in Figure 3.22. Find the partitions of 6 and draw the Ferrers diagram.

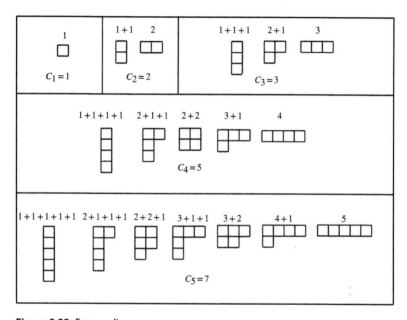

Figure 3.22 Ferrers diagram.

An asymptotic estimate for $p(n)$ was found in 1918 by G.H. Hardy and Ramanujan to be

$$p(n) \sim \frac{\exp\left(\pi\sqrt{2n/3}\right)}{4n\sqrt{3}} \quad \text{as } n \to \infty$$

Use this formula to estimate $p(100)$, $p(200)$, $p(1000)$.

24. **Internet Research**
    There is a wealth of information related to topics introduced in this section just waiting for curious minds. Try aiming your favorite search engine toward *ordered structures in math, ordered sets*.

3.3

Equivalence Relations

Purpose of Section   To introduce the concept of an **equivalence relation** and show how it partitions a set into disjoint subsets. We also introduce the idea of the congruence of integers and modular arithmetic.

3.3.1   Introduction

As you well know every fraction has many equivalent forms. For example

$$\frac{1}{2}, \frac{2}{4}, \frac{5}{10}, \frac{-1}{-2}, \frac{-15}{-30}, \cdots$$

are different ways to represent the same number. They may appear different and are called different names, but they are all equal. The idea of grouping things together that appear different, but from a certain perspective are the same which is the fundamental idea behind equivalence relations.

An equivalence relation is a relation that holds between two elements that relaxes the sometimes over-restrictive "equals relation" and replaces it by "equals from a certain point of view." This allows one to partition sets into groups called equivalence classes which share common properties. For example, we might say two integers as the same if they have the same remainder when divided by a certain number. For example, from some points of view, we may consider the integers ... $-5, -2, 1, 4, 7, \ldots$ "equal" since they all have a remainder of $+1$ when divided by 3.

Definition   An **equivalence relation** on a set A, denoted by "$\equiv$" (or sometimes by "$\sim$") is a relation on A such that for all x, y, z in A, it is **reflexive, symmetric,** and **transitive**. That is

Reflexive: $x \equiv x$
Symmetric: if $x \equiv y$, then $y \equiv x$
Transitive: if $x \equiv y$ and $y \equiv z$, then $x \equiv z$.

Example 1   Equivalence Relations
Some examples of equivalence relations are the following.

a) $x \equiv y$ means $x = y$ for real numbers x, y.
b) $x \equiv y$ means x is congruent to y for triangles x, y.
c) $x \equiv y$ means $x \Leftrightarrow y$ for logical sentences x, y.
d) $x \equiv y$ means "x has the same birthday as y."
e) $x \equiv y$ means x differs from y by a multiple of 5.
f) $A \equiv B$ means sets A, B have the same cardinality.

Example 2   Nonequivalence Relations
The following relations are not equivalence relations. Tell why one of the properties that defines an equivalence relation fails.
a) $x \equiv y$ means "x is in love with y" on the set of all people.
Ans: The relation is not symmetric for at least one couple.

b) $x \equiv y$ means $x \leq y$ on the real numbers.
Ans: The relation is not symmetric since $2 \leq 3$ does not imply $3 \leq 2$.

c) $x \equiv y$ means integers x, y have a common factor greater than 1.
Ans: The relation is not transitive; 2 and 6 have a common factor, 6 and 3 have a common factor, but 2 and 3 have no common factors.

d) $x \equiv y$ means $x \subseteq y$ on a family of sets.
Ans: The relation is not symmetric since $A \subseteq B$ does not imply $B \subseteq A$.

3.3.2   Partition of a Set

A partition of a set is a grouping of the members of the set into nonempty sets in such a way that each element is included in one and only one of the subsets. This leads us to the formal definition of a partition of a set.

Definition   A **partition** of a set A (See Figure 3.23) is a (finite or infinite) family of nonempty subsets $E = \{A_1, A_2, ...\}$ of A that satisfy

i) $\bigcup_{k \in E} A_k = A$

ii) $A_i \cap A_j = \varnothing$ for every pair A_i and A_j.

The following theorem reveals the reason equivalences relation play an important role in mathematics.

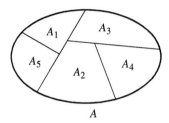

Figure 3.23 Set partition.

3.3.3 The Partitioning Property of the Equivalence Relation

We now see how an equivalence relation on a set allows one to create a partition of the set. If the equivalence relation was the equals relation "=," then the sets in the partition would consist of a single element, but for other equivalence relations the size of the partitions vary.

Theorem 1 Equivalence Classes
If R is a relation defined on a set A, then

$$R \text{ is an equivalence relation on } A \Leftrightarrow R \text{ induces a partition of } A$$

Proof
($\Rightarrow$) We begin by defining the concept of an equivalence class. The set

$$[x] = \{y \in A, y \equiv x\}$$

consisting of members of A that are equivalent to a fixed $x \in A$ is called the **equivalence class** of x, where x is called the class representative or representative of the class.

To show that an equivalence relation "$\equiv$" induces a partition of A, note that the reflexive property of an equivalence relation tells us $x \equiv x$ for all $x \in A$, which in turn tells us that every equivalence class is nonempty and that the union of all equivalence classes is the whole set A. Hence, the only remaining thing to show is that distinct equivalence classes do not overlap. In other words, if $[x] \cap [y] \neq \varnothing$, then $[x] = [y]$. So we assume $[x] \cap [y] \neq \varnothing$ and prove $[x] = [y]$. We begin by doing a little "background" work by picking an element $s \in [x] \cap [y]$ and using properties of an equivalence relation we have $x \equiv s$ and $y \equiv s$. But $\equiv$ is symmetric so we also have $y \equiv s$, and by transitivity $x \equiv y$ and by symmetry $y \equiv x$. We are now ready to start the proof of $[x] = [y]$ by first showing $[x] \subseteq [y]$. We select an arbitrary $d \in [x]$ which implies $d \equiv x$, but we have seen $x \equiv y$ and so by transitivity we have $d \equiv y$ which means $d \in [y]$. Hence, $[x] \subseteq [y]$. A similar

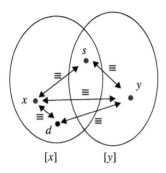

Figure 3.24 Disjoint equivalence classes.

argument shows that $[y] \subseteq [x]$ and so $[x] = [y]$. Figure 3.24 gives a broad idea of the players in the proof.

($\Leftarrow$) If we define members of the set A as equivalent if they belong to the same equivalence class, this defines an equivalence relation on A. ∎

3.3.4 Counting Partitions

Given a set with n members, how many ways are there to subdivide the set into disjoint subsets? The total number of partitions of a set of size n is called the **Bell number** B_n of the set, and the first few Bell numbers for sets of size $n = 0, 1, 2, \dots$ are

$$1, 2, 5, 15, 52, 203, 677, 4\,140, 21\,147, 115\,975\dots$$

The set $\{a, b, c\}$ of three members has a Bell number $B_3 = 5$ and the five partitions of $\{a, b, c\}$ are drawn in Figure 3.25.

a	a b	a	a	a b
b $\mid$ c	c	c $\mid$ b	b c	c
$a \equiv a$	$a \equiv b$	$a \equiv c$	$a \equiv a$	$a \equiv b$
$a \equiv b$	$c \equiv c$	$b \equiv b$	$b \equiv c$	$a \equiv c$
$a \equiv c$				$b \equiv c$

Figure 3.25 Partitions of $\{a, b, c\}$ and their induced equivalence relations.

Note how this partition gives rise to an equivalence relation on $\{a, b, c\}$. We say that two elements of the set are equivalent if they belong to the same set in the partition.

3.3.5   Modular Arithmetic

Two integers $x, y \in \mathbb{Z}$ are said to be **congruent modulo** N, denoted by[1]

$$x \equiv y \pmod{N}$$

if they have the same remainder when divided by the integer N. Dividing two congruent integers x, y by N, we have

$$\frac{x}{N} = Q_1 + \frac{r}{N}, \quad \frac{y}{N} = Q_2 + \frac{r}{N}$$

where Q_1, Q_2 are their respective quotients and r their common remainder. Subtracting the two equations gives

$$\frac{x}{N} - \frac{y}{N} = (Q_1 - Q_2) \quad \text{or} \quad x - y = (Q_1 - Q_2)N$$

which says if x, y are congruent modulo N, then their difference is divisible by N. In other words,

$$x \equiv y \pmod{N} \Leftrightarrow (\exists k \in \mathbb{Z})(x - y = kN)$$

We now show that the congruence relation is an equivalence relation.

Theorem 2   Congruence Is an Equivalence Relation on $\mathbb{Z}$

Proof
We show the congruence relation $\equiv$ is reflexive, symmetric, and transitive.

- **Reflexive:** $x \equiv x \pmod{N}$ since N divides $x - x = 0$.
- **Symmetric:** If $x \equiv y \pmod{N}$, then N divides $x - y$. Hence, there exists an integer k such that

$$x - y = kN \text{ or } y - x = -kN = N(-k)$$

  which means N divides $y - x$. Hence, $y \equiv x \pmod{N}$ which means $\equiv$ is a symmetric relation.

- **Transitive:** For integers x, y, z assume $x \equiv y \pmod{N}$ and $y \equiv z \pmod{N}$. Hence,

$$\begin{cases} x \equiv y \pmod{N} \\ y \equiv z \pmod{N} \end{cases} \Rightarrow \begin{cases} (\exists k_1 \in \mathbb{Z})(x - y = k_1 N) \\ (\exists k_2 \in \mathbb{Z})(y - z = k_2 N) \end{cases}$$

---

1 We use the notation "$\equiv$" here for integers being congruent since it is an equivalence relation.

Adding these equations gives

$$(x-y) + (y-z) = k_1 N + k_2 N$$

or

$$x - z = (k_1 + k_2)N$$

which shows N divides $x - z$ or $x \equiv z \pmod{N}$. Hence, $\equiv$ is a transitive relation. ∎

The congruence relation "$\equiv$" on $\mathbb{Z}$ partitions the integers into **congruence classes** called **residue classes**, where integers in each residue class have the same remainders when divided by N. For example if $N = 5$ the residue classes are denoted by $[0]_5$, $[1]_5$, $[2]_5$, $[3]_5$, $[4]_5$, which are listed in Table 3.3.

Table 3.3 Residue classes mod 5.

Residue classes for $\mathbb{Z}$ modulo (5)
$[0]_5 = \{5n : n \in \mathbb{Z}\} \quad = \{\cdots -10, -5, \underline{0}, 5, 10 \cdots\}$
$[1]_5 = \{5n+1 : n \in \mathbb{Z}\} \quad = \{\cdots -9, -4, \underline{1}, 6, 11 \cdots\}$
$[2]_5 = \{5n+2 : n \in \mathbb{Z}\} \quad = \{\cdots -8, -3, \underline{2}, 7, 12 \cdots\}$
$[3]_5 = \{5n+3 : n \in \mathbb{Z}\} \quad = \{\cdots -7, -2, \underline{3}, 8, 13 \cdots\}$
$[4]_5 = \{5n+4 : n \in \mathbb{Z}\} \quad = \{\cdots -6, -1, \underline{4}, 9, 14 \cdots\}$

Note that the residue classes partition the integers into five disjoint sets:

$$\mathbb{Z} = [0]_5 \cup [1]_5 \cup [2]_5 \cup [3]_5 \cup [4]_5$$

The collection of partitions is called the **quotient set** of $\mathbb{Z}$ modulo 5, and denoted by $\mathbb{Z}/5\mathbb{Z}$. In other words

$$\mathbb{Z}/5\mathbb{Z} = \{[0]_5, [1]_5, [2]_5, [3]_5, [4]_5\}$$

Modular Arithmetic Modular arithmetic (also called clock arithmetic) is a system of arithmetic whose numbers "wrap around" after they reach a certain value, called the **modulus**. Modular arithmetic was introduced by Carl Friedrich Gauss at the age of 24 in 1801 in his seminal book on number theory *Disquisitiones Arithmeticae* (Latin: discourse into arithmetic).

Table 3.4 shows the properties of some common relations.

Table 3.4 Common mathematical relations.

	Reflexive	Symmetric	Transitive	Antisymmetric
$\perp$	No	Yes	Yes	Yes
$=$	Yes	Yes	Yes	Yes
$\leq$	Yes	No	Yes	Yes
$<$	No	No	Yes	Yes
$\parallel$	Yes	Yes	Yes	No
$\perp$	No	Yes	No	No
$\subseteq$	Yes	No	Yes	Yes
$\equiv \bmod(n)$	Yes	Yes	Yes	No
$\cong$	Yes	Yes	Yes	No

> **Important Note** Do you understand why the remainder of the fraction $-3/5$ is 2? Remainders are defined as nonnegative integers, so $-3/5 = (-5 + 2)/5 = -1 + 2/5$.

Example 3 Equivalence Classes in the Plane

The Cartesian product $\mathbb{N} \times \mathbb{N}$ defines the grid points in the first quadrant of the Cartesian plane (i.e. points with positive integer coordinates). The relation

$$(a,b) \sim (c,d) \Leftrightarrow a + d = b + c.$$

between two points is an equivalence relation. We leave this proof to the reader. See Problem 20.

The equivalence classes resulting from this equivalence relation are illustrated in Figure 3.26, where each equivalence class consists of the grid point on a 45° line

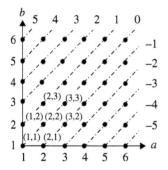

Figure 3.26 Equivalence classes as grid points on lines $y = x + n$.

in the first quadrant. We will see in Section 4.1 that there is a one-to-one correspondence between these equivalence classes and the integers $\mathbb{Z}$, thus allowing us to *define* the integers as equivalence classes of pairs of natural numbers.

> **Important Note** The significance of equivalence relations and equivalence classes in mathematics and other sciences lies in the fact that sometimes when a theorem is proven or a scientific result is established for a single "thing," the result follows for other "things" without having to establish the result for other things, being that the other "things" belong to the same equivalence class of which its member share common properties. This is what happens in all areas of science, one carries out an experiment and discovers some new phenomenon and then interpolates the result to other situations that are believed to be equivalent.

Problems

1. **Testing Relations**
    Let A denote the student body at a university and individual students by x and y. Determine if the following relations are equivalence relations on A.
    a) x is related to y iff x and y have the same major.
    b) x is related to y iff x and y have the same GPA.
    c) x is related to y iff x and y are from the same country.
    d) x is related to y iff x and y have the same major.

2. **Equivalence Relations**
    Which of the following relations R are equivalence relations on the given set A. For those relations that are equivalence relations, find the equivalence classes.
    a) xRy if and only if $y = x^2$. $(A = \mathbb{R})$
    b) mRn if and only if m is a factor of n. $(A = \mathbb{N})$
    c) xRy if and only if x and y have the same remainder when divided by 5. $(A = \mathbb{N})$
    d) xRy if and only if $|x - y| \le 1$. $(A = \mathbb{R})$
    e) $(a, b)R(c, d)$ if and only if $a^2 + b^2 = c^2 + d^2$. $(A = \mathbb{R}^2)$

3. **Not Equivalence Relations**
    Determine if the following relations are equivalent relations and if not, which condition: reflexive, symmetric, or transitive fails?
    a) The relation "$\le$" on the real numbers.
    b) The empty relation on an empty set (i.e. xRy never true)

c) Relation "⊂" of being a proper subset on a family of sets.
d) Relation of being perpendicular on lines in the plane.

4. **Finding the Equivalence Relation**
Partition the set $A = \{a, b, c, d, e\}$ into the equivalence classes $\{\{a, c\}, \{b, e\}, \{d\}\}$. Find the equivalence relation induced by this partition.

5. **Finding the Quotient Set**
Show that the relation

$$R = \{(1,1),(2,2),(3,3),(4,4),(5,5),(1,2),(2,1)\}$$

is an equivalence relation on $A = \{1, 2, 3, 4, 5\}$. What is the partition of A induced by this relation?

6. **Finding Equivalence Classes**
The set $\{1, 2, 3, 4\}$ is partitioned into $\{\{1, 2\}, \{3, 4\}\}$ by an equivalence relation R. Find the following:
a) [1]
b) [2]
c) [3]
d) [4]

7. **Hmmmmmmmmmm**
If an equivalence relation R on a set A has only one equivalence class, what is the relation?

8. **Unusual Equivalence Relation**
Define the relation $\equiv$ on $\mathbb{Z}$ by $m \equiv n$ if and only if 3 divides $m + 2n$.
a) Show that $\equiv$ is an equivalence relation
b) Find the equivalence classes?

9. **Equivalence Relation in Calculus**
Given the set of continuous functions $C[0, 1]$ defined on the closed interval $[0, 1]$, define $R \in C[0, 1] \times C[0, 1]$ by

$$f\,Rg \text{ if and only if } \int_0^1 f(x)dx = \int_0^1 g(x)dx$$

a) Show that R is an equivalence relation.
b) Find $g \in C\,[0, 1]$ equivalent to $f(x) = x$, but $f \neq g$.

10. **Equivalence Relations in Analysis**
Let $A = [-1, 1]$ and define an equivalence relation R on A by xRy if and only if $x^2 = y^2$, $x, y \in [-1, 1]$. Find the equivalence classes.

11. **Equivalence Sets of Polynomials**
    The set P consists of all polynomials defined on the real line, while $I \subseteq P$ are those polynomials that satisfy $p(0) = 0$. For $f, g \in P$ show that $f \equiv g \Leftrightarrow f - g \in I$ is an equivalence relation.[2]

12. **Modular Arithmetic**
    If $x, y \in \mathbb{Z}$ we say $x \equiv y \pmod{n}$ if n divides $x - y$ for a positive integer n. Show the relation $\equiv$ is an equivalence relation.

13. **An Old Favorite**
    The equals relation "=" is the most familiar equivalence relation. What are the equivalence classes on the set $A = \{1, 2, 3, 4, 5\}$ induced by this relation?

14. **Equivalence Classes in Logic**
    Define an equivalence relation on logical sentences by saying two sentences are equivalent if they have the same truth value. Place the following sentences in their proper equivalence class.
    a) $1 + 2 = 3$
    b) $3 < 5$
    c) $2 \mid 7$
    d) $x^2 < 0$ for some real number.
    e) $\sin^2 x + \cos^2 x = 1$
    f) Georg Cantor was born in 1845.
    g) Leopold Kronecker was a big fan of Cantor.
    h) Cantor's theorem guarantees larger and larger infinite sets.

15. **Similar Matrices**
    Two square matrices A, B are equivalent if there is an invertible matrix M that satisfies $MAM^{-1} = B$. Show this relation between matrices defines an equivalence relation.

16. **Counting Equivalence Relations**
    a) Count the number of equivalence relations on $A = \{1, 2\}$.
    b) Count the number of equivalence relations on $A = \{1, 2, 3\}$.

---

2 In the language of abstract algebra, the set $P(x)$ is a polynomial ring and the subset I a vanishing ideal in the ring.

17. **Arithmetic in Modular Arithmetic**
    Suppose

    $$a \equiv c \pmod 5$$
    $$b \equiv d \pmod 5$$

    Show that
    a) $a + b \equiv c + d \pmod 5$
    b) $a - b \equiv c - d \pmod 5$
    c) $ab \equiv cd \pmod 5$

18. **Mapping into the Equivalence Class**
    Let X denote student body at your college or university and define the equivalence relation on the student body as "being in the same class," class referring to freshman, sophomore, junior or senior. Define the mapping $f : x \to [x]$ that sends each student $x \in X$ into his or her equivalence class $[x]$. Is this a well-defined function? What is your own value under this mapping?

19. **Equivalence Classes as Directed Graphs**
    Inasmuch as equivalence relations are binary relations, they can be represented by digraphs. Draw a digraph that represents the equivalence classes of the set $\{0, 1, 2, 3, 4, 5, 6, 7\}$ if two elements are equivalent when they have the same remainder when divided by 3.

20. **Defining Integers from Natural Numbers**
    Example 3 shows how the integers can be defined in terms of pairs of positive integers by means of the equivalence relation.

    $$(a, b) \sim (c, d) \text{ if and only if } a + d + b + c$$

    Show this relation is an equivalence relation on $\mathbb{N} \times \mathbb{N}$ and describe the different equivalence classes. Observe that the equivalence classes can be placed in a one-to-one correspondence with the integers, thus allowing one to define the integers in terms of pairs of natural numbers.

21. **Counting Partitions**
    Find the different partitions of the sets
    a) $A = \{1, 2\}$
    b) $A = \{1, 2, 3\}$

22. **Interesting Equivalence Relation**
    Define a relation R on the nonnegative integers

    $$A = \{0, 1, 2, 3, \ldots, 29, 30\}$$

by

$mRn \Leftrightarrow$ (product of the digits of m = product of the digits of n).

For example 16R23, 4R14....

a) Show that R is an equivalence relation on A.
b) Find the equivalence classes of the relation.
c) The equivalence classes are listed in Table 3.5.

Table 3.5 Equivalence classes.

Product	Integers
0	0,10,20,30
1	1,11
2	2,12,21
3	3,13
4	4,14,22
5	5,15
6	6,16,23
7	7,17
8	8,18,24
9	9,19
10	25
12	26
14	27
16	28
18	29

23. **Internet Research**

There is a wealth of information related to topics introduced in this section just waiting for curious minds. Try aiming your favorite search engine toward *important equivalence relations in math, important equivalence classes in math, important partitions of a set.*

3.4

The Function Relation

> **Purpose of Section** We introduce the concept of the **function**, both as a "rule" that assigns a unique value to every member of a set, and from the relation viewpoint, as a subset of a Cartesian product. We discuss again the important concepts of injections, surjections, and bijections, which were introduced earlier in our study of the cardinality of sets.

3.4.1 Introduction

No doubt the concept of a function covers familiar territory for many readers of this book.[1] Normally, in the beginning of mathematics books, a function $f: A \rightarrow B$ is defined as a rule that assigns to each value $x \in A$ a unique value $y \in B$. This is the definition proposed by German mathematician Peter Lejeune Dirichlet (1805–1859) in the 1830s. When we write an algebraic formula like

$$y = f(x) = \sin x$$

where x is taken as a real number, the rule is clearly understood, it assigns to each x the value sin x. We denote the function by the letter f and the value of the function at x by $f(x)$. This motivates the Dirichlet definition of a function.

---

1 A fascinating reference for functions is *Atlas for Computing Mathematical Functions* by William Jackson Thompson, which gives analytical, visual, and descriptive properties of over 150 special functions useful in pure and applied mathematics.

Advanced Mathematics: A Transitional Reference, First Edition. Stanley J. Farlow.
© 2020 John Wiley & Sons, Inc. Published 2020 by John Wiley & Sons, Inc.
Companion website: www.wiley.com/go/farlow/advanced-mathematics

Peter Gustav Lejeune Dirichlet (1805–1859)

Dirichlet Definition of a Function   Let *A* and *B* denote sets. A **function** *f* from *A* to *B*, denoted $f : A \rightarrow B$, is a *rule* that assigns to each element $x \in A$, a unique element $f(x)$ in *B*. The set *A* is called the **domain** of the function, written dom(*f*), and *B* is the **codomain** of the function. For $x \in A$, the assigned value in *B* is called the **image** of *x* under *f* and denoted by $f(x)$, which is read as *the value of f at x*. See Figure 3.27.

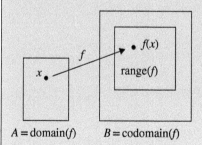

$A = \text{domain}(f)$ $B = \text{codomain}(f)$

Figure 3.27 Illustration of a function.

The **range** of *f*, denoted by range (*f*), or *f*(*A*), is the set of "outputs" of the function, or

$$\text{range}(f) = \{f(x) : x \in A\} \subseteq B$$

The **graph** of a function *f* is the set

$$\text{graph}(f) = \{(x, f(x)) : x \in A\} \subseteq A \times B$$

There are several synonyms for the word "function." The words *mapping* (or *map*), *transformation*, and *operator* are often used depending on the context as well as the domain and codomain of the function.

Example 1   Functions

The following are examples of functions with different domains, codomains, ranges, and graphs.

a) Define $f: [-1, 1] \rightarrow \mathbb{R}$ by the rule $f(x) = \sqrt{1-x^2}$. Here
   - domain$(f) = [-1, 1]$
   - codomain$(f) = \mathbb{R}$
   - range$(f) = [0, 1] \subseteq \mathbb{R}$
   - graph$(f) = \left\{ \left(x, \sqrt{1-x^2} \right) : x \in [-1, 1] \right\}$

b) Define $f: \mathbb{N} \rightarrow \mathbb{R}$ by the rule $f(n) = \sin n, \quad n = 1, 2, \ldots$
   - domain$(f) = \mathbb{N}$
   - codomain$(f) = \mathbb{R}$
   - range$(f) = \{\sin n, n = 1, 2, \ldots\}$
   - graph$(f) = \{(n, \sin n) : n \in \mathbb{N}\} \subseteq \mathbb{N} \times \mathbb{R}$

c) Define $f: [0, \infty) \rightarrow \mathbb{R}^3$ by the rule $f(t) = (\cos t, \sin t, t)$.
   - domain$(f) = [0, \infty)$
   - codomain$(f) = \mathbb{R}^3$
   - range$(f) = \{(\cos t, \sin t, t) : 0 \le t < \infty\} \subseteq \mathbb{R}^3$
   - graph$(f) = \{(t, (\cos t, \sin t, t)) : 0 \le t < \infty\} \subseteq [0, \infty) \times \mathbb{R}^3$
   We cannot view the graph of this function since it lies in a four-dimensional space. We can view the image of the function, however, which is a helix in three-dimensional space.

d) Define $f: \mathbb{R}^2 \rightarrow \mathbb{R}$ defined by the rule $f(x_1, x_2) = 3x_1 + 2x_2$.
   - domain$(f) = \mathbb{R}^2$
   - codomain$(f) = \mathbb{R}$
   - range$(f) = \{3x_1 + 2x_2 : (x_1, x_2) \in \mathbb{R}^2\} = \mathbb{R}$
   - graph$(f) = \{((x_1, x_2), 3x_1 + 2x_2) : (x_1, x_2) \in \mathbb{R}^2\} \subseteq \mathbb{R}^3$

A Brief History of the Function   Although the concept of a function is one of the most important concepts in all of mathematics, its history is relatively short. The mathematician and writer of math history, Morris Kline, credits Galileo (1564–1642) with the first statement of dependency of one quantity on another.[2] In 1673, the German mathematician Gottfried Leibniz (1646–1716) used the word "function" to mean any quantity that varies from point to point along a curve. One of the first formal definition of a function is due to the Swiss mathematician Leonhard Euler (1707–1783), who defined a function as

> *Quantities dependent on others, such that as the second changes, so does the first, are said to be functions.*

Euler and other leading mathematicians of the times, such as the French mathematician Joseph Fourier (1768–1830), thought of functions in terms of equations, such as $y = x^2$ or $y = \sin x$. For mathematicians of the time, an expression like

$$f(x) = \begin{cases} 0 & x < 0 \\ 1 & x \geq 0 \end{cases}$$

was *not* considered a function since it is not an equation, only a "rule" for assigning values to a variable. Finally in 1837, the German mathematician Peter Lejeune Dirichlet expanded the definition of a function to the meaning we accept today, when he wrote:

> *A variable quantity y is said to be a function of a variable quantity x, if to each value of x there corresponds a uniquely determined value of the quantity y.*

3.4.2   Relation Definition of a Function

In addition to defining a function as a rule (ala Dirichlet), we can also think of a function in terms of relations.

Definition   A relation between two sets A and B, called the domain and codomain of the relation, respectively, such that each member in the domain is assigned exactly one element in the codomain, is called a **function** from A to B.

---

2 The statement is 'The time of descent along inclined planes of the same height, but of different slopes, are to each other as the length of these slopes."

An Important Classification of Functions

1) **Algebraic Functions:** An algebraic function is a function that can be constructed, starting from x, using only a finite number of operations of addition, subtraction, multiplication, division, and root extraction. Typical algebraic functions are

$$f(x) = 1/x, \ f(x) = \sqrt{x}, f(x) = \frac{\sqrt[4]{1 + x^{4/5}}}{\sqrt{x} - (3 + x^3)}$$

2) **Transcendental Functions:** Functions that are not algebraic are called transcendental functions. Important transcendental functions are $\sin x$, $\cos x$ and e^x.

Important Note In 1890, the Italian mathematician Giuseppe Peano shocked the mathematical world with a construction of a continuous space-filling curve; a curve in the plane, defined by two continuous functions $x = f(t), y = g(t)$, such that as t varies over [0, 1], the point $(x(t), y(t))$ passes through *every point* in the unit square $[0, 1] \times [0, 1]$.

3.4.3 Composition of Functions

The function $f(x) = \sin x^2$ can be interpreted as assigning $x \to \sin x^2$. However, it can also be interpreted as a combination or composition of two functions: the first assigning $x \to x^2$, the second assigning $x^2 \to \sin x^2$, which leads us to the following definition.

Composition of Two Functions Given sets A, B, C and functions

$$g : A \to B, f : B \to C$$

we define the **composition** of f and g, denoted by $f \circ g$, and read "f circle g" as the function that sends the point $x \in A$ into

$$(f \circ g)(x) = f(g(x)) \in C.$$

The domain of $f \circ g$ consists of the following points of A:

$$\mathrm{dom}(f \circ g) = \{x \in A : x \in \mathrm{dom}(g)\} \cap \{x \in A : g(x) \in \mathrm{dom}(f)\}$$

where $\mathrm{dom}(f)$, $\mathrm{dom}(g)$ are the domains of f, g, respectively. See Figure 3.28.

In other words, the domain of a composition $f(g(x))$ consists of those x in the domain of g whose values $g(x)$ lie in the domain of f.

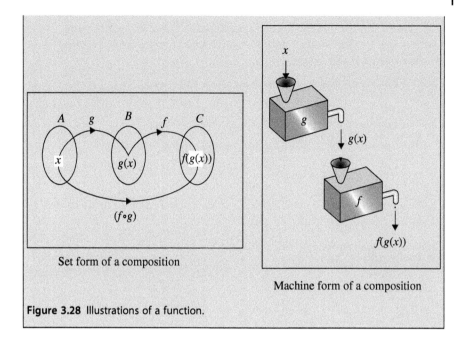

Set form of a composition

Machine form of a composition

Figure 3.28 Illustrations of a function.

Example 2   Composition of Real-Valued Functions

Find the compositions $f \circ g$ and $g \circ f$ of the functions

- $f(x) = x^2, \; -\infty < x < \infty$
- $g(x) = \sqrt{x}, \; 0 \le x < \infty$

Solution

- $(f \circ g)(x) = f(g(x)) = f(\sqrt{x}) = (\sqrt{x})^2 = x \;\; x \in [0, \infty)$
- $(g \circ f)(x) = g(f(x)) = g(x^2) = \sqrt{x^2} = |x| \;\; x \in \mathbb{R}$

Note that $f \circ g \neq g \circ f$. Their domains are also different.

Domain of a Composition   Here is a nice way to think about it. Person g and person f want to move several bags of cement 200 yards. Person g carries the bags the first 100 yards and gives them to person f, whose goal is to carry them the last 100 yards. Person g is able to carry eight bags (domain of g) after which he gives them to person f. Person f, however, is only able to carry only five bags (domain of person f), thus only five bags make it from the start to the end, and hence the domain of the composition $f(g(x))$ is five bags of cement.

1–1, Onto, and One-to-One Correspondence   Three important types of functions are **1–1 functions** (injections), **onto functions** (surjections), and **1–1 correspondences** (bijections).

- **1–1:** A function $f : A \rightarrow B$ is **one-to-one** (or an **injection**) if

  direct form: $(\forall a,\ b \in A)[a \neq b \Rightarrow f(a) \neq f(b)]$
  contrapositive form: $(\forall a,\ b \in A)[f(a) = f(b) \Rightarrow a = b]$

- **Onto:** A function $f : A \rightarrow B$ is **onto** (or a **surjection**) if

  $$(\forall b \in B)(\exists a \in A)(f(a) = b)$$

- **1–1 correspondence:** A function $f : A \rightarrow B$ is a **one-to-one correspondence** (or a **bijection**) if it is both 1–1 and onto.

These three types of functions are illustrated in Figure 3.29.

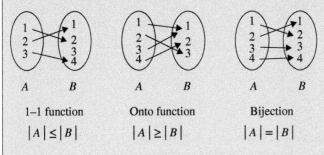

1–1 function	Onto function	Bijection
$|A| \le |B|$	$|A| \ge |B|$	$|A| = |B|$

Figure 3.29 Three major types of functions.

Example 3   1–1, Onto, One-to-One Correspondence
The graphs in Figure 3.30 illustrate typical functions from $\mathbb{R}$ to $\mathbb{R}$. Note that the graph of a $1-1$ function intersects any horizontal line at most once.

Example 4   Injection
Show that the function $f : \mathbb{N} \rightarrow \mathbb{N}$ defined by $f(n) = n^2$ is 1–1.[3]

Proof
If one simply lines up the natural numbers against their squares as shown in Table 3.6, it is easy to see that the function maps different values into different values and thus is a 1–1 mapping.

---

3 The domain of a function can influence whether a function is one-to-one. The function $f(x) = x^2$ with domain the real numbers is not one-to-one, but the same function defined on the nonnegative real numbers is one-to-one.

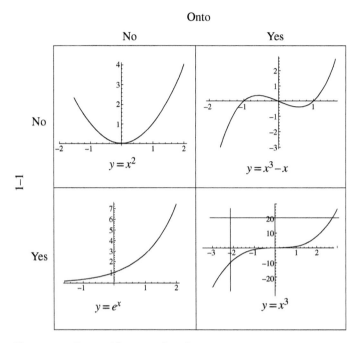

Figure 3.30 Types of functions $\mathbb{R} \to \mathbb{R}$.

Table 3.6 An injection.

n	1	2	3	4	$\cdots$	n	$\cdots$
$f(n) = n^2$	1	4	9	16	$\cdots$	n^2	$\cdots$

To actually prove f is 1–1, we show that for any $m, n \in \mathbb{N}$, where $m \neq n$, we have $f(m) \neq f(n)$, and do this by proving its contrapositive

$$f(m) = f(n) \Rightarrow m = n.$$

We have

$$f(m) = f(n) \Rightarrow m^2 = n^2$$
$$\Rightarrow m^2 - n^2 = 0$$
$$\Rightarrow (m-n)(m+n) = 0$$
$$\Rightarrow m = n \text{ or } m = -n$$

But $m = -n$ is not possible since m, n are positive numbers. Hence, we conclude $m = n$ verifying that f is 1 – 1 on $\mathbb{N}$.

Example 5   Surjection

If $f: \mathbb{R} \to \mathbb{R}$ is defined by $f(x) = x^3 + 1$, show that f maps $\mathbb{R}$ onto $\mathbb{R}$. That is show f is a surjection or onto mapping.

Proof

For any $y \in \mathbb{R}$, we must show there exists an $x \in \mathbb{R}$ that satisfies $y = x^3 + 1$. Choosing

$$x = \sqrt[3]{y-1} \in \mathbb{R}$$

we see

$$f(x) = \left(\sqrt[3]{y-1} \right)^3 + 1 = (y-1) + 1 = y.$$

Hence, f maps $\mathbb{R}$ onto $\mathbb{R}$.

Example 6   Counterexample

Is $f: \mathbb{R} \to \mathbb{R}$ defined by $y = x^2 + 2x$ a surjection?

Solution

The function is not a surjection since the number $y = -2$ in the codomain has no preimage in the domain since the equation $x^2 + 2x = -2$ has only complex solutions.

3.4.4   Inverse Functions

In arithmetic, some numbers have inverses. For example -3 is the additive inverse of $+3$ since $3 + (-3) = 0$. Some functions also have inverses in the sense that the inverse "undoes" the operation of the function.

Historical Note   The study of functions changed qualitatively with ideas of the Italian mathematician Vito Volterra (1860–1940), who introduced the idea of **functions of functions**, functions whose arguments were themselves functions. The French mathematician Jacques Hadamard (1865–1963) named these types of functions **functionals** and Paul Lévy (1886–1971) gave the name **functional analysis** to the study of functions interpreted as points in some abstract space, not unlike points in the plane.

Definition   Function Inverse If $f: A \to B$ is 1–1, then for each $y \in f(A)$ in the range of A, the equation $f(x) = y$ has a *unique* solution $x \in A$. This yields a new function $f^{-1}: \text{range}(f) \to A$, defined by

$$x = f^{-1}(y).$$

This function is called the **inverse** of f. See Figure 3.31.

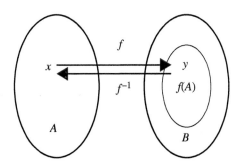

Figure 3.31 Inverse function.

Example 7   Inverse Function
The function $f: [0, \infty) \to [1, \infty)$ defined by

$$f(x) = 1 + x^2, \ x \geq 0$$

is a 1–1 function from $[0, \infty)$ onto $[1, \infty)$ and, hence, has an inverse f^{-1}: $[1, \infty) \to [0, \infty)$. Find and draw the graph of this inverse.

Solution
Solving the equation

$$y = 1 + x^2$$

for $x \geq 0$ in terms of y, we find the unique value

$$x = \sqrt{y - 1}, \ y \geq 1$$

or

$$f^{-1}(y) = \sqrt{y - 1}, \ y \geq 1.$$

At this stage, one often renames the variables and writes the inverse as

$$f^{-1}(x) = \sqrt{x - 1}, \ x \geq 1.$$

The graphs of f and f^{-1} are drawn in Figure 3.32. Note that the graph of f^{-1} is the reflection of the graph of f through the 45° line $y = x$.

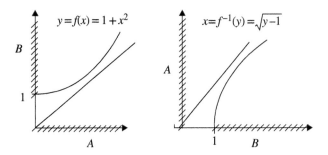

Figure 3.32 A function and its inverse.

Some common inverses of 1–1 functions defined of given domains are listed in Table 3.7.

Problems

1. **Testing Relations**
   Determine which of the following relations are functions. For functions, what is the domain and range of the function?
   a) $R = \{(1, 3), (3, 4), (4, 1), (2, 1)\}$
   b) $R = \{(1, 3), (1, 4), (1, 2), (3, 1)\}$
   c) $R = \{(1, 3), (3, 4), (1, 1)\}$
   d) $R = \{(1, 2), (2, 2), (3, 2), (2, 3)\}$

2. **Graphing Relations and Functions**
   Graph each of the following relations on $\mathbb{R}$ and tell which relations are functions.
   a) $R = \{(x, y) : y = x^2\}$
   b) $R = \{(x,y) : y = \pm\sqrt{x}\}$
   c) $R = \left\{(x,y) : y = \dfrac{1}{x+1}\right\}$
   d) $R = \{(x, y) : x = |y|\}$
   e) $R = \{(x, y) : |x| + |y| = 1\}$

3. **Find the Mystery Function**
   Find a function that "tears" the interval $[0, 1]$ into two parts at its midpoint and then "stretches" each part uniformly to twice its length.

4. **Compositions**

Find $f \circ g$ and $g \circ f$ and their domains for the following functions f and g. We assume the domains of the functions are all values for which the function is well defined.

a) $\begin{cases} f(x) = \{(-2, 3), (-1, 1), (0, 0), (1, -1), (2, -3)\} \\ g(x) = \{(3, 1), (0, 2), (-1, -2), (2, 0), (-3, 1)\} \end{cases}$

b) $f(x) = 2x + 3, \quad g(x) = -x^2 + 5$

c) $f(x) = \dfrac{1}{x^2 + 1}, \quad g(x) = x^2$

d) $f(x) = |x|, \quad g(x) = |x|$

e) $f(x) = \sqrt{x}, \quad g(x) = x - 2$

f) $f(x) = \sqrt{1 - x^2}, \quad g(x) = \sqrt{x^2 - 1}$

5. **Composition of Three Functions**

For each function f, g, h below that maps $\{1, 2, 3, 4\}$ to itself, find the composition $f \circ (g \circ h)$.

$f = \{(1, 3), (2, 4), (3, 1), (4, 2)\}$

$g = \{(1, 2), (2, 2), (3, 4), (4, 3)\}$.

$h = \{(1, 4), (2, 4), (3, 1), (4, 3)\}$

6. **Backwards Compositions**

One can sometimes interpret a function h as a composition of two functions. For the given function h given below determine f, g such that $h = f \circ g$.

a) $h(x) = (x - 1)^2 + (x - 1) + 3$

b) $h(x) = \sin(1/x)$

c) $h(x) = x^2 + x + 1$

d) $h(x) = e^{x^2} + 1$

7. **Decomposing a Function as a Composition**

Write the function $h(x) = x^2 + 1$ as a composition $h = f \circ g$ of two functions in an infinite number of different ways.

8. **Classroom Function**

Let A be the set of students in your Intro to Abstract Math Class and B be the natural numbers from 1 to 100.

a) To each student, assign the student's age. Is this a function from A to B?

b) To each natural number $n \in B$, assign students in the class whose age is n. Is this a function from B to A?

9. **More Compositions**
   Given functions f, g illustrated in Figure 3.33, both having domains and codomains $A = \{1, 2, 3, 4\}$, find the following.
   a) $f \circ g$
   b) $g \circ f$
   c) $f \circ f$
   d) $g \circ g$

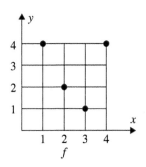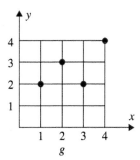

Figure 3.33 Compositions.

10. **Shifting Domain of a Composition**
    Given the function defined by

    $$f(x) = \frac{1}{1-x}, \quad x \neq 1$$

    whose domain is the real numbers, except 1, find the domain of $f \circ f$.

11. **Graphing a Composition**
    Draw the graph for two arbitrary real-valued functions f, g of a real variable. Then select an arbitrary real number x and use the graphs to find the location of $(f \circ g)(x)$.

12. **Compositions**
    Find $f \circ g$ if
    $f : \mathbb{R} \to \mathbb{R}^3$, $f(t) = (t, t^2, t^3)$
    $g : \mathbb{R} \to \mathbb{R}$, $g(t) = \sin t$

13. **Composition of Operators**
    Given the differential operators
    - $L_1(f) = x\, f(x) + 1$

    - $L_2(f) = x^2 \dfrac{df}{dx}$

find
a) $L_1 \circ L_2$
b) $L_2 \circ L_1$

14. **Recursive Functions**

A recursive function is one that is defined in terms of itself, normally defined over a restricted subset of its domain. For example, the factorial function $n! = n(n-1)(n-2)\cdots(2)(1)$ can be defined **recursively** as

$$n! = \begin{cases} 1 & n = 1 \\ n(n-1)! & n > 1 \end{cases}$$

Another example of a recursively defined function is the greatest common divisor of two positive integers m and n, which is defined as the largest positive integer that divides both m and n. For $0 < n \le m$, we can define the greatest common divisor of m and n by

$$\gcd(m,n) = \begin{cases} n & \text{if } n \text{ divides } m \\ \gcd(n, \text{remainder of } m/n) & \text{otherwise} \end{cases}$$

Use this recursive definition to find the greatest common divisor of the following numbers.
a) $m = 25, \quad n = 5$
b) $m = 101, \quad n = 13$
c) $m = 37, \quad n = 3$

15. **Functional Equation**

A **functional equation** is an equation which expresses the value of the function at a point in terms of the value of the function at another point or points. Below, are listed four well-known functional equations. Find a function or functions that satisfies the given functional equation.
a) $f(x + y) = f(x) + f(y)$
b) $f(x + y) = f(x)f(y)$
c) $f(xy) = f(x) + f(y), \quad x, y > 0$
d) $f(xy) = f(x)f(y), \quad x, y > 0$

16. **Injections, Surjections, Bijections**

Give examples of the following functions f_1, f_2, f_3, f_4 from $\mathbb{N}$ to $\mathbb{N}$ that satisfy the following properties.
a) f_1 is neither 1–1 or onto.
b) f_2 is 1–1, but not onto.
c) f_3 is onto, but not 1–1.
d) f_4 is both 1–1 and onto.

17. **Find the Function**
    Find a function f that satisfies the following properties.
    a) f maps $\mathbb{R}$ to $\{1, 2, 3\}$
    b) f maps $\mathbb{N}$ to $\mathbb{R}$
    c) f maps $\mathbb{R} \times \mathbb{R}$ to $\mathbb{R}$
    d) f maps $\mathbb{R}$ to $\mathbb{R} \times \mathbb{R}$
    e) f maps $\{a, b, c\}$ to $[0, 1]$

18. **Injections, Surjections, and Bijections**
    Which of the following functions $f: \mathbb{R} \to \mathbb{R}$ are injective, surjective, bijective, or none of the three. Assume the domains of the functions are subsets of $\mathbb{R}$ for which the function is well-defined.
    a) $f(x) = x^3 - 2x + 1$
    b) $f(x) = \sin(1/x)$
    c) $f(x) = \begin{cases} x^2 & x \le 0 \\ x + 1 & x > 0 \end{cases}$
    d) $f(x) = e^{-x}$

19. **Interesting Function**
    Let $f: \mathbb{N} \to \mathbb{N}$ be the function defined by

    even numbers $2n \to n$
    odd numbers $2n - 1 \to n$

    a) Draw part of the graph of this function.
    b) Is this function 1–1?
    c) Is the function an onto function?

20. **Inverse Function**
    Given the function defined by

    $$f(x) = \sqrt{x-2}, \quad x \ge 2$$

    a) Draw the graph of f
    b) Find the domain and range of f.
    c) Prove that the function is 1–1.
    d) Find the inverse of the function.
    e) Find the domain and range of the inverse function.
    f) Draw the graph of the inverse function.

21. **Function as Ordered Pairs**
    For $f: \{1, 2, 3\} \to \mathbb{N}$ defined by the ordered pairs $f = \{(1, 3), (2, 5), (3, 1)\}$:
    a) Is f 1–1?
    b) Is f onto?
    c) What is the range of f?

22. **1–1 But Not Onto**
    Give an example of a function $f: \mathbb{R} \to \mathbb{R}$ that is 1–1, but not onto.

23. **Hmmmmmmmmm**
    For what value of the exponent $n \in \mathbb{N}$ is the function $f(x) = x^n$ a 1–1 function?

24. **Counting Functions I**
    Let $A = \{1, 2\}$, $B = \{a, b, c\}$. Answer the following questions. Hint: It may help by drawing a simple picture.
    a) How many functions are there from A to B?
    b) How many 1–1 functions are there from A to B?
    c) How many onto functions are there from A to B?
    d) How many one-to-one correspondences are there from A to B?

25. **Counting Functions II**
    Let $A = \{1, 2, 3\}$, $B = \{a, b\}$. Answer the following questions. Hint: It may help by drawing a simple picture.
    a) How many functions are there from A to B?
    b) How many 1–1 functions are there from A to B?
    c) How many onto functions are there from A to B?
    d) How many one-to-one functions are there from A to B?

26. **Finding Injections and Surjections**
    a) Find a function $f: \mathbb{N} \to \mathbb{N}$ that is 1–1 but not onto.
    b) Find a function $f: \mathbb{N} \to \mathbb{N}$ that is onto but not 1–1.

27. **Composition of Onto Maps**
    Prove that if

    $$g: X \to Y \text{ (onto)}$$
    $$f: Y \to Z \text{ (onto)}$$

    then the composition $f \circ g$ is an onto function from X to Z. In short, the composition of surjections is a surjection.

28. **Composition of 1–1 Functions**
    Prove that if g is a 1–1 mapping from X to Y, and f is a 1–1 mapping from Y to Z, then the composition $f \circ g$ is a 1–1 mapping from X to Z. In other words, the composition of injections is an injection.

29. **Hmmmmmmmmmm**
    If

    $$g: A \to B, f: B \to C, f \circ g: A \to C.$$

Find examples of the following:
a) A 1–1 composition $f \circ g$, where f is not 1–1.
b) An onto composition $f \circ g : A \to C$, where $g : A \to B$ is not onto.
c) A one-to-one correspondence $f \circ g$, where g is not onto and f is onto.

30. **More Counting Functions**
   Let $S = \{1, 2, 3\}$.
   a) How many functions are there from S to S?
   b) How many onto functions are there from S to S?
   c) How many 1–1 functions are there from S to S?
   d) How many bijections are there from S to S?

31. **Counting Functions in General**
   If a set A has m elements and B has n elements, how many functions of different types map A into B?
   a) All functions
   b) All 1–1 functions
   c) All bijections

32. **Euler Totient Function**
   In number theory, the Euler totient function $\phi(n)$ (or phi function) is a function

   $$\phi : \mathbb{N} \to \mathbb{N}$$

   defined on the natural numbers that gives the number of natural numbers less than n that are **coprime** with n, where a number is coprime with another if the greatest common divisor of the two numbers is 1. For example $\phi(6) = 2$ since 1 and 5 are coprime with 6, but 2, 3, and 4 are not. Verify the following special cases of some important properties of the Euler totient function.
   a) $\phi(17) = 16$, general theorem $\phi(p) = p - 1$, p prime
   b) $\phi(1) + \phi(2) + \phi(4) + \phi(8) = 8$, general theorem $\sum_{k|n} \phi(k) = n$
   c) $\phi(15) = \phi(3)\phi(5)$, general theorem $\phi(mn) = \phi(m)\phi(n)$, m, n coprime
   d) $\phi(5^3) = (5 - 1)5^2$, general theorem $\phi(p^k) = (p - 1)p^{k-1}$

33. **Carmichael's Totient Function Conjecture**
   An open question in number theory is the Carmichael Totient Function Conjecture, which states that for every natural number n, there is at least one other natural number m that satisfies $\phi(m) = \phi(n)$. In other words, both have the same number of coprimes. As of 2019, it is unknown whether the conjecture is true or false.

34. Internet Research

There is a wealth of information related to topics introduced in this section just waiting for curious minds. Try aiming your favorite search engine toward *strange functions, history of the function, and list of mathematical functions.*

3.5

Image of a Set

Purpose of Section   To introduce the concept of the image and inverse image of a set. We show that unions of sets are preserved under a mapping, whereas intersections are preserved under one-to-one functions. On the other hand, the inverse function preserves both unions and intersections.

3.5.1   Introduction

In many areas of mathematics, be it topology, measure theory, real analysis, and others, one seeks to find the image of a *set A* under the action of a function f: $A \rightarrow B$, yielding the image set

$$f(A) = \{f(x) : x \in A\}$$

In medical imaging, such as CT scans, MRI imaging, X-rays, one is not interested in images of points, but of images of sets. Although the image viewed by the medical professionals is not the kind of function we have studied thus far, it is a function nevertheless.

Advanced Mathematics: A Transitional Reference, First Edition. Stanley J. Farlow.
© 2020 John Wiley & Sons, Inc. Published 2020 by John Wiley & Sons, Inc.
Companion website: www.wiley.com/go/farlow/advanced-mathematics

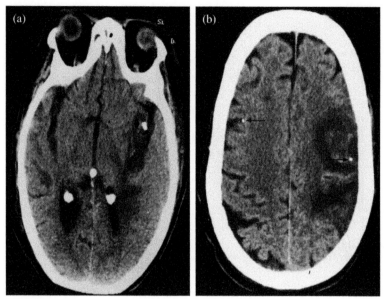

Medical imagings are images of sets under some function.

This discussion motivates the image and inverse image of a set.

Definition   Image and Inverse Image of a Set
Given sets X, Y let $f: X \to Y$ and $A \subseteq X$. As x takes on all values in the set $A \subseteq X$, the set of values $f(x) \subseteq Y$ defines the **image of** A:

$$f(A) = \{f(x) : x \in A\} \subseteq Y$$

Also, for $B \subseteq Y$ we can define the **inverse image of** B as the set

$$f^{-1}(B) = \{x \in X : f(x) \in B\}.$$

Note that $f^{-1}(B)$ is a well-defined set regardless of whether the function f is 1–1 and has an inverse. See Figure 3.34.

Example 1   Images of Sets
Let $X = \{1, 2, 3, 4\}$, $A = \{1, 2, 3\} \subseteq X$, and $Y = \{a, b, c\}$, and a function $f: A \to f(A) \subseteq Y$ defined by

$$f(1) = a, \quad f(2) = a, \quad f(3) = c.$$

Here

$$f(A) = \{a, c\}$$
$$f^{-1}(\{b, c\}) = \{3\}$$

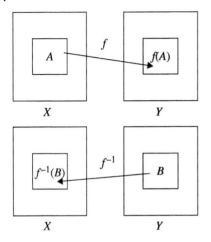

Figure 3.34 Image and inverse image of a set.

$$f^{-1}(\{a,c\}) = \{1,2,3\}$$
$$f^{-1}(\{b\}) = \varnothing$$

We illustrate these ideas visually in Figure 3.35.

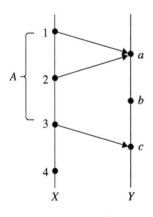

Figure 3.35 Image of a set.

Example 2 Image of a Set

Let $f: X \rightarrow Y$ be defined by

$$f(x) = 1 + x^2.$$

as drawn in Figure 3.36.

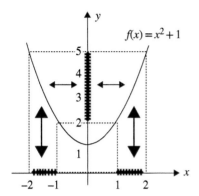

Figure 3.36 $f^{-1}([2, 5]) = [-2, -1] \cup [1, 2]$.

Then we have

a) $f(\{-1, 1\}) = \{2\}$ since both $f(-1) = f(1) = 2$.
b) $f([-2, 2]) = [1, 5]$
c) $f([-2, 3]) = [1, 10]$
d) $f^{-1}(\{1, 5, 10\}) = \{-3, -2, 0, 2, 3\}$
e) $f^{-1}([0, 1]) = \{0\}$
f) $f^{-1}([2, 5]) = [-2, -1] \cup [1, 2]$

Important Note It is often important to know if certain properties of sets are preserved under certain types of mappings. For instance, if X is a connected set and f a continuous function, then it can be proven that $f(X)$ is also connected.

3.5.2 Images of Intersections and Unions

The following theorem gives an important property for the image of the intersection of two sets.

Theorem 1 If $f: X \to Y$ and $A \subseteq X$, $B \subseteq X$, then the images of intersections satisfy

$$f(A \cap B) \subseteq f(A) \cap f(B)$$

See Figure 3.37.

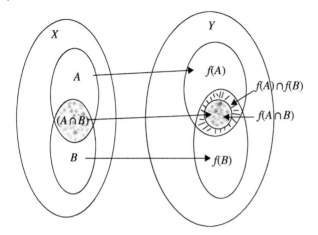

Figure 3.37 Image of an intersection.

Proof

If $y \in f(A \cap B)$, then there exists an $x \in A \cap B$ such that $f(x) = y$ which means $x \in A$ and $x \in B$. But

$$x \in A \Rightarrow y = f(x) \in f(A)$$
$$x \in B \Rightarrow y = f(x) \in f(B)$$

and so $y = f(x) \in f(A) \cap f(B)$, which proves the result.

To show the converse does not hold. That is

$$f(A \cap B) \not\supseteq f(A) \cap f(B)$$

we take $A = [-1, 0]$, $B = [0, 1]$, $f(x) = x^2$. Hence

$$f(A \cap B) = f(\{0\}) = \{0\}$$
$$f(A) \cap f(B) = [0,1] \cap [0,1] = [0,1]$$

which shows $f(A \cap B) \not\supseteq f(A) \cap f(B)$. We illustrate this in Figure 3.38. ∎

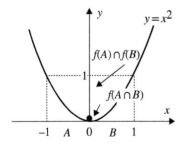

Figure 3.38 Counterexample to show $f(A) \cap f(B) \not\subseteq f(A \cap B)$.

Theorem 1 tells us that set intersections are not preserved under the image of a function. However, if a function is 1–1 then intersections are preserved.

Theorem 2   If a function $f: X \to Y$ is 1–1 and $A \subseteq X$, $B \subseteq X$, then intersections are preserved. That is

$$f(A \cap B) = f(A) \cap f(B).$$

Proof
($\subseteq$) To show $f(A) \cap f(B) \subseteq f(A \cap B)$ we let $y \in f(A) \cap f(B)$ from which we conclude $y \in f(A)$ and $y \in f(B)$. Hence

$$\left| \begin{matrix} \exists x_1 \in A \text{ such that } f(x_1) = y \\ \exists x_2 \in B \text{ such that } f(x_2) = y \end{matrix} \right. \Rightarrow f(x_1) = f(x_2)$$

But f is 1–1 so $x_1 = x_2$ and so

$$x_1 \in A \text{ and } x_1 \in B \text{ and thus } x_1 \in A \cap B$$

Hence, $y = f(x_1) \in f(A \cap B)$, which proves the result.

The proof of the set inclusion in the opposite direction ($\supseteq$) was proven in Theorem 1 without the 1–1 hypothesis. ∎

We now see that in contrast to intersections, unions are always preserved under set mappings.

Theorem 3   Unions Preserved
Given

$$f: X \to Y, \quad A \subseteq X, \quad B \subseteq X,$$

the union of two sets is preserved. That is

$$f(A \cup B) = f(A) \cup f(B)$$

Proof

a) ($\subseteq$) We begin by showing

$$f(A \cup B) \subseteq f(A) \cup f(B)$$

The argument goes

$$y \in f(A \cup B) \Rightarrow (\exists x \in A \cup B, \ f(x) = y)$$

hence, $x \in A$ or $x \in B$ which implies $f(x) \in f(A)$ or $f(x) \in f(B)$. Hence $f(x) \in f(A) \cup f(B)$ and so

$$f(A \cup B) \subseteq f(A) \cup f(B).$$

b) ($\supseteq$) The verification that $f(A \cup B) \supseteq f(A) \cup f(B)$ follows along similar lines and is left for the reader.

Although set intersections are not always preserved under the mapping of a function, both intersections and unions are preserved under inverse images of a function. ∎

Theorem 4   Unions, Intersections Preserved Under f^{-1}

Let $f:X \to Y$, $C \subseteq Y$, $D \subseteq Y$. The inverse image of both intersections and unions are preserved under the inverse image f^{-1}. That is.

a) $f^{-1}(C \cap D) = f^{-1}(C) \cap f^{-1}(D)$
b) $f^{-1}(C \cup D) = f^{-1}(C) \cup f^{-1}(D)$

Proof

We prove a) by the series of if-and-only-if statements:

$$x \in f^{-1}(C \cap D) \Leftrightarrow f(x) \in C \cap D$$
$$\Leftrightarrow f(x) \in C \text{ and} f(x) \in D$$
$$\Leftrightarrow x \in f^{-1}(C) \text{ and } x = f^{-1}(D)$$
$$\Leftrightarrow x \in f^{-1}(C) \cap f^{-1}(D)$$
∎

The proof of b) is left to the reader. See Problem 8.

3.5.2.1   Summary

Given a function $f: X \to Y$, where A, B are subsets of X and C, D are subsets of Y, the following properties hold.

1) $f(A \cup B) = f(A) \cup f(B)$
2) $f(A \cap B) \subseteq f(A) \cap f(B)$ (= if f is 1 - 1)
3) $f^{-1}(C \cup D) = f^{-1}(C) \cup f^{-1}(D)$
4) $f^{-1}(C \cap D) = f^{-1}(C) \cap f^{-1}(D)$
5) $f(f^{-1}(C)) \subseteq C$ (= if f is onto)
6) $A \subseteq f^{-1}(f(A))$ (= if f is 1 - 1)
7) $A \subseteq B \Rightarrow f(A) \subseteq f(B)$
8) $C \subseteq D \Rightarrow f^{-1}(C) \subseteq f^{-1}(D)$
9) $\overline{f^{-1}(A)} = f^{-1}(\bar{A})$
10) $f^{-1}(C - D) = f^{-1}(C) - f^{-1}(D)$

11) $f\left(\bigcup_{i \in I} A_i\right) = \bigcup_{i \in I} f(A_i)$

12) $f\left(\bigcap_{i \in I} A_i\right) \subseteq \bigcap_{i \in I} f(A_i)$ (= if f is 1-1)

13) $f^{-1}\left(\bigcup_{i \in I} A_i\right) = \bigcup_{i \in I} f^{-1}(A_i)$

14) $f^{-1}\left(\bigcap_{i \in I} A_i\right) = \bigcap_{i \in I} f^{-1}(A_i)$

Example 3   Inverse Images in Topology

In point-set topology, one is very much interested in inverse images of sets. It can be shown that the inverse image of an open interval (a, b) is always an open interval if the function is continuous.[1] For example, show that for the real-valued continuous function $f(x) = x^3$ of the real variable x, the inverse image of any open interval $(c, d) \subseteq \mathbb{R}$ is an open interval.

Solution

For any interval $(c, d) \subseteq \mathbb{R}$ on the y-axis, as drawn in Figure 3.39, its inverse image under $f(x) = x^3$ is

$$f^{-1}((c,d)) = \left(\sqrt[3]{c}, \sqrt[3]{d}\right)$$

which is an open interval. Typical examples are

$$f^{-1}(1,8) = (1,2),$$
$$f^{-1}(-8, -1) = (-2, -1).$$
$$f^{-1}(-8,8) = (-2,2)$$

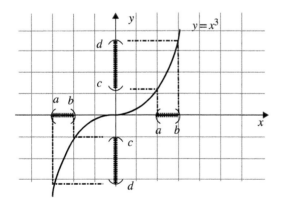

Figure 3.39 Inverse images of open intervals.

---

1 In Section 5.4, we will study open sets, but for now we restrict ourselves to open intervals.

In measure theory, one defines a measurable function f as a function whose inverse image $f^{-1}(A)$ of a measurable set A is a measurable set. In areas like measure theory, topology, and probability, one is more interested in images and inverse images of sets than images of points.

Note Although the inverse image of an open interval is preserved for continuous functions, open intervals are not always preserved for continuous functions. For example the constant function $f(x) = 1$ maps the open interval $(0, 1)$ into a set with one element, namely $\{1\}$, which is not an open interval.

Problems

1. **Party Time**
   We are having a party with possible desserts

   $$B = \{\text{cake, ice cream, pie}\}$$

   where the possible guests are among the group $A = \{a, b, c, d, e\}$. Each guest's favorite dessert is indicted by the function $f : A \rightarrow B$ where

   $$f(a) = \text{pie}, \ f(b) = \text{ice cream}, \ f(c) = \text{pie}, \ f(d) = \text{ice cream}, \ f(e) = \text{cake}$$

   What types of desserts will be required if the following groups of guests are invited to the party?
   a) $f(\{a,c\})$
   b) $f(\{b,d,e\})$
   c) $f(\{b\})$
   d) $f(\emptyset)$
   e) $f(\{a, b, c, d, e\})$

2. **Images of Sets**
   Given the sets $A = \{1, 2, 3, 4\}$, $B = \{a, b, c, d\}$ and the function $f : A \rightarrow B$ defined by

   $$f(1) = b, \ f(2) = a, \ f(3) = d, \ f(4) = c$$

   find the following:
   a) $f(\{1,3\})$
   b) $f(\{2,3,4\})$
   c) $f(\{2\})$
   d) $f(\{1,2,3,4\})$
   e) $f^{-1}(\{a,c\})$
   f) $f^{-1}(\{a,b,c\})$

3. **Interpretation of Images**

Translate the following statements into English. For example $y \in f(A)$ means there exists an $x \in A$ such that $y = f(x)$.

a) $y \in f(A \cup B)$
b) $y \in f(A) \cup f(B)$
c) $y \in f(A \cap B)$
d) $y \in f(A) \cap f(B)$

e) $y \in f\left(\bigcup_{i \in I} A_i\right)$

f) $y \in \bigcup_{i \in I} f(A_i)$

g) $y \in f\left(\bigcap_{i \in I} A_i\right)$

h) $y \in \bigcap_{i \in I} f(A_i)$

4. **Images of Sets**

Given the function $f: \mathbb{R} \to \mathbb{R}$ defined by $f(x) = x^2 + 2$. Find the images of the given sets under the mapping f.

a) $f(\{-1, 1, 3\})$
b) $f(\emptyset)$
c) $f([0, 2])$
d) $f([-1, 2] \cup [3, 5])$
e) $f^{-1}([-1, 2])$
f) $f^{-1}([0, 2])$
g) $f^{-1}([6, 11])$
h) $f^{-1}([-1, 6])$

5. **Continuous Images of Intervals**

Given the continuous function $f: \mathbb{R} \to \mathbb{R}$ defined by $f(x) = |x| + 1$, find the images.

a) $f([-2, -1))$
b) $f([-2, 3])$
c) $f([-2, 2])$
d) $f^{-1}([0, 4])$
e) $f^{-1}([-2, 0])$
f) $f^{-1}(\{1, 2, 3\})$

6. **Identity or Falsehood?**

True or false

$$A \subseteq B \Rightarrow f(A) \subseteq f(B)?$$

7. **Image of a Union**
   Show $f(A \cup B) \subseteq f(A) \cup f(B)$.

8. **Inverse of Union**
   Show $f^{-1}(A \cup B) = f^{-1}(A) \cup f^{-1}(B)$.

9. **Complement Identity**
   Show $f^{-1}(\bar{A}) = \overline{f^{-1}(A)}$

10. **Composition of a Function with Its Inverse**
    Prove the following and give examples to show that equality does not hold.
    a) $f[f^{-1}(A)] \subseteq A$
    b) $A \subseteq f^{-1}[f(A)]$

11. **Inverse Images**
    Let $f: \mathbb{N} \to \mathbb{R}$ be a function defined by $f(n) = 1/n$. Find
    a) $f^{-1}\left(\left[\frac{1}{10}, 1\right]\right)$
    b) $f^{-1}\left(\left[\frac{1}{100}, \frac{1}{2}\right]\right)$
    c) $f^{-1}\left(\left[0, \frac{1}{10}\right]\right)$

12. **Inverse Image of an Open Interval**
    In topology, a function f is defined as a continuous function if the inverse image of every open set in the range is an open set in the domain. Show that for the function $f: \mathbb{R} \to \mathbb{R}$ defined by $f(x) = x^2$, the inverse image of the following open intervals[2] is an open interval or the union of open intervals, and thus $f(x) = x^2$ is a continuous function.
    a) $f^{-1}((-1, 1))$
    b) $f^{-1}((0, 4))$
    c) $f^{-1}(\mathbb{R})$
    d) $f^{-1}((4, 16))$

13. **Dirichlet's Function**
    Given Dirichlet's (shotgun) function[3] $f: [0, 1] \to R$ is defined by

    $$f(x) = \begin{cases} 1 & x \text{ is rational} \\ 0 & x \text{ is irrational} \end{cases} \quad 0 \le x \le 1$$

    ---

    2 Open intervals and union of open intervals are special cases of open sets and the real number system is a special topological space.
    3 Sometimes called the "shotgun" function since it is full of holes.

Find

a) $f^{-1}\left(\left[\dfrac{1}{2},1\right]\right)$

b) $f^{-1}\left(\left[0,\dfrac{1}{2}\right]\right)$

14. **Image of a Singleton**
    Let $f: X \to Y$. Show that for $x \in X$ one has

    $$f(\{x\}) = \{f(x)\}.$$

15. **Connected Sets**
    It can be proven that the continuous image of a connected set is connected.[4] Find the image of the connected set $[-1, 1]$ under the continuous functions.
    a) $f(x) = x^3$ **Ans:** $f([-1, 1]) = [-1, 1]$
    b) $f(x) = e^x$
    c) $f(x) = 2x + 1$

16. **Function of Functions**
    Define $C[0, 1]$ to be the set of continuous functions defined on $[0, 1]$.[5]
    Define

    $$I: C[0,1] \to \mathbb{R} \text{ by } I(f) = \int_0^1 f(x)\,dx.$$

    a) Find $f, g \in C[0, 1]$ so $I(f) = 1$, $I(g) = 0.5$, $I(h) = -4$.
    b) Express the following integral property in terms of I

    $$\int_0^1 [f(x) + g(x)]\,dx = \int_0^1 f(x)\,dx + \int_0^1 g(x)\,dx.$$

    c) Is the function I 1–1?
    d) Is the function I onto $\mathbb{R}$?

17. **Internet Research**
    There is a wealth of information related to topics introduced in this section just waiting for curious minds. Try aiming your favorite search engine toward *properties of images of sets*, *images*, and *inverse images of a set*.

---

4 Connectedness is a precise topological concept we will not go into here. Use your intuition of what it might mean for a set to be connected.
5 A function of a function is called a functional.

Chapter 4

The Real and Complex Number Systems

4.1

Construction of the Real Numbers

Purpose of Section   Starting with the natural numbers, we logically construct the integers as equivalence classes of pairs of natural numbers. We then logically construct the rational numbers as equivalence classes of pairs of integers. When we arrive at the real numbers, we use the Dedekind cut to define the real numbers in terms of rational numbers.

4.1.1   Introduction

No doubt, most readers of this book think of real numbers as values on a number line, which has long been accepted by scientists and engineers as a model for measurements of length, mass, and time.

Although there is nothing wrong with this intuitive interpretation, it is the goal of this section to show how the real numbers can be logically created from more primitive number systems like the natural numbers, as well as introducing aspects of the real numbers decimal expansions, the least upper bound property, types of real numbers like rational, irrational, algebraic, and transcendental, and completeness properties.

Without going into the history of how numbers went from 1, 2, 3, ... to the real numbers, there are two fundamental approaches to how to define the real numbers. First, we can state axioms that we believe characterize our interpretation of the real numbers. This is the *here they are, the real numbers*. This approach is called the *synthetic* approach, whereby axioms hopefully embody what we believe a "continuum" should be.

On the other hand, we can "construct" the real numbers, much like a carpenter builds a house. In this approach, we begin with what our distant ancestors gave us, the natural numbers 1, 2, 3, ... We then take these numbers and doing some "mathematical carpentry" replacing hammers and nails with a little set

Advanced Mathematics: A Transitional Reference, First Edition. Stanley J. Farlow.
© 2020 John Wiley & Sons, Inc. Published 2020 by John Wiley & Sons, Inc.
Companion website: www.wiley.com/go/farlow/advanced-mathematics

theory and predicate logic, to logically construct, step-by-step, the real numbers. This is the approach taken in this section, starting with the natural numbers we construct the integers, then the rational numbers, all the way to real numbers. The synthetic axiomatic approach of *here they are* will be left to the next section.

4.1.2 The Building of the Real Numbers

The construction of the real numbers begins with the simplest numbers, the natural numbers

$$\mathbb{N} = \{1,2,3,...\}$$

then, by a series of steps we construct the integers

$$\mathbb{Z} = \{...-3,-2,-1,0,1,2,3,...\}$$

followed by the rational numbers

$$\mathbb{Q} = \{p/q : p,q \in \mathbb{Z}, q \neq 0\}$$

and finally, the real numbers $\mathbb{R}$.

Important Note The English mathematician/philosopher Bertrand Russell once said, it must have taken many ages for humans to realize that a pair of pheasants and a couple of days were both instances of the concept of "two."

4.1.3 Construction of the Integers: $\mathbb{N} \rightarrow \mathbb{Z}$

The way we construct the integers $0, \pm 1, \pm 2, ...$ from the natural numbers 1, 2, 3, ... is to define integers as *pairs* of nonnegative integers, where we interpret each pair (m, n) as the difference $m - n$. Thus, the pair $(2, 5)$ of natural numbers is defined as -3, the pair $(7, 1)$ as 6, and the pair $(4, 4)$ as zero. If we then define addition, subtraction, and multiplication of these pairs (m, n) consistent with the arithmetic of the natural numbers, we have successfully defined the integers in terms of the natural numbers.

To carry out this program, we begin with the Cartesian product

$$\mathbb{N} \times \mathbb{N} = \{(m,n) : m,n = 1,2,...\}$$

of pairs of natural numbers, called *grid* points, which are illustrated in Figure 4.1.

We now define an equivalence relation "$\equiv$" between pairs of natural numbers by

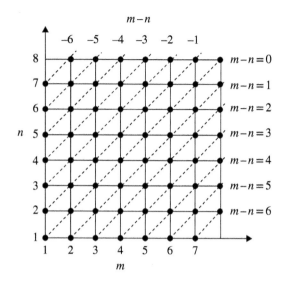

Figure 4.1 Partitioning $\mathbb{N} \times \mathbb{N}$ into equivalence classes.

$$(m,n) \equiv (m',n') \Leftrightarrow m + n' = m' + n$$

This relation can easily be verified to be an equivalence relation and its verification is left to the reader. See Problem 1. We saw in Section 3.3 that equivalence relations on a given set partitions the set into disjoint equivalence classes. In this case, the equivalence classes consist of grid points on 45° lines as drawn in Figure 4.1. A few of these equivalence classes are listed in Table 4.1, designated by $\overline{()}$ which shows a representative member of each equivalence class. Each one of these equivalence classes will define an integer. For example, the equivalence class $\overline{(0,0)}$ defines zero, $\overline{(1,2)}$ defines -1, $\overline{(2,1)}$ defines $+1$.

Table 4.1 Five equivalence classes.

Equivalence class	Integer definition
$\overline{(1,3)} = \{(1,3),(2,4),(3,5)...\}$	-2
$\overline{(1,2)} = \{(1,2),(2,3),(3,4)...\}$	-1
$\overline{(1,1)} = \{(1,1),(2,2),(3,3)...\}$	0
$\overline{(2,1)} = \{(2,1),(3,2),(4,3)...\}$	1
$\overline{(3,1)} = \{(3,1),(4,2),(5,3)...\}$	2

We now define the integers $\mathbb{Z}$ as the collection of all these equivalences classes. In other words

$$\mathbb{Z} = \left\{ ...\overline{(1,4)}, \overline{(1,3)}, \overline{(1,2)}, \overline{(1,1)}, \overline{(2,1)}, \overline{(3,1)}, \overline{(4,1)}, ... \right\}$$

which we write as

$$\mathbb{Z} = \{ ...-3, -2, -1, 0, 1, 2, 3, ... \}$$

We can also define the nonnegative and negative integers as

$$\text{Nonnegative integers} = \left\{ \overline{(k,1)} : k \in \mathbb{N} \right\}$$

$$\text{Negative integers} = \left\{ \overline{(0,k)} : k \in \mathbb{N} \right\}$$

We now must define addition, subtraction, and multiplication for the integers, consistent with the rules of the natural numbers. We define for $p, q, r, s \in \mathbb{N}$ addition $\oplus$, subtraction $\ominus$, and multiplication $\otimes$ of integers (represented by pairs of natural numbers) by:

- **Addition:** $\overline{(p,r)} \oplus \overline{(q,s)} = \overline{(p+q, r+s)}$
- **Subtraction:** $\overline{(p,q)} \ominus \overline{(r,s)} = \overline{(p+s, q+r)}$
- **Multiplication:** $\overline{(p,q)} \otimes \overline{(r,s)} = \overline{(pr+qs, ps+qr)}$

For example

- **Addition:** $\overline{(3,5)} \oplus \overline{(1,4)} = \overline{(4,9)}$ or -5
- **Subtraction:** $\overline{(3,6)} \ominus \overline{(2,7)} = \overline{(10,8)}$ or $+2$
- **Multiplication:** $\overline{(1,3)} \otimes \overline{(7,2)} = \overline{(13,23)}$ or -10

When $p > r$ and $q > s$, the pairs (p, r) and (q, s) correspond to $p - q > 0$ and $r - s > 0$, hence, the above definition of addition, subtraction, and multiplication of pairs of natural numbers reduces to identities of the natural numbers.

4.1.4 Construction of the Rationals: $\mathbb{Z} \to \mathbb{Q}$

We now move up the number chain and construct the rational numbers $\mathbb{Q}$ from the integers $\mathbb{Z}$ using a strategy similar to what we did in constructing the integers as pairs of natural numbers.

To carry this out, define an equivalence relation on $\mathbb{Z} \times (\mathbb{Z} - \{0\})$ by

$$(m,n) \equiv (m',n') \Leftrightarrow mn' = m'n$$

where (m, n) and (m', n') are pairs of integers and we "interpret" them as m/n and m'/n', respectively. This equivalence relation partitions the grid points in $\mathbb{Z} \times (\mathbb{Z} - \{0\})$ into distinct equivalence classes as drawn in Figure 4.2. Observing the following equivalences

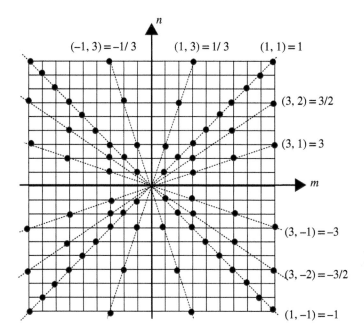

Figure 4.2 Equivalence classes defining rational numbers.

$$(1,2) \equiv (2,4) \equiv (-3,-6) \equiv (13,26) \equiv (-5,-10)$$

we see that the equivalence classes consist of the grid points on straight-lines passing through the origin, where the equivalence class with representative (m, n) is assigned the rational number m/n. For example

$$(2,3) = 2/3, \quad (-3,4) = -3/4, \quad (0,3) = 0/3 = 0$$

A few equivalence classes with an arbitrary designated representative are shown in Table 4.2.

Table 4.2 Five equivalence classes in $\mathbb{Z} \times (\mathbb{Z} - \{0\})$.

Equivalence class	Rational correspondence
$\overline{(1,2)} = \{(1,2),(2,4),(3,6)...\}$	$\dfrac{1}{2}$
$\overline{(1,-1)} = \{(1,-1),(2,-2),(3,-3)...\}$	-1
$\overline{(3,-5)} = \{(3,-5),(-3,5),(6,-10)...\}$	$-\dfrac{3}{5}$
$\overline{(0,1)} = \{(0,1)(0,-1),(0,2)...\}$	0

We now define the rational number p/q as

$$\frac{p}{q} = \overline{(p,q)}, \quad p, q \in \mathbb{Z}, \quad q \neq 0.$$

We now define the arithmetic operations on the newly formed rational numbers.

- **Addition:** $\overline{(p,q)} \oplus \overline{(r,s)} = \overline{(ps+qr,qs)}$
- **Subtraction:** $\overline{(p,q)} \ominus \overline{(r,s)} = \overline{(ps-qr,qs)}$
- **Multiplication:** $\overline{(p,q)} \otimes \overline{(r,s)} = \overline{(pr,qs)}$

For example

- **Addition:** $\overline{(1,2)} \oplus \overline{(2,10)} = \overline{(14,20)}$ or $\dfrac{14}{20}$

- **Subtraction:** $\overline{(3,6)} \ominus \overline{(1,4)} = \overline{(6,24)}$ or $\dfrac{6}{24}$

Interesting Note   In a recent seventh-grade textbook it was written that to add and multiply fractions, one should use the rules given in the book, whereas to add and multiply *real* numbers (guess the book is talking about irrational numbers like π and $\sqrt{2}$) one should resort to your calculator.

4.1.5   How to Define Real Numbers

We now come to a fork in the road. There are several ways to define the real numbers and each has its merits and demerits. On the one hand, we could define real numbers in the way they were described to you in grade school, as decimal expansions. This is nice since all numbers, rational and irrational are more or less treated in the same way, and we normally perform arithmetic when numbers are written in this form, like $2.35 + 4.91$. However, on the negative side, when real numbers are represented by expressions of the form

$$\pm a_1 a_2 a_3 \cdots a_m . b_1 b_2 b_3 \cdots$$

where the as and bs are decimal digits, it is the dot, dot, dot at the end of the expansion that causes problems theoretically. Another problem is that certain real numbers have more than one decimal expansion, like

$$2.318 = 2.317\,999\ldots$$

Like we said, there is that nasty dot, dot, dot again.

Still another demerit for the decimal expansion interpretation of real numbers is that they do not relate visually to points on a continuous number line, which is how most people like to think of real numbers.

However, there are two more approaches for defining the real numbers, one due to Cantor and the other due to his close friend Richard Dedekind. Both of these approaches are based on defining the real numbers in terms of the rational numbers, much like we defined the rational numbers in terms of the integers. Cantor's approach defines real numbers limits of sequences of rational numbers, like in the definition of the irrational real number π being the limit of the sequence

$$3, \quad 3.1, \quad 3.14, \quad 3.141, \quad 3.1415, \quad 3.14159, \quad 3.141592,\ldots \to \pi$$

Although this approach has intuitive appeal, it demands that the reader develop a background in sequences, convergence, null sequences, and other concepts from real analysis, not introduced in this book. Hence, we follow the second approach, that due to Richard Dedekind.

> **Historical Note**   The German mathematician **Richard Dedekind** (1831–1916) was one of the major mathematicians of the nineteenth century. He made major contributions to number theory and abstract algebra. His invention of ideals in ring theory as well as his contributions to algebraic numbers, fields, modules, lattice, and so on, were crucial in the development of abstract algebra. His book *Was sind and was sollen die Zahlen?* (What are numbers and what should they be?) laid the foundation for the real number system and was a milestone in the history of mathematics.

Although rational numbers have many desirable properties, they have from the perspective of calculus, the undesirable property that they have "gaps" like the gaps at $\sqrt{2}$ and π. In fact, there are an infinite number of gaps, and as we have seen in Section 2.5, an uncountable number of gaps. The idea is to "fill in" these gaps, and arriving at a new number system called the real numbers.

4.1.6   How Dedekind Cuts Define the Real Numbers

Dedekind's 1872 idea appeals to our intuitive grasp of the rational numbers aligned on a line. His idea was to partition the rational numbers into two disjoint sets L and U satisfying the following two conditions.

Definition   A **Dedekind cut**, denoted by (L, U), is a partition of the rational numbers into two disjoint, nonempty subsets $L \subseteq \mathbb{Q}$ and $U \subseteq \mathbb{Q}$ such that all members of $\mathbb{Q}$ in the so-called **lower set** L are less than all members of $\mathbb{Q}$ in the **upper set** U. Stated analytically, a Dedekind cut (L, U) is a pair of nonempty subsets of $\mathbb{Q}$ satisfying

- $L \cup U = \mathbb{Q}$
- $L \cap U = \varnothing$
- $[l \in L \text{ and } u \in U] \Rightarrow l < u$

The above definition of a Dedekind cut does not uniquely define a partition of the rational numbers, but three different partitions called Dedekind cuts. They are as follows.

Type 1 Dedekind Cut (L, U)   The first type of partition of $\mathbb{Q}$ arises when the lower set L does not have a largest element, but the upper set U has a smallest element, this smallest element being a rational number r. We call this a type of Dedekind cut a rational cut at r, where the lower and upper sets L, U can be expressed as follows:

$$L = \{l \in \mathbb{Q} : l < r\}, \quad U = \{u \in \mathbb{Q} : r \leq u\}$$

which is visualized in Figure 4.3.

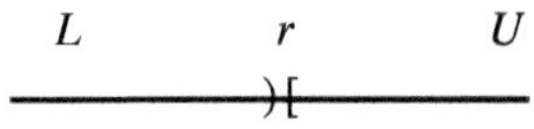

Figure 4.3 Type 1 Dedekind cut.

Type 2 Dedekind Cut (L, U)   The second type of partition of $\mathbb{Q}$ arises when the upper set U does not have a smallest element, but the lower set L has a largest element, this largest element being a rational number r. We call this a type of Dedekind cut a rational cut at r, where the lower and upper sets L and U can be expressed as follows:

$$L = \{l \in \mathbb{Q} : l \leq r\}, \quad U = \{u \in \mathbb{Q} : r < u\}$$

which is visualized in Figure 4.4.

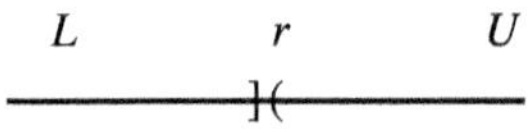

Figure 4.4 Type 2 Dedekind cut.

Type 3 Dedekind Cut (L, U)   The third type of Dedekind cut arises when the lower set L does not have a largest element and upper set U does not have a smallest element. Since the Dedekind cut partitions the rational numbers there is no rational number between the lower and upper sets. This type of Dedekind cut is expressed as follows:

$$L = \{l \in \mathbb{Q} : l < r\}, \quad U = \{u \in \mathbb{Q} : r < u\}$$

and is visualized in Figure 4.5. Dedekind cuts of this type identify gaps in the rational numbers and each gap is identified with a nonrational number called an **irrational number**.

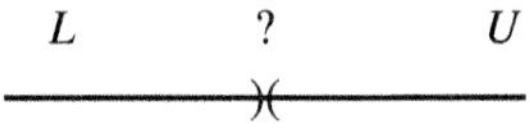

Figure 4.5 Type 3 Dedekind cut.

This Dedekind cut is visualized in Figure 4.5.

Each Dedekind cut of Type 1 or Type 2 corresponds to a rational number, whereas a Dedekind cut of Type 3 correspond to a number that is not rational. These are numbers that fill all the gaps between rational numbers. These are called **irrational numbers**. This leads us to the Dedekind cut definition of the real numbers.

Definition of the Real Numbers   The set of real numbers, denoted by $\mathbb{R}$, is defined as union of all Dedekind cuts of Type 1 and Type 3, or equivalently of Type 2 and Type 3.

Example 1   Dedekind Cuts Defining Rational and Irrational Numbers

a) The Dedekind cut (L, U) for the rational number three is

- $L = \{l \in \mathbb{Q} : l < 3\}$
- $U = \{u \in \mathbb{Q} : 3 \le u\}$

defines the rational number three.

b) The Dedekind cut for $\sqrt{2}$ is slightly more complicated since we cannot use $\sqrt{2}$ in the definition, else we would be using $\sqrt{2}$ to define $\sqrt{2}$.

- $L = \{l \in \mathbb{Q} : l < 0\} \cup \{l \in \mathbb{Q} : l^2 < 2\}$
- $U = \{u \in \mathbb{Q} : u \ge 0\} \cap \{u \in \mathbb{Q} : u^2 > 2\}$

The reader can carefully exam the rational numbers in L and see that they are less than any rational number in U, and both lower and upper sets of rational numbers have the form $L = (-\infty, x)$ and $U = (x, \infty)$.[1] Remember, the lower and upper

---

1 The number x is the square root of 2, but we are not allowed to write it at this stage.

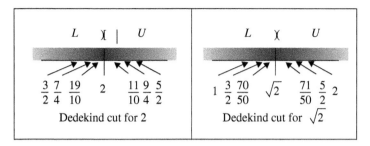

Figure 4.6 Dedekind cuts for 2 and $\sqrt{2}$.

sets of a Dedekind cut must partition the rational number into two disjoint sets whose union gives the rational numbers. Figure 4.6 illustrates typical cuts.

4.1.7 Arithmetic of the Real Numbers

Now that we have defined the real numbers, the next job is to Dedekind cuts to define addition, subtraction, multiplication, and division of the real numbers (as well as defining order relations like < and ≤) consistent with those of the rational numbers. Although we can define arithmetic operations of real numbers in this way, they are rather complicated, and we will not include them here. If we had to carry out even the easiest arithmetic, like adding $\pi + \sqrt{2}$ using the Dedekind cut definition, it would a difficult task. As we all know, the way we do arithmetic involving irrational numbers, we replace them by a decimal digits, $\pi \doteq 3.14, \sqrt{2} \doteq 1.41$ and find the approximate value $\pi + \sqrt{2} \doteq 4.55$. Of course, we cannot be 100% accurate like we are when we restrict arithmetic to the rational numbers, like

$$\frac{1}{3} + \frac{4}{5} = \frac{17}{15}$$

but that is the price of enlarging our number system to the real numbers.

Problems

1. **Equivalence Relation I**
   Show that the relation ≡ on ℕ defined by

   $(m,n) \equiv (m',n')$ if and only if $m + n' = m' + n$

   between pairs of natural numbers (m, n) and (m', n') is an equivalence relation.

2. **Equivalence Relation II**
   Show that the relation $\equiv$ on $\mathbb{Z}$ defined by

   $$(m,n) \equiv (m',n') \Leftrightarrow mn' = m'n$$

   between pairs of integers (m, n) and (m', n') is an equivalence relation.

3. **Arithmetic in $\mathbb{Z}$**
   We defined the integers $\mathbb{Z}$ in terms of pairs of natural numbers (m, n) belonging to an equivalence class. Perform the following arithmetic for integers as defined in the book. What does each operation represent in the language of integers?
   a) $\overline{(1,5)} \oplus \overline{(3,2)}$
   b) $\overline{(1,5)} \ominus \overline{(3,2)}$
   c) $\overline{(1,5)} \otimes \overline{(3,2)}$

4. **Arithmetic in $\mathbb{Q}$**
   We defined the rational numbers $\mathbb{Q}$ in terms of pairs of integers (m, n) belonging to an equivalence class. Perform the following arithmetic for rational numbers using the definition in the book. What does each operation represent in the language of rational numbers?
   d) $\overline{(1,5)} \oplus \overline{(3,2)}$
   e) $\overline{(1,5)} \ominus \overline{(3,2)}$
   f) $\overline{(1,5)} \otimes \overline{(3,2)}$

5. **Decimal to Fractions**
   Find the fraction for each of the following real numbers in decimal form, where the bar over numbers means the numbers are repeated.
   a) $0.9999....$ $\left(0.\bar{9}\right)$
   b) $0.232\,323\,23....$ $\left(0.\overline{23}\right)$
   c) $0.012\,312\,312\,3....$ $\left(0.0\overline{123}\right)$
   d) $0.001\,111....$ $\left(0.00\bar{1}\right)$

6. **Two Decimal Representations**
   The real numbers that have two different decimal representations agree up to some point, but then one continues with $a999...$, while the other continues with $b000...$, where the digit a is one more than the digit b. The dot, dot, dot at the end of the expression refers to a never-ending list of digits. Write several instances of real numbers that have two decimal representations.

7. **Irrationals Everywhere**
   Show that between any two rational numbers, there is an irrational number.
   Hint: For two rational numbers $r_1 < r_2$, there exists a natural number n that

satisfies $n(r_2 - r_1) > \sqrt{2}$. Using this inequality, we can construct a number between r_1 and r_2. The only remaining task is to show this number is irrational.

8. **Rationals Everywhere**
    Show that between every two irrational numbers there is a rational number. Hint: Let i_1 and i_2 be two irrational numbers with $i_1 < i_2$. Hence, there exists a natural number n satisfying

    $$n > \frac{1}{i_2 - i_1} \Rightarrow n(i_2 - i_1) > 1 \Rightarrow ni_2 - ni_1 > 1$$

    Use the last result to construct a rational number m/n that lies between i_1 and i_2.

9. **Internet Research**
    There is a wealth of information related to topics introduced in this section just waiting for curious minds. Try aiming your favorite search engine toward *construction of the integers from the natural numbers, construction of the rational numbers from the integers, Dedekind cut,* and *Richard Dedekind.*

4.2

The Complete Ordered Field

The Real Numbers

> **Purpose of Section** We present the axiomatic definition of the real numbers as a **complete ordered field.** We carry out this three-step process by first defining an **algebraic field**, then introduce an order on the algebraic field, yielding an **ordered field**, and then the **completeness axiom** that allows us to define the complete ordered field, of which there is only one, the real numbers.

4.2.1   Introduction

Advances in function theory in the nineteenth century demanded a deeper understanding of the real numbers, which led to a "rigorization" of analysis by such mathematical greats as Cauchy, Abel, Dedekind, Dirichlet, Weierstrass, Bolzano, Frege, Cantor, and others. A deeper understanding of functions required precise proofs which in turn required the real number system be placed on solid mathematical ground.

Although we generally think of real numbers as points on a continuous line that extends endlessly in both directions, the goal of this chapter is to strip away everything you know about the real numbers and start afresh. This is not easy since all knowledge and mental imagery of the real numbers created over a lifetime is firmly entrenched in our minds. But if the reader is willing to wipe the slate clean and start anew, we will introduce you to a *new* mathematical entity, known by mathematicians as the *complete, ordered field,* which, for lack of another name, we call $\mathbb{R}$. By building the axioms of the real numbers, you will have a deeper understanding of them than simply as "points on a very long line."

There are three types of axioms required to define the real numbers. First, there are the *arithmetic* axioms, called the **field axioms**, which provide the rules for adding, subtracting, multiplying and dividing. Second, there are the **order axioms**, which allow us to compare sizes of real numbers like $2 < 3$, $4 > 0$,

Advanced Mathematics: A Transitional Reference, First Edition. Stanley J. Farlow.
© 2020 John Wiley & Sons, Inc. Published 2020 by John Wiley & Sons, Inc.
Companion website: www.wiley.com/go/farlow/advanced-mathematics

and $-3 < 0$, and so on. And last, there is an axiom, called the completeness **axiom**, which gives the real numbers that special quality which allows us to think of real numbers as "flowing" continuously with no gaps.

So let us begin our quest to define the holy grail of real analysis.

> **Important Note** At this point in the book, we should stop and apologize to female readers. You may have noticed that basically every named mathematicians in the book is male, which is due to the fact that the great mathematicians of the nineteenth century and before were mostly male. As any student of history knows, women were kept away from all academic studies, even the Queen of the Sciences. Today, however, things are changing rapidly and women are making contributions at the highest level in every area of mathematics.

4.2.2 Arithmetic Axioms for Real Numbers

We begin by defining a set $\mathbb{R}$, but do not think of $\mathbb{R}$ as the real numbers *yet*. We begin by defining two binary functions from $\mathbb{R} \times \mathbb{R} \rightarrow \mathbb{R}$, one called the *addition* function and the *multiplication* function. The addition function assigns to each pair (a, b) of numbers in $\mathbb{R}$ a new element of $\mathbb{R}$ called the sum of a and b and denoted by $a + b$. The multiplication function assigns to each pair of elements in $\mathbb{R}$ a new element in $\mathbb{R}$ called the *product* of a and b and denoted by $a \times b$ or more often simply ab. These operations are called **closed operations** since when a, $b \in \mathbb{R}$ so are $a + b$ and ab.

These axioms have passed the test of time and are now chiseled in stone in the laws of mathematics and form an algebraic system called a **field**[1] (or an **algebraic field**), which is summarized as follows.

Table 4.3 Field axioms.

Field axioms

A **field** is a set, which we call $\mathbb{R}$, with two binary operations, called $+$ and $\times$, where for all a, b and c in $\mathbb{R}$, the following axioms hold.[a]

Addition axioms	Name of axiom
$(\forall a, b \in \mathbb{R})[a + (b + c) = (a + b) + c]$	Associativity of addition
$(\forall a, b \in \mathbb{R})(a + b = b + a)$	Addition commutes
$(\exists ! 0 \in \mathbb{R})(\forall a \in \mathbb{R})(a + 0 = 0 + a = a)$	Unique additive identity
$(\forall a \in \mathbb{R})(\exists !(-a) \in \mathbb{R})[a + (-a) = (-a) + a = 0]$	Unique additive inverse

---

1 Modern algebra or abstract algebra, which is distinct from elementary algebra as taught in schools, is a branch of mathematics that studies algebraic structures, such as groups, rings, fields, modules, vector spaces and other algebraic structures.

Table 4.3 (Continued)

Multiplication axioms[b]	Name of axiom
$(\forall a,b,c \in \mathbb{R})[a(bc) = (ab)c]$	Associativity of multiplication
$(\forall a,b \in \mathbb{R})(ab = ba)$	Multiplication commutes
$(\exists!1 \in \mathbb{R})(\forall a \in \mathbb{R})(a{\cdot}1 = 1{\cdot}a = a)$	Unique multiplicative identity
$(\forall a \in \mathbb{R})[a \neq 0 \Rightarrow (\exists!a^{-1} \in \mathbb{R})(a{\cdot}a^{-1} = a^{-1}{\cdot}a = 1)]$	Unique multiplicative inverse

Distributive axiom	Name of axiom
$(\forall a, b, c \in \mathbb{R})[a(b + c) = ab + ac]$	Multiplication distributes over addition

[a] We call the field $\mathbb{R}$ since we are concentrating on the real numbers, but keep in mind there are many examples of an algebraic field. It is assumed in the axioms for a field that the additive identity 0 and the multiplicative identity 1 are not equal.
[b] We often drop the multiplication symbol "$\cdot$" and denote multiplication of two elements as $a \cdot b = ab$.

4.2.3 Conventions and Notation

In addition to the above axioms, we make the following conventions;

1) The associative axioms for both addition and multiplication tell us it does not matter where the parentheses are placed. In other words, we can write $a + b + c$ for $a + (b + c)$ or $(a + b) + c$. The same associative law holds for multiplication, which allows us to write $abc = a(bc) = (ab)c$.
2) The unique additive inverse of an element a is denoted by $-a$. Hence, we have $a + (-a) = 0$. The multiplicative inverse of a is denoted by a^{-1} and often written $1/a$. Hence, $aa^{-1} = a(1/a) = 1$.

Two other operations of *subtraction* and *division* can be defined directly from addition and multiplication by

$$\text{Subtraction}: a - b = a + (-b) \quad (\text{read } a \text{ minus } b)$$
$$\text{Division}: a/b = ab^{-1} \quad (\text{for } b \neq 0)(\text{read } a \text{ divided by } b)$$

> **Important Note** A field is an algebraic system where you can add, subtract, multiply, and divide (except by 0) in the same manner you did as a child. As a child, you were taught these were "properties" of numbers. But they are not properties, they are the definition or rules of engagement of the real numbers. A subtle, but important point.

We know what you are thinking: you have known all this since third–grade. If your argument is that the axioms of arithmetic are simple and elementary, that is no argument at all. Axioms are supposed to be simple and elementary. The question you should ask is what kind of results can be proven from the axioms. The answer is an algebraic field can give rise to many deep results. Ask yourself if these are the simplest axioms that give rise to a system of arithmetic? Do you need any more axioms? Can you get by with less? These are not trivial questions and their answers are even less so. There are other systems of axioms that allow you to perform "arithmetic" operations on elements of a set, such as groups and rings that we will learn about in Chapter 6 when we study abstract algebra.

4.2.4 Fields Other than $\mathbb{R}$

1) **Boolean Field:** Let $F_2 = \{0, 1\}$ and define the binary operations of addition and multiplication by Table 4.4.

Table 4.4 Boolean field.

+	0	1	×	0	1
0	0	1	0	0	0
1	1	0	1	0	1

The set A with these arithmetic operations is an algebraic field. We leave it to the reader to check all the properties a field must possess.

2) **Complex Numbers:** The complex numbers $a + bi$, where a, b are real numbers and $i = \sqrt{-1}$, where addition and multiplication are defined in the usual manner.

3) **Rational Numbers $\mathbb{Q}$:** The rational numbers where addition and multiplication are defined in the usual way.

4) **Rational Functions F:** The set of all rational functions

$$f(x) = \frac{p(x)}{q(x)}$$

where $p(x)$, $q(x) \neq 0$ are polynomials with real coefficients, where addition and multiplication are defined in the usual way and 0 and 1 are the standard additive and multiplicative identities.

There are many other examples of fields studied by mathematicians, including the Galois finite fields, p-adic number fields, and fields of functions, such as meromorphic and entire functions.

We now come to the second group of the three types of axioms required to describe the real numbers.

> **Important Note** The real numbers are ineptly named. They are no more or less real from a mathematical point of view any other number system like the integers or rational numbers. The name "real numbers" is one of those chronicled lapses, not unlike the name "Pythagorean Theorem": which was known long before Pythagoras was born.

4.2.5 Ordered Fields

Although an algebraic field allows us to carry out arithmetic on a set, what it cannot do is compare sizes of members of the set. The job now is to include "order" on the field. To do this, we split the field into two disjoint sets, P and N, called the **negative** and **positive** members of the field. These two sets mimic the properties of the positive and negative real numbers. This motivates the general definition of an ordered (algebraic) field.

Definition: Ordered Field
An algebraic field F is said to be ordered if its **nonzero members** can be split into two **disjoint subsets**, $F = P \cup N$ called respectively, the **negative** (N) and **positive** (P) members in such a way that

- $x, y \in P \Rightarrow x + y \in P$
- $x, y \in P \Rightarrow xy \in P$
- $x \in P \Leftrightarrow -x \in N$

These properties allow us to define a **strict (total) order** $<$ on F by

$$x < y \Leftrightarrow y - x \in P$$

which is a total order on F since

- irreflexive since $x - x = 0 \notin P \Rightarrow x \not< x$
- asymmetric $x < y \Rightarrow \sim (y < x)$
- transitive since $[(x < y) \wedge (y < z)] \Rightarrow x < z$

For convention, we say

- $x > 0$ when $-x < 0$
- $x \geq 0$ when $x > 0$ or $x = 0$
- $x \leq 0$ when $x < 0$ or $x = 0$

We can now prove all the usual properties of inequalities of real numbers.

> **Important Note** One does not just make rules or axioms willy-nilly hoping good things will follow. In fact, it is just the opposite. Knowing what is desirable, one designs axioms from which the desirable theorems will follow. As the mathematician Oswald Veblen once said, "The test of a good axiom system lies in the theorems it produces."

> **Historical Note**   The concept of an ordered field was introduced by Austrian/
> American mathematician Emil Artin (1898–1962) in 1927. Artin was one of the
> leading algebraists of the twentieth century who emigrated to the United States
> in 1937 and spent many years at Indiana University and Princeton University.

4.2.6   The Completeness Axiom

If we were to stop with ordered fields, we would be neglecting that special ingre-
dient that defines the real numbers as a *continuum*. There are many examples of
ordered fields that are not the real numbers, and all those algebraic systems have
"gaps" between their elements. The set of rational numbers is an ordered field,
which as we all know, has an uncountable number of gaps between its members,
two gaps being the solutions of $x^2 = 2$, which are $x = \pm\sqrt{2}$, which was proven long
ago to be not a rational number. What we need is an axiom that "fills in" these
gaps and this is where the *completeness* (or *continuum*) axiom comes into play.
   An interesting aspect of this axiom is that over the years mathematicians have
found several completeness axioms that are logically equivalent. Thus, it is possible to introduce any one of them as the "completeness" axiom. In this book, we
have chosen the "version" of the completeness axiom as the **least upper bound**
axiom, our reason being many interesting concepts can be deduced by working
with it. Then, there is the other benefit, it is easy to understand. Before stating
the axiom, however, we review a few important ideas about the least upper
bound of a set introduced in Section 3.2.

4.2.7   Least Upper Bound and Greatest Lower Bounds

We use the four intervals in Figure 4.7 as a prop for reviewing the concepts of
the least upper bound (lub) and the greatest lower bound (glb) introduced in our
study of orders in Section 3.2.

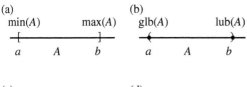

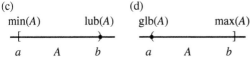

Figure 4.7 Max, min, lub, glb.

The intervals (a, b), $[a, b]$, $[a, b)$, and $(a, b]$ are all bounded, both above and below. Bounded above simply means there is at least one number greater than or equal to the elements in the set. Likewise, a lower bound for a set is a number less than or equal to the elements in the set. Of course, not all sets are bounded; the set $[1, \infty)$ is bounded below but not above, and $(-\infty, \infty)$ is bounded neither above nor below. The intervals $[a, b]$ and $(a, b]$ each have a maximum value of b, whereas the intervals (a, b) and $[a, b)$ do not have a maximum value. The same arguments hold for minimum values. The intervals $[a, b]$, $[a, b)$ each has a minimum value of a, but the intervals (a, b) or $(a, b]$ do not have minimum values.

So what is the meaning lub (A) and glb (A) in Figure 4.7? Note that two of the intervals contain their maximum value and two do not. However, and this is the important part, for *each* of the four intervals $[a, b]$, $[a, b)$, (a, b), and $(a, b]$, the set of upper bounds is the *same*, namely $[b, \infty)$, and note that this set of upper bounds contains its minimum of b. For the intervals (a, b), $[a, b)$ where b does not belong to the interval, we call b the **least upper bound** (or **supremum**) of the set since it is the least of the upper bounds of the set. We denote this value by lub(A). For the two sets $[a, b]$ and $(a, b]$ that *have* a maximum value, the least upper bound of the set is the same as the maximum of the set. For the sets (a, b) and $[a, b)$ that do not have maximum values, the least upper bound b is a kind of "surrogate" for the maximum.

The same principle holds for lower bounds. The set of lower bounds for the four intervals is the same, namely $(-\infty, a]$. The number a is the greatest of all these lower bounds and is called the glb(A) for each of the four intervals. Any set that is bounded below may or may not have a *minimum* value, but the set of lower bounds will always have a *maximum* value, and that maximum value is called the **greatest lower bound** (or **infimum**) of the set and denoted by glb(A).

Definition   Let A be a set in an ordered field that is bounded above. The number lub(A) is the **least upper bound** or **supremum** of A if

- lub(A) is an upper bound of A, i.e. $x \le$ lub(A) for all $x \in A$.
- If u is any upper bound for A, then lub$(A) \le u$.

See Figure 4.8.

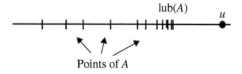

Figure 4.8 Least upper bound.

Likewise the number glb(A) is the **greatest lower bound** or **infimum** of A if

- glb(A) is a lower bound of A, i.e. glb(A) ≤ x *for all* x ∈ A.
- If l is any lower bound for A, then l ≤ glb(A).

See Figure 4.9

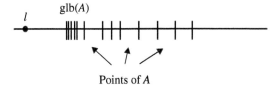

Points of A

Figure 4.9 Greatest lower bound.

This leads us to the completeness axiom for ℝ, which up to now, we endowed with only field and order axioms. The last set of axioms we assign to ℝ (actually only one axiom) is called the completeness axiom.

Completeness Axiom: least Upper Bound Axiom[2] In an ordered field, if every nonempty set that is bounded above has a least upper bound, then the ordered field satisfies the **completeness axiom.**

We are now (finally) ready to define the real numbers.

Definition of the Real Numbers The **real number system** ℝ is a **complete ordered field**, that is, an ordered field that satisfies the completeness axiom. Stated another way, it is a set ℝ that satisfies the axioms of an **algebraic field**, the **order axioms**, and the **completeness axiom.**

The least upper bound axiom is necessary since there are ordered fields that do not "look like" the real numbers and the reason is that they do not satisfy the least upper bound axiom. Of the ordered fields that do not satisfy the completeness axiom, the rational numbers are the most well-known. By including the completeness axiom with an ordered field, the ordered field behaves like the real numbers.

When we refer to the real numbers as a complete ordered field, we always say *the* complete ordered field since all complete ordered fields are *isomorphic*. We say that two abstract structures are isomorphic if they have exactly the same

---

2 In Section 4.1, we defined the real numbers in terms of rational numbers by Dedekind cuts. Although we did not prove it at that time, the real numbers defined by Dedekind cuts satisfies the completeness axiom.

mathematical structure and differ only in the symbols used to represent various objects and operations in the system.

Example 1   Rational Numbers and the Completeness Axiom

Do the rational numbers satisfy the completeness axiom?

Solution

In order for the rational numbers to satisfy the completeness axiom, every set of rational numbers that is bounded above must have a least member. The rational numbers fail the completeness axiom since the set

$$A = \left\{ q \in \mathbb{Q} : 0 < q < \sqrt{2} \right\}$$

of rational numbers is bounded above (5 is an upper bound), but it has no least upper bound since the set of upper bounds

$$\text{Set of upper bounds} = \left[\sqrt{2}, \infty \right)$$

has no smallest (rational) member.

Historical Note   Czech mathematician Bernard Bolzano (1781–1848) conceptualized the least upper bound property of the real numbers in an 1817 paper in which he gave the first *analytic* proof of the Intermediate Value Theorem. He realized the proof must depend on deep properties of the real numbers.

Important Note   In the previous Section 4.1, when we defined the real numbers in terms of Dedekind cuts of rational numbers, we did not show how an ordering could be defined on the real numbers in terms of Dedekind cuts, nor did we show the real numbers were complete using Dedekind cuts. This could have been done, but it would have been a long and tedious task and so it was generously omitted.

Problems

1. **True or False**
    a) The natural numbers $\mathbb{N}$ with operations of addition and multiplication is an ordered field.
    b) $\mathbb{Q}$ and $\mathbb{R}$ are ordered fields but $\mathbb{C}$ is not.
    c) For A, B bounded sets of real numbers, the identity

$$\sup(A-B) = \sup(A) - \sup(B)$$

holds, where $A - B = \{a - b : a \in A, b \in B\}$.

d) The integers $\mathbb{Z}$ constitute an ordered field.

e) All finite sets of real numbers have a least upper bound.

f) The rational numbers less than 1 have a least upper bound.

g) If a subset of the real numbers has an upper bound, then it has exactly one least upper bound.

h) $\sup(\mathbb{Z}) = \infty$

i) Every finite set can be ordered.

j) The set of linear functions $f(x) = ax + b$ with the usual addition and multiplication of functions is an algebraic field.

k) In plain English, the completeness axiom ensures there are no "holes" in the real numbers.

2. **Solving a Middle School Equation**

   Show that $(\forall a, b \in \mathbb{R})$ the equation $a + x = b$ has exactly one solution, which is $x = b + (-a)$.

3. **Glb, Lub, Max, and Min**

   If they exist, find $\max(A)$, $\min(A)$, $\text{lub}(A)$, and $\text{glb}(A)$ for the following sets.

   a) $A = \{1, 3, 9, 4, 0\}$

   b) $A = [0, \infty)$

   c) $A = \{x \in \mathbb{Q} : 0 \le x < 1\}$

   d) $A = [-1, 3]$

   e) $A = \{x : x^2 - 1 = 0\}$

   f) $A = \{n \in \mathbb{N} : n \text{ divides } 100\}$

   g) $A = \{x \in \mathbb{R} : x^2 < 2\}$

   h) $A = (-\infty, \infty)$

   i) $A = \left\{1, \dfrac{1}{2}, \dfrac{1}{3}, \dfrac{1}{4}, \ldots\right\}$

4. **More Difficult Sup and Inf**

   If they exist, find the least upper bound and greatest lower bound of the set

   $$A = \left\{\frac{1}{n} - \frac{1}{m} : m, n \in \mathbb{N}\right\}.$$

5. **Algebraic Field**

   Show that the rational numbers with the operations of addition and multiplication form an algebraic field.

6. **Boolean Field**
   Show that the set $F_2 = \{0, 1\}$ consisting of two elements and following arithmetic operations forms an algebraic field as shown in Table 4.5.

Table 4.5 Algebraic field.

+	0	1	×	0	1
0	0	1	0	0	0
1	1	0	1	0	1

7. **Ordered Field**
   Show that the rational numbers with the operations of addition and multiplication and usual ordering relation "less than" "<" form an ordered field.

8. **Not an Ordered Field**
   Show that the complex numbers is an algebraic field but not an ordered field.

9. **Well-Ordering Principle**
   The **well-ordering principle**[3] states that every (nonempty) subset $A \subseteq \mathbb{N}$ contains a smallest element under the usual ordering $\leq$. Does this principle hold for all subsets $A \subseteq \mathbb{Z}$?

10. **Well-Ordering Theorem**
    A partial order "$\preceq$" on a set X is called a **well-ordering** (and the set X is called **well-ordered**) if every nonempty subset $S \subseteq X$ has a least element. The **Well-Ordering Theorem**[4] states that every set can be well-ordered by some partial order. Are the following sets well ordered by the usual "less than or equal to" order "$\leq$"?
    a) $\mathbb{N}$
    b) $\{3, 4, 5\}$
    c) $\mathbb{Z}$
    d) $\left\{\frac{1}{n} : n \in \mathbb{N}\right\}$

---

3 This principle is equivalent to the Principle of Mathematical Induction.
4 The Well-Ordering Theorem is equivalent to the Axiom of Choice and was proven by the German mathematician Ernst Zermelo (1871–1953). Although the theorem says the real numbers $\mathbb{R}$ are well-ordered, no one has ever found a well-ordering.

11. **Internet Research**

    There is a wealth of information related to topics introduced in this section just waiting for curious minds. Try aiming your favorite search engine toward *least upper bound principle, least upper bound, greatest lower bound, complete ordered field,* and *definition of the real numbers.*

4.3

Complex Numbers

Purpose of Section   To introduce the field $(\mathbb{C}, +, \times)$ of complex numbers and their **Cartesian** and **polar representations** in the complex plane. We also show how **Euler's theorem** connects the **complex exponential** to real trigonometric functions.

4.3.1   An Introductory Tale

The study of numbers generally begins with children and counting numbers 1, 2, 3, ... then progressing to negative numbers, and then to fractions, and finally to the real numbers. To most students of mathematics, the complex numbers come last, if at all. Throughout history, every enlargement of the meaning of number had practical motivation. Negative numbers were required to solve $x + 3 = 1$, rational numbers were required to solve $3x = 5$, and real numbers were a response to the equation $x^2 = 2$. Finally, complex numbers came about when people wanted to solve equations like $x^2 + 1 = 0$.

Square roots of negative numbers first appeared in **Ars Magna** (1545) by the Italian mathematician Gerolamo Cardano (1501–1576) in his solution of the simultaneous equations

$$x + y = 10, \quad xy = 40$$

getting the solution

$$x = 5 + \sqrt{-15}, \quad y = 5 - \sqrt{-15}$$

Cardano did not give any interpretation of the square root of negative numbers, although he did say that they obeyed the usual rules of algebra and that

Advanced Mathematics: A Transitional Reference, First Edition. Stanley J. Farlow.
© 2020 John Wiley & Sons, Inc. Published 2020 by John Wiley & Sons, Inc.
Companion website: www.wiley.com/go/farlow/advanced-mathematics

solutions containing them could be verified. Cardano and other mathematicians at the time would go to great lengths to avoid both negative and complex numbers, referring to negative solutions as "false[1]" solutions and complex solutions as "useless." Even the great mathematician Carl Gauss said, as late as 1825, "the true metaphysics of $\sqrt{-1}$ is illusive." He overcame his doubts, however, by 1831 when he applied complex numbers to number theory. His acceptance of complex numbers provided a great boost to the acceptance of complex numbers in the mathematical community.

Today, complex numbers are crucial in the study of many areas of mathematics, including harmonic analysis, ordinary and partial differential equations, analytic number theory, analytic function theory, as well as being applied in many areas of engineering and science, including theoretical physics, where analytic function theory constitutes much of the foundations of quantum mechanics.

Someone once argued that real numbers are more "natural" than complex numbers since real numbers measure things we can all see and feel, like a person's height or weight, whereas no one can physically "experience" complex numbers. The person who makes such a claim just does not know where to look. Every engineer, physicist, and student of differential equations knows it is complex numbers that allows for the description of oscillatory motion. The next time you hear an orchestra tuning their instruments to the standard A above Middle C, you are hearing the complex number $440\,i$, the complex number that describes 440 oscillations per second.

4.3.2 Complex Numbers

There are a number of ways to introduce complex numbers, each of which has its merits and demerits. We choose to define complex numbers formally as pairs of real numbers and then introduce a "more friendly" notation.

Definition We define the **complex numbers** $\mathbb{C}$ as the set of pairs of real numbers

$$\mathbb{C} = \{(x,y) : x \in \mathbb{R}, y \in \mathbb{R}\}$$

with the binary arithmetic operations of addition[2]

---

1 The French mathematician Rene Descartes referred to square roots of negative numbers as *imaginary* numbers, a word which unfortunately is still used by some people today. Descartes also called negative solutions as *false* solutions, which fortunately did not stick.
2 We use + to denote addition for both complex numbers and real numbers, and the dot · to denote multiplication for complex numbers.

$$(a,b) + (c,d) = (a + c, b + d)$$

and multiplication

$$(a,b) \cdot (c,d) = (ac - bd, ad + bc)$$

The question we ask is do these arithmetic operations result in a legitimate arithmetic system having properties similar to those of the real numbers? We have reason to believe this to be true since we can think of the complex numbers $\mathbb{C}$ as an extension of the real numbers $\mathbb{R}$ in the sense that complex numbers of the form $(a, 0)$ behave exactly like the real numbers:

$$(a,0) + (b,0) = (a + b, 0) \quad (a,0) \cdot (b,0) = (ab, 0)$$

That is, we can think of the real numbers as those complex numbers where the second coordinate is zero.

4.3.3 Complex Numbers as an Algebraic Field

The complex numbers along with the above binary operations of addition and multiplication forms an algebraic field similar to the real numbers, with $(0, 0)$ being an additive identity and $(1, 0)$ the multiplicative identity. We will not list all the field properties for the complex numbers, but a few of them you might recall are

a) $(a, b), (c, d) \in \mathbb{C} \Rightarrow (a, b) + (c, d) \in \mathbb{C}$
b) $(a, b), (c, d) \in \mathbb{C} \Rightarrow (a, b) \cdot (c, d) \in \mathbb{C}$
c) $(a, b), (c, d) \in \mathbb{C} \Rightarrow (a, b) + (c, d) = (c, d) + (a, b)$
d) $(a, b) \in \mathbb{C} \Rightarrow (a, b) + (0, 0) = (a, b)$
e) $(a, b) \in \mathbb{C} \Rightarrow (a, b) + (-a, -b) = (0, 0)$
f) $(a, b), (c, d) \in \mathbb{C} \Rightarrow (a, b) \cdot (c, d) = (c, d) \cdot (a, b)$
g) $(a,b) \in \mathbb{C} \setminus \{(0,0)\} \Rightarrow (a,b) \cdot \left(\dfrac{a}{a^2 + b^2}, \dfrac{-b}{a^2 + b^2} \right) = (1,0)$
h) $(0, 1) \cdot (0, 1) = (-1, 0)$

The last identity (h) introduces the special complex number $(0, 1)$, which we denote by the letter i, and since we identify the complex number $(-1, 0)$ with the real number -1, we can interpret this statement as $i \cdot i = -1$ or $i^2 = -1$. For this reason, one often calls the complex number i as the square root of -1 and often written $i = \sqrt{-1}$. If we give the complex number $(0, 1)$ the special name i, then we can write the general complex number (a, b) as

$$(a,b) = (a,0) + (0,b)$$
$$= (a,0) + (b,0) \cdot (0,1)$$
$$= (a,0) + (b,0) \cdot i$$

which normally is expressed as $a + bi$. Using this notation to express complex numbers, we can express addition and multiplication of complex numbers as

Complex addition : $(a + bi) + (c + di) = (a + c) + (c + d)i$

Complex multiplication : $(a + bi)(c + di) = (ac - bd) + (ad + bc)i$

The real number a is called the **real part** of the complex number $a + bi$, and b is called the **imaginary part**, often denoted by

$\text{Re}(a + bi) = a \quad \text{Im}(a + bi) = b$

4.3.4 Imaginary Numbers and Two Dimensions

A good way of thinking about complex numbers is to think of them as a two-dimensional vector of real numbers, and when plotting these vectors we call the x-axis the **real axis** and the y-axis the **imaginary axis**.

We can give thanks to nonmathematicians Casper Wessel and Jean Robert Argand for their insight in representing complex numbers as points in the plane (see Figure 4.10).

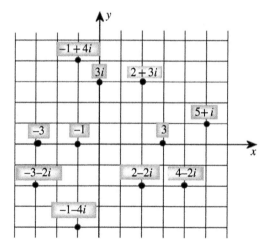

Figure 4.10 Complex plane representation of complex numbers.

Important Note The "+" in $x + iy$ is unlike any "+" you have seen before. It is not addition in the traditional sense, but a "placeholder" which separates the real and pure imaginary parts of the complex number something else.

The **absolute value** (or **modulus**) of a complex number $z = x + iy$ is defined as the nonnegative real number

$$|z| \equiv \sqrt{x^2 + y^2}$$

which is the length of the line segment from the origin to z in the complex plane (see Figure 4.11). The **conjugate** of a complex number $z = x + iy$ is defined to be $\bar{z} = x - iy$, which geometrically is the reflection of z through the real axis. The absolute value of a complex number can be written in terms of its conjugate by $|z| = \sqrt{z\bar{z}}$.

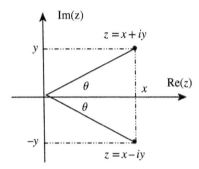

Figure 4.11 Magnitude and complex conjugate of a complex number.

The complex numbers do not form an *ordered* field like the real numbers, but do form what is called an **algebraically closed field**,[3] which means that all polynomial equations with complex coefficients have complex roots (real roots being special cases of complex roots). This property contrasts with the real numbers where polynomial equations like $z^2 + 1 = 0$ with real coefficients, but roots outside the real number system, i.e. $x = \pm i$. Algebraists would say that the complex numbers are an **algebraically closed field**, whereas the real numbers are not algebraically closed.

4.3.5 Polar Coordinates

Recall that a point (x, y) in the Cartesian plane can be written in terms of polar coordinates (r, θ), where the relationship between them is given by $x = r \cos \theta$,

---

3 Algebraically closed is different from the topological completeness we defined when defining the real numbers as a complete ordered field.

$y = r \sin \theta$. Hence, any complex number can be written in terms of polar coordinates as

$$z = x + iy = r\cos\theta + ir\sin\theta = r(\cos\theta + i\sin\theta)$$

where $r = |z| = \sqrt{x^2 + y^2}$ is the absolute value of z, and θ is the **argument** of z, written $\theta = \arg(z)$, which measures the angle between the positive real axis and the line segment from 0 to z (see Figure 4.12).

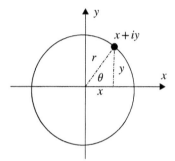

Figure 4.12 Polar form of a complex number.

Since the argument θ can wrap around the origin several times, either clockwise or counterclockwise, the **principle argument** of a complex number is the unique argument that lies in the interval $(-\pi, \pi]$. Thus, the complex number i has argument $\pi/2$, and -1 has argument π, and $-i$ has argument $-\pi/2$ (although sometimes we say $3\pi/2$).

4.3.6 Complex Exponential and Euler's Theorem

Any student of calculus knows that the exponential function e^x bears little relationship to the trigonometric functions $\sin x$ and $\cos x$. The exponential function e^x grows without bound as x gets large, while the trigonometric functions oscillate between plus and minus 1. In one of the most important discoveries in mathematics, Swiss mathematician Leonhard Euler showed in 1748 that although real exponential functions may be unrelated to trigonometric functions, complex exponentials and trigonometric functions have an intimate relationship. Euler does this by replacing the θ in the Taylor series expansion of e^θ with the complex number $i\theta$, thus defining a *new function* $e^{i\theta}$, called the **complex exponential**. Euler shows that this exponential had an interesting relationship with the trigonometric functions.

Euler's Theorem

If θ is an real number, then

$$e^{i\theta} = \cos\theta + i\sin\theta$$

where $e^{i\theta}$ is the result of replacing θ by $i\theta$ in the exponential e^{θ}. Replacing θ by $-\theta$ gives the reciprocal

$$e^{-i\theta} = \cos\theta - i\sin\theta.$$

Proof

The Taylor series expansions of e^{θ}, $\sin\theta$, $\cos\theta$ about the point $\theta = 0$, which converge for all real values θ, are

$$\sin\theta = \theta - \frac{\theta^3}{3!} + \frac{\theta^5}{5!} - \frac{\theta^7}{7!}\cdots$$

$$\cos\theta = 1 - \frac{\theta^2}{2!} + \frac{\theta^4}{4!} - \frac{\theta^6}{6!} + \cdots$$

$$e^{\theta} = 1 + \theta + \frac{\theta^2}{2!} + \frac{\theta^3}{3!} + \frac{\theta^4}{4!} + \cdots$$

If we replace θ by $i\,\theta$ in e^{θ} and using the fact that $i^2 = -1$, $i^3 = -i$, $i^4 = 1$, $i^5 = i$, ..., we arrive at the complex exponential[4]

$$\boxed{\begin{aligned} e^{i\theta} &= 1 + i\theta + \frac{(i\theta)^2}{2!} + \frac{(i\theta)^3}{3!} + \frac{(i\theta)^4}{4!} + \cdots \\ &= \left(1 - \frac{\theta^2}{2!} + \frac{\theta^4}{4!} - \frac{\theta^6}{6!} + \cdots\right) + i\left(\theta - \frac{\theta^3}{3!} + \frac{\theta^5}{5!} - \frac{\theta^7}{7!}\cdots\right) \\ &= \cos\theta + i\sin\theta \end{aligned}}$$

∎

Euler's theorem allows one to work with exponentials and all their wonderful manipulative properties like

- $e^{i\theta_1} e^{i\theta_2} = e^{i(\theta_1 + \theta_2)}$
- $\left(e^{\theta}\right)^n = e^{n\theta}$

Euler's equation also allows us to write the important identity

$$e^z = e^{x+iy} = e^x e^{iy} = e^x(\cos y + i\sin y)$$

---

4 The rearrangement of terms is allowed since the Taylor series are *absolutely convergent*.

Example 1   Complex Exponentials

a) $e^{-x + 3y} = e^{-x}(\cos 3y + i \sin 3y)$
b) $e^{(-3 + 2i)t} = e^{-3t}(\cos 2t + i \sin 2t)$
c) $e^{-ix} = \cos x - i \sin x$

4.3.7   Complex Variables in Polar Form

Euler's equation allows us to write complex numbers either in Cartesian or **polar form** by equation:

$$z = x + iy = r(\cos\theta + i\sin\theta) = re^{i\theta}$$

Note that

$$\left|e^{i\theta}\right| = \left|\cos\theta + i\sin\theta\right| = \sqrt{\cos^2\theta + \sin^2\theta} = 1$$

We can visualize the complex exponential $e^{i\theta}$ as a point in the complex plane[5] with argument θ lying on the unit circle. As θ moves from 0 to 2π, the exponential $e^{i\theta}$ moves counterclockwise around the unit circle (see Figure 4.13).[6] For example[7]

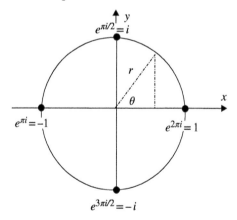

Figure 4.13 The complex exponential $e^{i\theta}$.

---

5 We get so used to representing complex numbers as points in the plane that we use them interchangeably.

6 We really should say θ goes from $-\pi$ to $+\pi$ since we have adopted the negative angle convention, but people are sloppy about this so we simply say 0 to 2π.

7 Note the convention of sometimes placing the "i" in front of the constants in the exponent and sometimes at the end. Also the argument of $-i$ is often represented interchangeably by either $3\pi/2$ or $-\pi/2$.

$$e^{\pi i/2} = i, e^{i\pi} = -1, e^{3\pi i/2} = e^{-i\pi/2} = -i, e^{2\pi i} = 1.$$

Important Note Euler's equation is a wealth of important information. If we write $e^{\pi i/2} = i$, and then raise both sides to the ith power, we get $i^i = e^{-\pi/2} \doteq 0.207\ 879\ 57....$

Example 2 Complex Numbers in Polar Form

There are always two ways to write a complex number, Cartesian or complex.

a) $1 + i = \sqrt{2}[\cos(\pi/4) + \sin(\pi/4)] = \sqrt{2}e^{i(\pi/4)}$

b) $i = \cos(\pi/2) + i\sin(\pi/2) = e^{i(\pi/2)}$

c) $1 + \sqrt{3}i = 2[\cos(\pi/3) + i\sin(\pi/3)]$

d) $-1 = \cos(\pi) + i\sin(\pi) = e^{i\pi}$

Important Note Four important number systems are the one-dimensional real numbers, the two-dimensional complex numbers, the four-dimensional quaternions, and the eight-dimensional octonians. Although addition and multiplication are defined for all four systems, multiplication is not commutative for quaternions as it is for real and complex numbers, and as for octonians, not only is multiplication not commutative, the associative law does not hold.

4.3.8 Basic Arithmetic of Complex Numbers

Complex numbers are an algebraic field just like the real numbers. They can be added, subtracted, multiplied, and divided and have interesting geometric interpretations in the complex plane.

Addition

$$z_1 + z_2 = (x_1 + iy_1) + (x_2 + iy_2) = (x_1 + x_2) + i(y_1 + y_2)$$

Two complex numbers are added by adding the real and complex parts of the two numbers. In the complex plane, the sum of two complex numbers corresponds to the point lying on the diagonal of a parallelogram whose sides are the two complex numbers (see Figure 4.14).

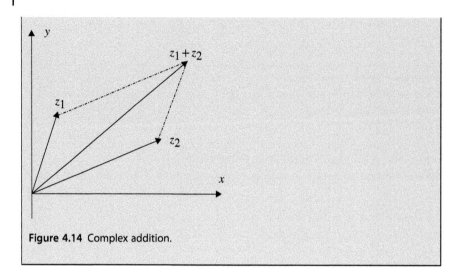

Figure 4.14 Complex addition.

Subtraction

$$z_1 - z_2 = (x_1 + iy_1) - (x_2 + iy_2) = (x_1 - x_2) + i(y_1 - y_2)$$

Subtraction is analogous to addition but real and complex parts are subtracted (see Figure 4.15).

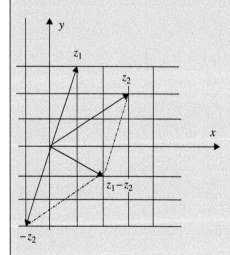

Figure 4.15 Complex subtraction.

Multiplication

$$z_1 z_2 = \left(r_1 e^{i\theta_1}\right)\left(r_2 e^{i\theta_2}\right) = r_1 r_2 e^{i(\theta_1 + \theta_2)}$$

Or

$$(x_1 + iy_1)(x_2 + iy_2) = (x_1 x_2 - y_1 y_2) + i(x_1 y_2 + x_2 y_1)$$

The product of two complex numbers is best interpreted in polar coordinates. In the complex plane, the product of two complex numbers is a complex number whose magnitude is the product of the magnitudes of the two numbers, and whose argument is the sum of the arguments of the two numbers (see Figure 4.16).

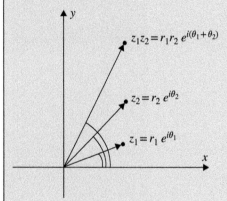

Figure 4.16 Complex multiplication.

Division

$$\frac{z_1}{z_2} = \frac{r_1 e^{i\theta_1}}{r_2 e^{i\theta_2}} = \frac{r_1}{r_2} e^{i(\theta_1 - \theta_2)}$$

$$\frac{x_1 + iy_1}{x_2 + iy_2} = \left(\frac{x_2 - iy_2}{x_2 - iy_2}\right)\left(\frac{x_1 + iy_1}{x_2 + iy_2}\right) = \left(\frac{x_1 x_2 + y_1 y_2}{x_2^2 + y_2^2}\right) + i\left(\frac{x_2 y_1 - x_1 y_2}{x_2^2 + y_2^2}\right)$$

Division of two complex numbers is also best interpreted in polar coordinates. In the complex plane, the quotient of two complex numbers is a complex number whose magnitude is the quotient of the magnitudes of the two numbers, and whose argument is the difference of the arguments of the two numbers.

If division is performed in Cartesian form, it is accomplished by the process of **rationalizing the denominator**, where one multiplies both numerator and denominator by the conjugate of the denominator and then collecting real and complex parts (see Figure 4.17).

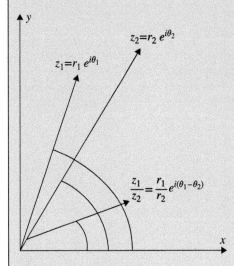

Figure 4.17 Complex division.

Example 3 Complex Arithmetic

a) $(2 + 3i)(3 + i) = 3 + 11i$

b) $2e^{i\pi} \cdot 3e^{i\pi/2} = 6e^{3\pi i/2}$

c) $\dfrac{2 + i}{1 - 3i} = \left(\dfrac{1 + 3i}{1 + 3i}\right)\left(\dfrac{2 + i}{1 - 3i}\right) = \left(\dfrac{2 - 3}{1^2 + 3^2}\right) + \left(\dfrac{1 + 6}{1^2 + 3^2}\right)i = -\dfrac{1}{10} + \dfrac{7i}{10}$

d) $\dfrac{6e^{i\pi}}{2e^{i\pi/2}} = 3e^{i\pi/2}$

e) $\dfrac{1}{z} = \dfrac{1}{r}e^{-i\theta} = \dfrac{1}{r}(\cos\theta - i\sin\theta)$

4.3.9   Roots and Powers of a Complex Number

One of the benefits of representing complex numbers in polar form is visualizing the roots of the equation $x^n = a$. Most beginning students of mathematics know there are two square roots of a positive number like four, knowing them to be ±2. However, if you answered two as the cube root of eight, you would be one-third correct, since there are two other two cube roots. And what about the cube roots of a negative number like minus one? Certainly, minus one is one cube root, but there are two more.

This leads to the following problem. Given a complex number $a \neq 0$ and an integer n greater or equal to two, find the roots of $z^n = a$? To solve this equation, we write z and a in polar form, getting

$$z^n = \left(re^{i\theta}\right)^n = r^n e^{in\theta} = |a|e^{i\phi}$$

where $|a| = r^n$. Taking the absolute value of each side of this equation yields $r^n = |a|$ from which we get $r = |a|^{1/n}$. Plugging this back into the equation, the equation reduces to $e^{in\theta} = e^{i\phi}$ from which we find $n\theta = \phi$ or $\theta = \phi/n$. Hence, putting all this together, the nth root of $z^n = a$ is

$$z = re^{i\theta} = |a|^{1/n} e^{i\phi/n} = |a|^{1/n}[\cos(\phi/n) + i\sin(\phi/n)].$$

However, this is not the only root since

$$z_k \equiv |a|^{1/n} e^{i(\phi + 2\pi k/n)}, \quad k = 0, 1, 2, \ldots, n-1$$

also satisfies $z^n = a$, which can be seen by direct computation

$$\left(|a|^{1/n} e^{i(\phi + 2\pi)/n}\right)^n = |a|e^{i(\phi + 2\pi)} = |a|e^{i\phi}e^{2\pi i} = |a|e^{i\phi} = a.$$

These results can be summarized as follows.

Roots of a Complex Number   For each nonzero complex number[8] $a = |a|e^{i\phi} \in \mathbb{C}$ there are n distinct nth roots. In the complex plane, they are the following n points on a circle of radius $|a|^{1/n}$.

$$a^{1/n} = |a|^{1/n} e^{[i(\phi + 2\pi k)/n]}$$

$$= |a|^{1/n}\left[\cos\left(\frac{\phi + 2\pi k}{n}\right) + i\sin\left(\frac{\phi + 2\pi k}{n}\right)\right], \quad k = 0, 1, 2, \ldots, n-1$$

The drawings in Figure 4.18 show the roots of the equation $z^n = 1$ for $n = 2, 3, 4, 5$.

---

8 When we say complex number, this includes real numbers as well.

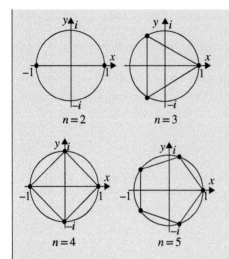

Figure 4.18 Roots of unity.

To find the *n*th roots of the complex number *a*, first find the **principal root**, whose absolute value is $|a|^{1/n}$ and whose argument is $\arg(a)/n$. The other $n-1$ roots are equally spaced points on the same circle whose angles between them is $2\pi/n$.

Figure 4.19 shows the six roots of the equation $z^6 = 1$ called the six roots of unity.

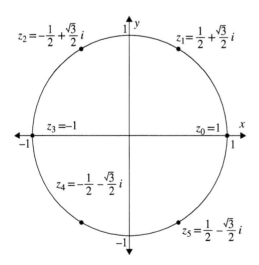

$z_2 = -\dfrac{1}{2} + \dfrac{\sqrt{3}}{2}i$

$z_1 = \dfrac{1}{2} + \dfrac{\sqrt{3}}{2}i$

$z_3 = -1$

$z_0 = 1$

$z_4 = -\dfrac{1}{2} - \dfrac{\sqrt{3}}{2}i$

$z_5 = \dfrac{1}{2} - \dfrac{\sqrt{3}}{2}i$

Figure 4.19 Six roots of unity.

Example 4   Roots

a) $\sqrt{4} = 2\left[\cos\left(\dfrac{0+2\pi k}{2}\right) + i\sin\left(\dfrac{0+2\pi k}{2}\right)\right]$

$= 2[\cos(k\pi) + i\sin(k\pi)], \quad k = 0,1$

$= \begin{cases} 2, & k = 0 \\ -2, & k = 1 \end{cases}$

b) $\sqrt{i} = \sqrt{1}\left[\cos\left(\dfrac{\pi/2+2\pi k}{2}\right) + i\sin\left(\dfrac{\pi/2+2\pi k}{2}\right)\right]$

$= \cos\left(\dfrac{\pi}{4} + k\pi\right) + i\sin\left(\dfrac{\pi}{4} + k\pi\right), \quad k = 0,1$

$= \begin{cases} \dfrac{1}{\sqrt{2}} + \dfrac{i}{\sqrt{2}}, & k = 0 \\[2mm] -\dfrac{1}{\sqrt{2}} - \dfrac{i}{\sqrt{2}}, & k = 1 \end{cases}$

c) $i^{1/3} = |1|^{1/3}\left[\cos\left(\dfrac{\pi/2+2\pi k}{3}\right) + i\sin\left(\dfrac{\pi/2+2\pi k}{3}\right)\right], \quad k = 0,1,2$

$= \begin{cases} \dfrac{\sqrt{2}}{2} + \dfrac{1}{2}i & k = 0 \\[2mm] -\dfrac{\sqrt{2}}{2} + \dfrac{1}{2}i & k = 1 \\[2mm] -i & k = 2 \end{cases}$

The three cube roots of the complex number i are shown in Figure 4.20.

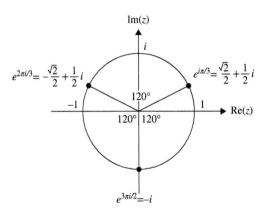

Figure 4.20 Three cube roots of i.

Problems

1. **Complex Numbers**
   Let
   a) $w = i$, $z = 1 + i$
   b) $w = 2$, $z = -i$
   c) $w = i$, $z = -i$
   d) $w = -1 - i$, $z = 1 + i$
   e) $w = 1 + i$, $z = -1 + i$

   Find the following values.

   - $|z|$
   - $\theta = \arg(z)$
   - $w + z$
   - wz
   - w/z

2. **Convert to Polar Form**
   Convert the following complex numbers to polar form.
   a) $2i$ **Ans:** $re^{i\theta} = 2e^{i(\pi/2)2}$
   b) $-1 + i$
   c) $-i$ **Ans:** $re^{i\theta} = e^{i(3\pi/2)}$
   d) $2 - 2i$

3. **Convert to Cartesian Form**
   Convert the following complex numbers to Cartesian form $x + iy$.
   a) $e^{3\pi i}$ **Ans:** $e^{3\pi i} = \cos 3\pi + i \sin 3\pi = -1$
   b) $2e^{i\pi/4}$
   c) $e^{2\pi i}$
   d) $e^{5\pi i}$
   e) $5e^{3\pi i/4}$

Important Note There used to be a company that sold T-shirts with "Mathematicians, we're Number $e^{2\pi i}$ written on them.

---

4. Evaluate $(1 + i)^{100}$.

5. **Useful Identity I**
   Show that the complex conjugate of the sum of two complex numbers is the sum of the conjugates; that is $\overline{(w + z)} = \bar{w} + \bar{z}$.

6. **Useful Identity II**
Verify the identity $|z|^2 = z\bar{z}$ for $z = 2 + 3i$.

7. **Taking a Complex Function Apart**
Find the real and imaginary parts of the following.
a) z^3 **Ans:** $\text{Re}(z^3) = x^3 - 3xy^2$, $\text{Im}(z^3) = 3x^2y - y^3$
b) $1/z$

8. Compute the following.
a) $\sqrt{i}$
b) $\sqrt{-i}$
c) $\sqrt{1+i}$
d) $\sqrt[3]{-1}$
e) $\sqrt[4]{-1}$

9. **de Moivre's Formula**
Use Euler's theorem to prove **de Moivre's formula**

$$(\cos\theta + i\sin\theta)^n = \cos n\theta + i\sin n\theta$$

for any positive integer n. Hint: Use induction.

10. **Primitive Roots of Unity**
The n roots the equation $z^n = 1$ are called the n roots of unity. Find and plot the roots when $n = 1, 2, 3, 4,$ and 8.

11. **Fractional Powers**
Find the following.
a) $i^{3/2}$
b) $(-1)^{3/4}$
c) $\sqrt{1+i}$
d) i^i

12. **Hmmmmmmmmmm**
Show $\dfrac{1}{i} = -i$

13. **Internet Research**
There is a wealth of information related to topics introduced in this section just waiting for curious minds. Try aiming your favorite search engine toward *applications of complex numbers, history of complex numbers,* and *roots of unity.*

Chapter 5

Topology

5.1

Introduction to Graph Theory

Purpose of Section   To introduce the basics of one of the most important areas of topology: graph theory and networks. We introduce Euler paths and cycles, Hamiltonian tours, minimum spanning trees, and the Euler Characteristic.

5.1.1   Introduction

The subject of graph theory and networks is an area of topology that is rapidly moving into mainstream mathematics. It has modern applications in artificial intelligence, data mining, large-scale communication networks, and industrial scheduling. Few subjects in mathematics can be traced back to such an exact and interesting beginnings as graph theory to the famous *Seven Bridges of Konigsberg Problem,* solved by the Swiss mathematician Leonhard Euler in 1736.

The town of Konigsberg, Prussia (now Kaliningrad, Russia), rests on the banks of the Pregel River, and as it flows through the town, it divides the town into four distinct regions connected by seven bridges, illustrated in the accompanying drawing. The town flourished in the seventeenth and eighteenth centuries, and the story goes that the people of Konigsberg spent their evenings strolling throughout the city, crossing the seven bridges that spanned the river. The question was asked whether it was possible to start at one of the four land areas, cross each bridge exactly once, and return to the starting point. The mathematician Euler learned of the problem and in a published paper[1] demonstrated that in

---

1 Euler published his findings in a paper titled *Solutio problematis ad geometriam situs pertinentis* (The solution of a problem relating to the geometry of position) in the *Proceedings of St. Petersburg Academy* in 1736.

Advanced Mathematics: A Transitional Reference, First Edition. Stanley J. Farlow.
© 2020 John Wiley & Sons, Inc. Published 2020 by John Wiley & Sons, Inc.
Companion website: www.wiley.com/go/farlow/advanced-mathematics

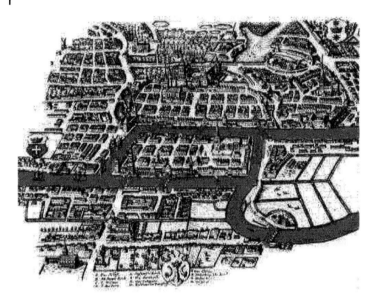

order for a person to cross each bridge exactly once and return to the starting point, each land mass must be connected by an even number (2, 4, 6, ...) of bridges. Since this was not the case for the Konigsberg bridges such a stroll was not possible.

Before showing how Euler solved this problem, we must introduce the concept of a graph.

Definition A graph $G = (V, E)$ is a finite set V of **vertices** (or **nodes**) and a finite set E of **edges** connecting pairs of vertices. The diagram below depicts a graph with five vertices and eight edges defined by

$$V = \{a, b, c, d, e\}$$
$$E = \{(a,b), (a,c), (a,d), (b,c), (b,d), (b,e), (c,d), (c,e)\}$$

See Figure 5.1.

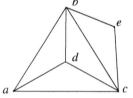

Figure 5.1 Typical graph.

5.1.2 Glossary of Important Concepts in Graph Theory

Some of the main ingredients of a graph are

- The **order** of a graph G is the number of vertices in the graph, denote by $|G|$.
- The **degree** of a vertex is the number of edges connecting the vertex.

- A **path** is a sequence of vertices connected by edges.
- A **cycle** is a path starting and ending at the same vertex.
- A graph is **connected** if it is possible to traverse from any vertex to any another vertex by a path.
- A vertex is **odd** (**even**) if it has an odd (even) number of edges adjacent to it.

In the graph illustrated in the definition, we have

- the order is $|G| = 5$.
- the degree of vertices a, b, c, d, e is 3, 4, 4, 3, 2.
- typical paths are *acdb* and *aba*.
- typical cycles are *adbeca* and *cdac*.
- the graph is connected.
- the vertices b, c, e are even, a, d are odd.

5.1.3   Euler Paths and Circuits

Two important concepts in the study of graphs are Euler paths and Euler circuits.

- **An Euler path** is a path that passes through each edge of the graph exactly once and ends at a *different* vertex.
- An **Euler cycle** (**or tour**) is a cycle that passes through each edge of the graph exactly once and starts and ends at the *same* vertex.

Example 1   Euler Paths and Cycles
Find, if any, Euler paths and cycles in the graph in Figure 5.2.

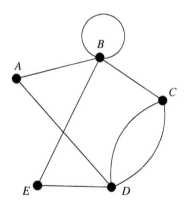

Figure 5.2 Typical graph.

Solution

Typical Euler paths and tours are

Euler paths: *BBADCDEBC, CDCBBADEB*
Euler tours: *CDCBBADEBC, CDEBBADC*

5.1.4   Return to Konigsberg

We now state and indicate the proof of Euler's famous 1736 theorem, effectively solving the *Konigsberg Bridge Problem*.

Euler's Original Graph Theorem   If all vertices of a graph are even (2, 4, 6, ...), then the graph contains a cycle that crosses every edge exactly once. Such a cycle through a graph is called an **Euler Tour**.[2]

Proof

Euler argued that since every vertex is even, a path arriving at a vertex other than the starting vertex can always leave that vertex.   ∎

Figure 5.3 shows a rough diagram of the Pregel River and four adjoining land masses along with the graphical representation drawn by Euler. Note that not all the vertices in the graph are even (in fact none are even), hence there does not exist an Euler Tour in the graph.

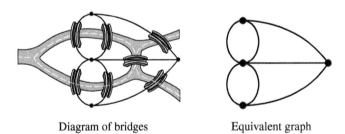

Diagram of bridges   Equivalent graph

Figure 5.3 Graphical representation.

The existence of Euler paths or cycles depends on the number of odd nodes: 0, 2, or more than 2.

---

2 Euler proved that if a connected graph has an "Euler" tour, then all the vertices must be even. The converse that a connected graph with an Euler tour must have all even vertices was not proven until 1873.

5.1.4.1 Euler Paths and Tours and Odd Nodes

The following implications are basic to Euler tours and paths.

- zero odd vertices ⇒ graph has an Euler tour.
- two odd vertices ⇒ graph has an Euler path starting at one odd vertex and ending at the other.
- more than two odd vertices ⇒ graph does not have an Euler path.

Example 2   Euler Paths and Cycles

Determine if the graphs in Figure 5.4 have an Euler tour, Euler path, or neither[3]

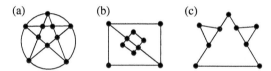

Figure 5.4 Finding Euler tours.

Solution

We leave the fun for the reader.

Important Note   Since the time Euler first realized that the Konigsberg Bridge problem could be solved using the "geometry of relations," others have acknowledged the power of graph theory. In the 1800s, the German physicist Gustav Kirchhoff observed that electric circuits could be analyzed by drawing them as graphs with wires as edges and junction points as vertices. Today, graph theory is an active area of mathematical research with applications in integrated circuits, genomics, scheduling problems, and all sorts of large-scale networks.

Historical Note   In a 1670 letter to Christian Huygens, the mathematician Gottfried Leibniz (1646–1716) wrote as follows: "I am not content with algebra, in that it yields neither the shortest proofs nor the most beautiful constructions of geometry. Consequently, in view of this, I consider that we need yet another kind of analysis, geometric or linear, which deals directly with position, as algebra deals with magnitude." What Leibniz thought needed was topology.

---

3 We thank the Math Forum at Drexel University for the use of these examples.

One characteristic of graph theory is the variety of problems that can be solved using its principles. One such fascinating problem that can be solved using the tools of graphs is the Handshaking Problem.

Handshaking Problem At any social gathering, the number of people who shake hands an odd number of times is an even number.[4] The problem illustrates the type of reasoning often used in graph theory.

Proof

We begin by drawing a graph with a given number of vertices and no edges as illustrated in Figure 5.5a) and note that each vertex has degree 0 and so the number of odd vertices is 0. We then simulate two persons shaking hands by drawing an edge between two vertices and observing that for each new edge the number of odd vertices goes up or down by 2, thus keeping the number of odd vertices an even number.

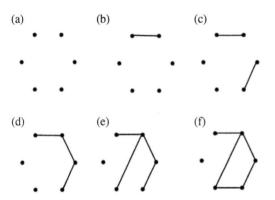

Figure 5.5 Handshaking problem. (a) Odd vertices = 0, (b) odd vertices = 2, (c) odd vertices = 4, (d) odd vertices = 2, (e) odd vertices = 4, and (f) odd vertices = 2.

One of the fascinating aspects of graph theory is that its applications are almost endless. Anything that involves objects and relations between them can be modeled as a graph. The nodes can be cities as vertices and airplane flights as the edges between them. But how do we model the cost of a flight? We simply use a weighted graph that assigns the costs of the flights to the edges. There are algorithms for finding cheapest paths in a network.

---

4 This theorem was also proven by Euler.

5.1.5   Weighted Graphs

In many applications in modern technology, it is useful to assign nonnegative numbers or weights to the edges of a graph.

Definition   A **weighted graph** is a graph whose edges are assigned nonnegative numbers called **costs** or **weights**.

One class of problems modeled by **weighted graphs**, which are becoming more and more important in network design. Network design ranges from the design of telephone and Internet landlines to electrical networks, gas lines, water mains, sewage lines, and on and on. The goal is to design networks that operate effectively at minimum cost.

Consider the weighted graph in Figure 5.6 consisting of eight cities, where the numbers on the edges represent distances between cities. A problem might be to design a high-speed train network connecting the cities and whose total length is minimum. This is a simple task for eight cities, but when designing communication networks connecting thousands of vertices, the problem is far from trivial.

From the perspective of graph theory, the goal is to find the minimum spanning tree in the graph. A **tree** is a graph that contains no cycles, and a **spanning tree** is a subtree that includes every vertex of the graph, and a **minimum spanning tree** is a spanning tree (there may be many) whose sum of its weights is a minimum.

There is a well-known algorithm for finding the minimum spanning tree in a graph called **Kruskal's Algorithm**. The idea behind the algorithm is very simple. One starts with only the vertices of the graph, then, one by one, start adding edges, starting with the edge with the smallest weight, then continue by adding edges with increasing weights, provided no cycle is created. If a cycle is formed, bypass that edge and continue on until all vertices are included in the tree. The result will be a minimum spanning tree in the graph.[5] Carrying out Kruskal's Algorithm for the eight cities in Figure 5.6, we find its minimum spanning tree that we draw in Figure 5.7. The total length of the minimum spanning tree is 4330 miles. This will give planners an estimate of the total cost of the new rail system.

---

5 Kruskal's minimum spanning tree algorithm is an example of a "greedy algorithm," where the strategy is to make the optimal choice at each stage. Most algorithms in graph theory are not greedy algorithms and involve more intricate strategies.

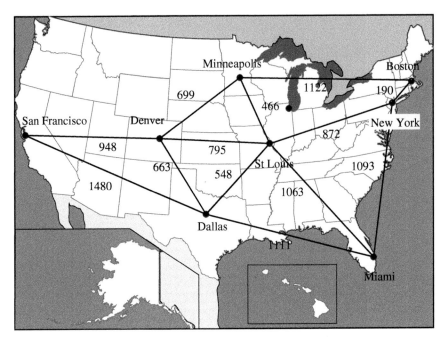

Figure 5.6 Weighted graph of cities and distances.

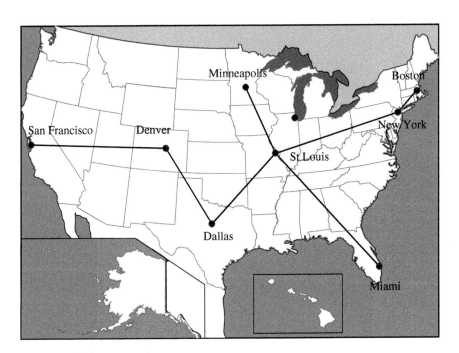

Figure 5.7 Minimum spanning tree.

5.1.6   Euler's Characteristic for Planar Graphs

In graph theory, a **planar graph** is one that can be drawn in the plane in a way that no two edges cross each other. It is sometimes called the "integrated circuit" graph. For example, the complete[6] graph K_4 of order four in Figure 5.8a) is a planar graph. At first glance, the graph may not appear planar since two edges cross, but keep in mind graph theory is not Euclidean geometry. Graph theory is the "geometry of connections," where shapes of edges are not important, only how the vertices are connected. Hence, the graph in Figure 5.8a can be redrawn as Figure 5.8b or Figure 5.8c that show no intersecting edges.

(a)   (b)   (c)

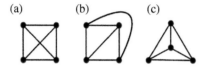

Figure 5.8 Equivalent drawings the planar graph K_4.

An important concept of planar graphs is that of the faces of graph, which simply are regions bounded by the edges of the graph. The graph in Figure 5.8 has four vertices, six edges, and four faces. The reader may have counted only three faces, but in our present analysis, we count the infinite region outside the graph as a face.

One property common to every connected planar graph is its Euler Characteristic. No matter how different the graphs may appear, they have something in common and that's their Euler Characteristic. This leads us to Euler's Characteristic Theorem. We will see in Section 5.3 how the Euler Characteristic is an important tool for distinguishing topological shapes in three dimension.

Euler's Characteristic Theorem   If G is a connected, planar graph with v vertices, e edges, and f faces, then

$$v - e + f = 2$$

Proof
First note that the "triangle graph" ($v = 3$, $e = 3$, $f = 2$) as drawn in Figure 5.9 has an Euler Characteristic of 2.[7]

---

6 A graph is called complete if every node is connected to every other node.
7 When counting the faces of a graph, it is an accepted practice to count the region of the plane that lies "outside" the graph as one of the faces.

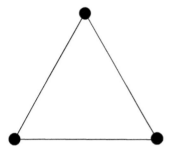

Figure 5.9 Euler characteristic.

Starting with an arbitrary connected, planar graph, the idea behind the proof is to continually **triangulate** the graph, throwing out "boundary triangles," until eventually arriving at a triangle graph, and to show at each step of this process the Euler Characteristic is unchanged. Hence, if the final graph has an Euler Characteristic of 2, so does the initial graph.

Although we will not prove the result for an arbitrary connected, planar graph, the general proof can be understood by working through the proof for the planar representation for a cube as drawn in Figure 5.10a. Starting with the graph in Figure 5.10a, we "triangulate" the five internal faces by drawing edges between different vertices. If we now count the number of vertices, edges, and faces in the triangulated graph in Figure 5.10b, we see that the number of vertices is unchanged; the new graph has five additional edges and five new faces, and so the Euler Characteristic $v - e + f$ remains the same.

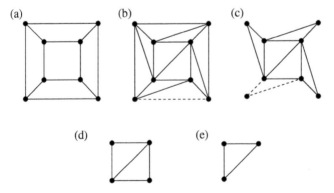

Figure 5.10 At each step $v - e + f$ is unchanged. (a) Start, (b) triangulate faces, (c) remove boundary triangles, (d) almost done, and (e) done.

The next step is to eliminate the triangles that have one or two edges on the boundary, such as four triangles in Figure 5.8b. We leave it to the reader to verify that the removal of these four boundary triangles, either the type in Figure 5.10b with one boundary edge, or the type in Figure 5.10c with two boundary edges, does not change the Euler Characteristic. Continuing this process of eliminating exterior triangles, we eventually arrive at the triangle graph whose Euler Characteristic is two. Therefore, the original graph also has Euler Characteristic of two. ∎

Problems

1. In the drawing in Figure 5.11 showing 10 bridges connecting 5 land masses, determine if it is possible to start and end at any land mass and traverse each bridge exactly once.

   For Problems 2–7, determine if the given graph has an Euler Tour and if so find one.

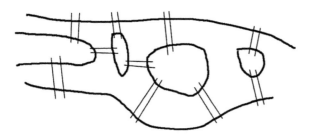

Figure 5.11 Find an Euler tour.

2.

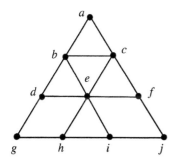

Figure 5.12 Find an Euler tour.

3.

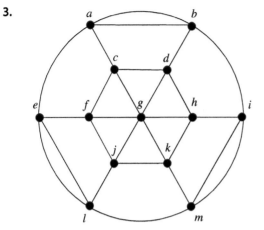

Figure 5.13 Find an Euler tour.

4.

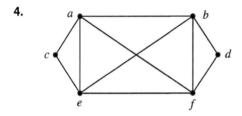

Figure 5.14 Find an Euler tour.

5.

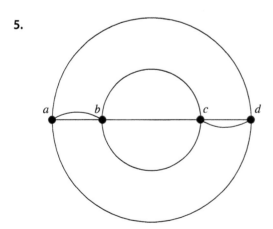

Figure 5.15 Find an Euler tour.

6.

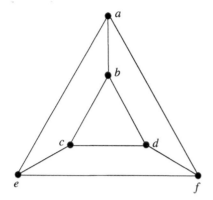

Figure 5.16 Find an Euler tour.

7.

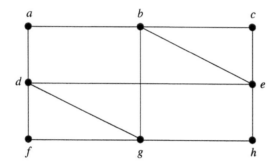

Figure 5.17 Find an Euler tour.

8. Euler Paths

If all nodes of a graph are even except two, then the graph has an **Euler path**. This means one starts at one of the odd nodes, traverses all the edges of the graph exactly once, and ends at the other odd node. Find the Euler paths in the graphs in Figure 5.18.

Hamiltonian Tour

Another type of path, or more accurately tour, through a graph is the **Hamiltonian tour**, which is a tour that starts at a given vertex, traverses each *vertex* (not edge) *exactly once*, and then returns to the starting vertex. A graph that contains a Hamiltonian tour is called a **Hamiltonian graph**. Unfortunately, unlike Euler tours, there is no simple test for determining if a graph has a Hamiltonian tour. For Problems, 9–12 find, if there is a Hamiltonian tour in the given graph.

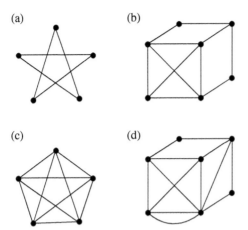

Figure 5.18 Find an Euler path.

9.

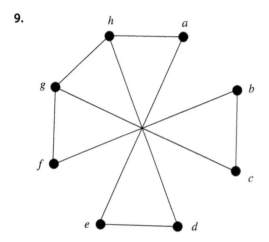

Figure 5.19 Find a Hamiltonian path.

10.

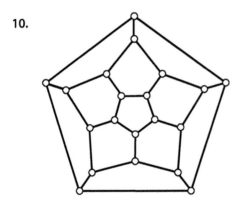

Figure 5.20 Find a Hamiltonian path.

11.

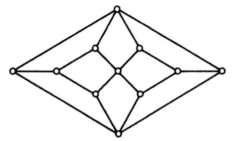

Figure 5.21 Find a Hamiltonian path.

12.

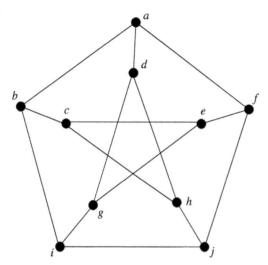

Figure 5.22 Find a Hamiltonian path.

13. Graphs of Order 3

Suppose the vertices of a graph are indistinguishable.
a) Draw all graphs with three vertices.
b) Draw all graphs with four vertices and three edges.

14. All Graphs of Order Four

Assuming the vertices of the graph are indistinguishable, draw all possible graphs with four vertices. Hint: There are 11 of them.

15. Hamilton's Famous Puzzle

In 1859, Irish mathematician William Rowan Hamilton (1805–1865) mar-keted a puzzle shaped as a regular dodecahedron, a solid with 12 faces where each face having the shape of a regular pentagon, as illustrated in Figure 5.23. The name of a city was assigned to each corner of the dodecahedron. The goal of the puzzle was to start at any city, find a route along the edges of the dodecahedron, visiting each city exactly once and returning to the starting city. Such a path is a Hamiltonian Tour. The planar representation of a dodecahedron is shown in the following figure. Can you find a Hamiltonian Tour through this dodecahedron?

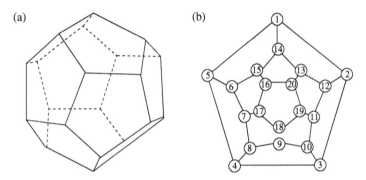

Figure 5.23 Planar representation of the regular dodachedron. (a) Regular dodachedron and (b) planar representation of dodachedron.

16. **Tours of Platonic Solids**
Graphs in Figure 5.24 are planar representations of the five Platonic solids; the tetrahedron, cube, octahedron, dodecahedron, and icosahedron. Tell if each has Euler and Hamiltonian Tours. If so, find one.

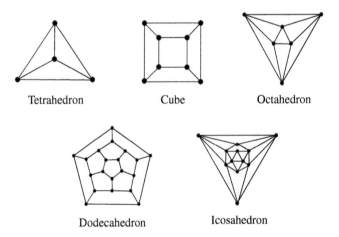

Figure 5.24 Planar representations of the Platonic solids.

17. **Knight's Tour**

A knight's tour on a chessboard in Figure 5.25 is a path on the board that visits each square exactly once, then returns to the starting square. The smallest chessboards for which a tour is possible are 5×6 board and 3×10 boards. A tour is not possible for 3×3 and 4×4 boards

a) Give an argument why a knight's tour is not possible for a 4×4 board. Hint: Any tour must pass through the upper left square 1 and the lower right square 16.

b) Draw a graph of 16 nodes representing a 4×4 board, where each vertex is connected by an edge if there is a knight's move between the vertices.

c) Show that a Hamiltonian tour, which is a cycle that passes through each vertex (not edge) exactly once and returns to the starting vertex, is not possible for the graph you drew in part (b).

1	2	3	4
5	6	7	8
9	10	11	12
13	14	15	16

Figure 5.25 Knight's tour.

18. **More Knight's Tour**

Place a chess knight (or a coin) on any one of the squares of the board drawn in Figure 5.26 and find a tour that lands on each square exactly

Figure 5.26 Find the knight's tour.

once and returns to the starting square. Such a tour would be a Hamiltonian tour if one interprets the squares of the board as the vertices of a graph.

19. **Regular Graphs**
   A regular graph is one where all vertices have the same degree. A k-regular graph is a regular graph where each vertex has degree k. Figure 5.27 below shows k-regular graphs with six vertices for k ranging from zero to three. Note that only the 3-regular graph is connected.
      Draw the following regular graphs
   a) 4-regular graph with 6 vertices
   b) 5-regular graph with 12 vertices.
   c) 3-regular graph with 8 vertices. These are called cubic graphs, and there are five of them. Draw as many as you can.

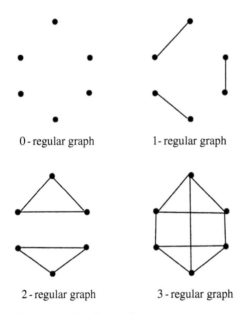

Figure 5.27 Regular graphs.

20. **Chromatic Number of a Graph**
   The chromatic number of a graph is the smallest number of colors needed to color a graph so that no two adjacent vertices have the same color. Find the chromatic numbers of the graphs in Figure 5.28.

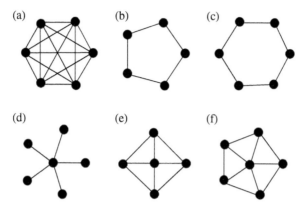

Figure 5.28 Find the chromatic number.

21. Moser Spindle

The chromatic number of the plane is the smallest number of colors required to color all the points in the plane so that no two points that are exactly one unit apart have the same color. The chromatic number is unknown but known to require *at least* five colors.[8] It is also known that any unit-distance graph, no matter how complicated, can be colored with seven colors. The Moser spindle is a unit-distance graph with 7 vertices and 11 edges all have the same length, and can be colored with four colors. Can you four-color the Moser spindle unit-distance graph drawn in Figure 5.29?

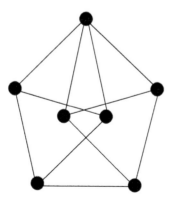

Figure 5.29 Moser Spindle.

---

8 In 2018 biologist Aubrey de Grey found a graph that could not be colored with four colors and thus raised the question to five colors.

22. **Internet Research**

There is a wealth of information related to topics introduced in this section just waiting for curious minds. Try aiming your favorite search engine toward *applications of graph theory, graph theory in computer science,* and *largest known graphs.*

5.2

Directed Graphs

> **Purpose of Section** We introduce the **directed graph** and its corresponding adjacency matrix and show how directed graphs arise in modern society, including social networks, tournaments, the Google Page Rank system, and the optimization technique of dynamic programming.

5.2.1   Introduction

Thus far in our discussion of graphs, we have not associated a "direction" with the edges of the graph. In many models, however, such as hyperlinks connecting web-pages, traffic flow problems, we impose a direction on the edges. A **directed graph** (or **digraph**) is defined as a graph whose edges are "directed" from one vertex to another. We call these types of edges **directed edges** (or **arcs**). If a directed edge goes from vertex u to vertex v, then u is called the **head** (or **source**) and v is called the **tail** (or **sink**) of the edge, and v is said to be the **direct successor** of u and u is the **direct predecessor** of v. See Figure 5.30.

The applications of directed graphs range from transportation problems in which traffic flow is restricted to one direction and one-way communication problems, to asymmetric social interactions, athletic tournaments, and even the World-Wide Web. Before talking about applications, however, we introduce the concept of the adjacency matrix of a directed graph.

Definition The **adjacency matrix** of a directed graph with n vertices is an $n \times n$ matrix $M = (m_{ij})$, where the m_{ij} entry of the matrix denotes

$$m_{ij} = \begin{cases} 1 & \text{if there is a directed edge from vertex } i \text{ to vertex } j \\ 0 & \text{if there is no directed edge from vertex } i \text{ to vertex } j \end{cases}$$

Advanced Mathematics: A Transitional Reference, First Edition. Stanley J. Farlow.
© 2020 John Wiley & Sons, Inc. Published 2020 by John Wiley & Sons, Inc.
Companion website: www.wiley.com/go/farlow/advanced-mathematics

For example, the adjacency matrix for the digraph in Figure 5.30 is

$$M = \begin{bmatrix} 0 & 1 & 0 & 0 & 0 & 0 \\ 0 & 0 & 0 & 1 & 0 & 0 \\ 0 & 1 & 0 & 0 & 0 & 0 \\ 1 & 0 & 1 & 0 & 1 & 0 \\ 0 & 1 & 0 & 1 & 0 & 1 \\ 1 & 0 & 0 & 0 & 0 & 0 \end{bmatrix}$$

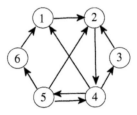

Figure 5.30 A directed graph with directed edges.

5.2.2 Tournament Graphs (Dominance Graphs)

A **tournament graph** (or **dominant graph**) is a directed graph where every pair of vertices is joined by *exactly one* directed edge. In other words, for every pair of vertices u and v, there is a directed edge from u to v or a directed edge from v to u. Figure 5.31 shows a tournament graph with five vertices.

Tournament graphs are aptly named since they model round-robin tournaments in tennis, baseball, and so on, where every team plays every other team exactly once, where we assume no ties.

The number of directed edges "leaving" a vertex u is called the **out-degree** of the vertex and denoted by

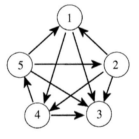

Figure 5.31 Tournament graph with five vertices.

$od(u)$. In the directed graph in Figure 5.31, we have $od(5) = 3$, $od(1) = 2$. If the vertices of a directed graph represent athletic teams, a directed edge from vertex u to vertex v means team u beats team v and the out-degree of u represents the number of wins for team u. On the other hand, the number of directed edges arriving at a vertex v is called **in-degree** of v and denoted by $id(v)$. In the graph in Figure 5.31, we have $id(5) = 1$, $id(1) = 2$. In connection with athletic tournaments, the in-degree of vertex v represents the number of losses for team v.

Tournament graphs are even used by sociologists, who call them dominance graphs, and are used to model social interactions.

5.2.3 Dominance Graphs in Social Networking

Many social situations involve people or groups of people (countries, cultures, universities, and so on) in which some individuals or groups "dominates" others. The word "dominate" can refer to a wide variety of dominance: cultural, physical, political, economic, and so on. Nowadays, online social networking services such as Facebook, Twitter, LinkedIn, and so on, bring people together and offer interesting dynamics between individuals.

Suppose a sociologist wishes to study dominance patterns in a close-knit group of college women. The group consists of five students: Amy (A), Betty (B), Carol (C), Denise (D), and Elaine (E). The sociologist begins by conducting interviews with each pair to determine their pair-wise dominance.[1] If it has been determined that person A dominates person B by some metric, we denote this by writing A → B. (We assume in this simple model that for each pair of students, one person dominates the other.) After conducting the interviews, the sociologist draws the dominance graph that represents pair-wise dominations of the entire group, which is shown in Figure 5.32. Note that Amy dominates Betty and that Denise dominates Carol.

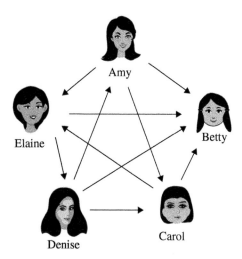

Figure 5.32 Dominance graph.

The adjacency matrix for the dominance graph is drawn in Figure 5.33.

The number of one's in each row is the out-degree of the row vertex and represents the number of **first-stage dominances** of that individual. Note that Amy and Denise each have a "score" of three, Carol and Elaine each have a score of two, and Betty has a score of zero. In other words, Amy and Denise are tied after the first round.

---

1 The determination of one-on-one dominance can be carried out by a series of questions, although it can be subjective in some instances.

$$M = \begin{array}{c|ccccc} & \text{Amy} & \text{Betty} & \text{Carol} & \text{Denise} & \text{Elaine} \\ \hline \text{Amy} & 0 & 1 & 1 & 0 & 1 \\ \text{Betty} & 0 & 0 & 0 & 0 & 0 \\ \text{Carol} & 0 & 1 & 0 & 0 & 1 \\ \text{Denise} & 1 & 1 & 1 & 0 & 0 \\ \text{Elaine} & 0 & 1 & 0 & 1 & 0 \end{array}$$

Figure 5.33 Adjacency matrix.

The goal is to find the **group leader**.[2] If one person dominates more people than all others in the first round, then that person is declared group leader. However, if two or more people tie in the first round, we resort to **second-stage dominances**. In our example, we have a tie between Amy and Denise, so we move on to the second round.

To understand second-stage dominances, note that Elaine dominates Denise and Denise dominates Amy. Hence, we say that Elaine has a second-stage dominance over Amy and denote this by Elaine → Denise → Amy. To find the number of second-stage dominances, we find the square[3] of the adjacency matrix M^2, which is given in Figure 5.34.

$$M^2 = \begin{array}{c|ccccc} & \text{Amy} & \text{Betty} & \text{Carol} & \text{Denise} & \text{Elaine} \\ \hline \text{Amy} & 0 & 2 & 0 & 1 & 1 \\ \text{Betty} & 0 & 0 & 0 & 0 & 0 \\ \text{Carol} & 0 & 1 & 0 & 1 & 0 \\ \text{Denise} & 0 & 2 & 1 & 0 & 2 \\ \text{Elaine} & 1 & 1 & 1 & 0 & 0 \end{array}$$

Figure 5.34 Second order dominances.

To interpret M^2, note that Amy → Elaine → Denise, which is the only second-stage dominance of Amy over Denise. This fact is indicated by the one in row Amy, column Denise of M^2. Amy has 2 second-stage dominances over Betty (Amy → Carol → Betty) and (Amy → Elaine → Betty), indicated with a two in row Amy, column Betty of M^2. In other words, the elements of M^2 give

---

2 If these were athletic teams playing in a round-robin tournament, we would want to know who should be declared the winner.

3 It is not necessary that the reader knows how to find the square of a matrix. One can simply look at the graph to count the second-order dominances.

the second-order dominances of row members over column members. From M^2 we see that Denise has a total of 5 second-state dominances (sum of the numbers in row Denise) while Amy has four (sum of the numbers in row Amy). Thus, we declare Denise the second-stage group leader.[4]

We could continue by finding third-order dominances, but we stop here and find the sum of the matrices $M + M^2$, whose elements give the total number of first-stage and second-stage dominances of each individual shown in Figure 5.35.

		Amy	Betty	Carol	Denise	Elaine
	Amy	0	3	1	1	2
$M + M^2 =$	Betty	0	0	0	0	0
	Carol	0	2	0	1	1
	Denise	1	3	2	0	2
	Elaine	1	2	1	1	0

Figure 5.35 First and second order dominances.

Here, Amy has a total of 7 first- and second-stage dominances over the members in the group while Denise has 8, so we call Denise the leader of the group.

It is interesting that a group leader in the social network depends on the out-degrees of the vertices, whereas for ranking webpages on the Internet, one focuses on the in-degrees or the number of links pointing to a webpage.

5.2.4 PageRank System

Google's search engine models the Internet as a directed graph where vertices are websites and the edges are links between websites. The strategy behind Google's PageRank system is based on counting the number and quality of links to a website as a way to measure the importance of the site. Consider the tiny Internet of four websites as drawn in Figure 5.36 with several individuals online. The numbers on the edges of the digraph estimate the probability that an individual in the column website will move to the row website.

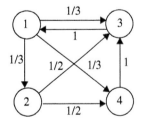

Figure 5.36 Web graph of a small internet.

---

4 Since Amy and Denise both have at least 3 second-stage dominances, we could compute $M^3(G)$, which would give the number of **third-stage dominances**. One suspects, however, that although we might define third-order dominances in theory, it is difficult to observe them in reality.

	Webpage 1	Webpage 2	Webpage 3	Webpage 4
Webpage 1	0	0	1	0
$T=$ Webpage 2	1/3	0	0	0
Webpage 3	1/3	1/2	0	1
Webpage 4	1/3	1/2	0	0

Figure 5.37 Transition matrix.

The digraph in Figure 5.35 gives rise to the **transition matrix** in Figure 5.37.
Note that an individual at website one will probably move to websites two,
three, or four with equal probabilities of 1/3, whereas an individual at website
three will probably move to website one.

Initially, we assume the fraction of individuals at each website is 0.25, which
we represent by the **PageRanks** of the four sites:

$$R_0 = (0.25, 0.25, 0.25, 0.25)$$

The idea is to simulate the movement of online individuals, using the transition
matrix, and estimate the fraction of individuals frequenting each website after a
period of time. To determine this fraction after one "click of the mouse," we compute the product of the transition matrix with the initial PageRank distribution, or

$$R_1 = TR_0 = \begin{bmatrix} 0 & 0 & 1 & 0 \\ 1/3 & 0 & 0 & 0 \\ 1/3 & 1/2 & 0 & 1 \\ 1/3 & 1/2 & 0 & 0 \end{bmatrix} \begin{bmatrix} 0.25 \\ 0.25 \\ 0.25 \\ 0.25 \end{bmatrix} = \begin{bmatrix} 0.25 \\ 0.08 \\ 0.45 \\ 0.21 \end{bmatrix}$$

This new PageRank tells us, probabilistically speaking, that 25% of online individuals will favor website one, 8% website two, 45% website three, and 21% favor
website four. These fractions come from multiplying the fraction of individuals
that prefer a given website times the probability people at that website will move
to the next website. The calculations that yield the values in R_1 are

fraction at website 1 $=(1)(0.25) = 0.25$
fraction at website 2 $=(1/3)(0.25) = 0.08$
fraction at website 3 $=(1/3)(0.25) + (1/2)(0.25) + (1)(0.25) = 0.45$
fraction at website 4 $=(1/3)(0.25) + (1/2)(0.25) = 0.21$

We now continue this process again and again, finding

$$R_2 = \begin{bmatrix} 0.45 \\ 0.08 \\ 0.33 \\ 0.12 \end{bmatrix}, R_3 = \begin{bmatrix} 0.33 \\ 0.16 \\ 0.32 \\ 0.19 \end{bmatrix}, R_4 = \begin{bmatrix} 0.32 \\ 0.12 \\ 0.37 \\ 0.19 \end{bmatrix}, \ldots, R_{100} = \begin{bmatrix} 0.35 \\ 0.13 \\ 0.35 \\ 0.17 \end{bmatrix}$$

which in the long run we interpret that 35% of individuals will be on webpage 1, 13% on webpage 2, 35% on webpage 3, and 17% on webpage 4. This ordering gives search engines a way of measuring the popularity and importance of website pages.

> **Important Note** As the reader might suspect, Google's actual PageRank system includes several bell and whistles in addition to this basic description. The mathematics behind Google's search engine is a **Markov Chain,** which is a probabilistic model that describes the movement of individuals through the internet. The states of the Markov Chain are the websites and Google wants to know the steady state of the Markov Chain, which determines the popularity of websites.

5.2.5   Dynamic Programming

Many applications of directed graphs relate to the movement of objects – be it cars, trucks, airplanes, or even Internet packets from one location to another, and the objects being moved involve costs. Figure 5.38 shows a directed graph which represents a collection of possible paths from the "start" to the "end" with numbers on the edges representing the distance of traversing the edge. The problem is to find the path that minimizes the total distance of going from the start to the end.

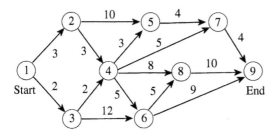

Figure 5.38 Minimum path problem.

Although a quick examination of this small graph will convince you the minimum path is 1 – 3 – 4 – 7 – 9, resulting in a minimum distance of 13, for larger graphs with maybe 1000 nodes a visual inspection of the graph would probably provide no useful information about the shortest path.

To solve this problem, we use a powerful technique called **dynamic programming**, introduced by Richard Bellman in the 1950s. The general philosophy of the technique is to subdivide complex problems into smaller parts, solve the component parts and use these results to solve the larger problem.

In the current problem, the strategy is to work backwards, from the end node to the starting node. We begin by introducing the quantity

s_i = shortest distance from vertex i to the end

In the graph in Figure 5.38, we have $s_9 = 0$, $s_7 = 4$, $s_6 = 9$, $s_8 = 10$. Our goal is to find s_1 We now let

d_{ij} = distance from vertex i to vertex j

which means $d_{12} = 3$, $d_{46} = 5$, and so on. Hence, $d_{ij} + s_j$ = distance from vertex i to vertex j plus the smallest distance from j to the end.

To find the shortest distance s_i from vertex i to the end, we compute the minimum of

$$s_i = \min_{j} \left\{ d_{ij} + s_j \right\}$$

taken over all vertices j connected to vertex i. For example, to find the shortest distance s_4 from vertex $i = 4$ to the end, we find the minimum of the following four distances

Table 5.1 Finding minimum distance to the end.

Vertex	s_i = minimum path to end from vertex i	Path
9	$s_9 = 0$	
8	$s_8 = d_{89} + s_9 = 10 + 0 = 10$	$8 \rightarrow 9$
7	$s_7 = d_{79} + s_9 = 4 + 0 = 4$	$7 \rightarrow 9$
6	$s_6 = \min \begin{cases} d_{68} + s_8 \\ d_{69} + s_9 \end{cases} = \min \begin{cases} 5 + 10 \\ 9 + 0 \end{cases} = 9$	$6 \rightarrow 9$
5	$s_5 = d_{57} + s_7 = 4 + 4 = 8$	$5 \rightarrow 7$
4	$s_4 = \min \begin{cases} d_{45} + s_5 = 3 + 8 = 11 \\ d_{46} + s_6 = 5 + 9 = 14 \\ d_{47} + s_7 = 5 + 4 = 9 \\ d_{48} + s_8 = 8 + 10 = 18 \end{cases} = 9$	$4 \rightarrow 7$
3	$s_3 = \min \begin{cases} d_{34} + s_4 \\ d_{36} + s_6 \end{cases} = \min \begin{cases} 2 + 9 = 11 \\ 12 + 9 = 21 \end{cases} = 11$	$3 \rightarrow 4$
2	$s_2 = \min \begin{cases} d_{24} + s_4 \\ d_{25} + s_5 \end{cases} = \min \begin{cases} 3 + 9 \\ 10 + 8 \end{cases} = 12$	$2 \rightarrow 4$
1	$s_1 = \min \begin{cases} d_{12} + s_2 \\ d_{13} + s_3 \end{cases} = \min \begin{cases} 3 + 12 \\ 2 + 11 \end{cases} = 13$	$1 \rightarrow 3$

$$4 \text{ to } 5 \text{ distance} + \min \text{distance home from } 5 = d_{45} + s_5 = 3 + 8 = 11$$
$$4 \text{ to } 6 \text{ distance} + \min \text{distance home from } 6 = d_{46} + s_6 = 5 + 9 = 14$$
$$4 \text{ to } 7 \text{ distance} + \min \text{distance home from } 7 = d_{47} + s_7 = 5 + 4 = 9$$
$$4 \text{ to } 8 \text{ distance} + \min \text{distance home from } 8 = d_{48} + s_8 = 8 + 10 = 18$$

Taking the minimum of these subproblems, the shortest distance from vertex four to the end vertex is $s_4 = 9$. We continue this process, moving backward in the graph and finding the minimum distances $s_8, s_7, \ldots, s_1$ from each vertex, although not necessarily descending in perfect order. We now advise the reader to get out pencil and paper compute the minimum distance to the end from each vertex, eventually finding s_1, the shortest distance from the starting vertex to the final vertex. The computations in Table 5.1 show the shortest distant s_i to the end from every vertex i in the graph.

Hence, the minimum distance from vertex 1 to vertex 9 is 13 and retracing the path from start to end we find path that gives the shortest distance is $1 \rightarrow 3 \rightarrow 4 \rightarrow 7 \rightarrow 9$.

Problems

1. Problems 1–6 find the adjacency matrix M of the given digraph. Then compute M^2 and $M + M^2$ and verify that the elements of these matrices agree with the number of dominations in the graphs.

Figure 5.39 Find the adjacency matrix.

2.

Figure 5.40 Find the adjacency matrix.

3.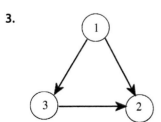

Figure 5.41 Find the adjacency matrix.

4.

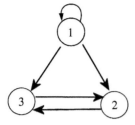

Figure 5.42 Find the adjacency matrix.

5.

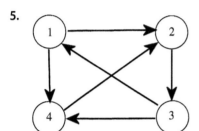

Figure 5.43 Find the adjacency matrix.

6.

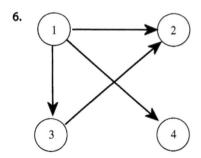

Figure 5.44 Find the adjacency matrix.

7. **Group Dominance**

The graph shown in Figure 5.45 shows the dominance of a group of four classmates: Amy, Betty, Charlie, and Denise.

a) Construct the adjacency matrix M for this graph.

b) Is there a first-stage dominance leader?

c) Compute M^2 and interpret its elements.

d) Who is the group leader?

e) Which person is dominated by the most other people?

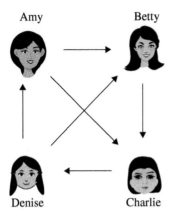

Amy   Betty

Denise   Charlie

Figure 5.45 Group dominance.

8. Round-Robin Tournaments
The graph in Figure 5.46 shows the results of a round-robin tournament for five baseball teams.

Round-robin tournament graph

a) Construct the adjacency matrix M for this graph.
b) Is there a consensus leader for the group?
c) Compute M^2 and interpret its elements.
d) Which team is the conference winner?

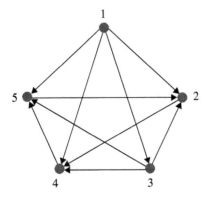

Figure 5.46 Round robin tournament.

9. **Landau's Theorem**

   A theorem by Landau states that if some vertex u in a dominance graph has a larger out-degree than all other vertices, then either u dominates all other vertices v, or if it does *not* dominate a given vertex v, then u dominates a third vertex w which in turn dominates v. What does the theorem say in the language of round-robin tournaments? Verify this theorem for the dominance graphs in Problems 1–6.

10. **Landau's Theorem in the Yankee Conference**

    Suppose the football teams in the Yankee Conference play every other team exactly once during the season. At the end of the season, Maine has won more games than any other team. However, Maine lost to Vermont. What does Landau's theorem say in the language of the Yankee Conference?

11. **Dynamic Programming**

    Use dynamic programming to find the shortest way to accomplish the project in Figure 5.47.

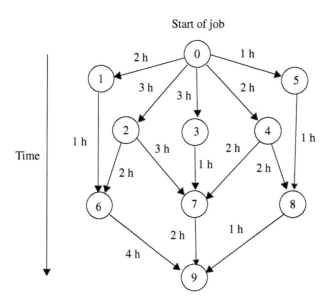

Figure 5.47 Shortest time problem.

12. Ranking Webpages

The dominance graph in Figure 5.48 illustrates a tiny Internet of four webpages where the vertices represent the webpages and the directed edges represent links from one webpage to another. Rank the webpages from first to last. Note that node 1 moves to nodes 2, 3, and 4, each with probability 1/3, whereas node 2 moves to nodes 3 and 4, each with probability 1/2.

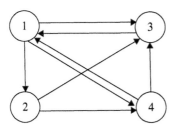

Figure 5.48 Ranking webpages.

13. Game Time on a Directed Graph

Sixteen objects are placed on a table. At each play of the game, each player selects either 1, 2, 3, or 4 objects from the pile. The players alternate turns and the person who takes the last object (or objects) from the table wins. Draw a directed graph of 17 vertices labeled from 0 to 16 illustrating the number of objects on the table and directed edges between vertices that give the best possible move at each vertex. The player who starts the game will always win if the right strategy is used. Find the different paths through the graph from vertex 16 to vertex 0 what will ensure the first player wins. What is the strategy?

14. Internet Research

There is a wealth of information related to topics introduced in this section just waiting for curious minds. Try aiming your favorite search engine toward *applications of directed graphs, Google's PageRank system, directed graphs in social rankings,* and *directed graphs in athletic rankings.*

5.3

Geometric Topology

Purpose of Section   To introduce the idea of topological equivalence between figures (i.e. curves, surfaces, …) and the idea of a homeomorphism between them. We show how the Euler characteristic, previously defined for connected, planar graphs and convex polyhedra, comes into its own as an aid in determining if two figures are topologically equivalent.

5.3.1   Introduction

It is instructive to compare a geometer with a topologist. For a geometer, the pattern of ridges on your fingers and palms are too amorphous for serious analysis. There are no lines, circles, or other geometric shapes which geometers love. On the other hand, a topologist is not restricted to the rigid shapes of Euclidean geometry, but studies more general patterns and classifies them according to specific topological rules. See Figure 5.49.

When a topologist looks at the ridges on your fingers and palms, the topologist sees nearly parallel curves, which is nature's way of preferring order and continuity, but when ridges collide things get complicated. Ridges come together in a variety of interesting, often unexpected ways, and it is the job of a topologist to classify the ways this can occur. You may not have had an interest in topology at this stage in your life, but the ridges on your fingers provide a lesson in basic topology.

The study of topology is not just about vertices and edges studied in the previous two sections on graphs, or even about one-dimensional fingerprints, but extends to surfaces in higher dimensions 2, 3, 4, …, where the standard rules of Euclidean geometry are relaxed, yielding an exciting new geometry, what some call "rubber-sheet" geometry.

Advanced Mathematics: A Transitional Reference, First Edition. Stanley J. Farlow.
© 2020 John Wiley & Sons, Inc. Published 2020 by John Wiley & Sons, Inc.
Companion website: www.wiley.com/go/farlow/advanced-mathematics

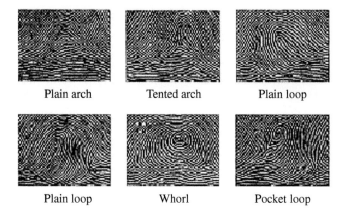

Plain arch	Tented arch	Plain loop
Plain loop	Whorl	Pocket loop

Figure 5.49 Topological fingerprints.

When we are told topology is concerned with properties of surfaces, curves, knots, and the intricacies of geometric objects, the first thing that comes to mind is *what*? Physical objects are so commonplace to our experience that it is hard to imagine anyone telling us something that is not familiar. That said, take a strip of paper and give one of the ends a half-twist, then tape the ends together creating a loop. Now, take a second strip, but now give the ends a complete twist rather than a half-twist. Now take a pencil and draw a line down the center of each of the two strips. Anyone not familiar with this simple experiment will be surprised with the result. The strip given a complete twist results in a simple loop as expected, but the strip given a half-twist does not result in a loop at all, but a one-sided surface called a Möbius strip. Now take a pair of scissors and cut the Möbius strip down the middle. Does your lifetime experiences with surfaces predict what happens? Are you left with two smaller Möbius strips? Or maybe you get one long two-sided Möbius strip. Now try cutting the Möbius strip, not down the middle of the strip, but one-third the way from an edge. Do you know what happens now? Each time it is different. You can carry out your own experiments to find out if it is not intuitively clear.

While the Möbius strip is mesmerizing, probably its greatest impact has been to spur interest in topological concepts. Imagine if you can a surface made from rubber that can be stretched, bent, twisted, and deformed in any way one pleases, but just do not tear it. Although the shape can change in many ways, there are properties that remain unchanged. In plane geometry, one is concerned with distance, curvature, angle, and so on, but in topology one forgoes those more rigid geometric properties with the aim of discovering what one could argue are more fundamental properties of an object. Properties that remain unchanged under a continuous deformation are called **topological properties**, and we say two objects are **topologically equivalent** if each object can be stretched, bend or continuously deformed into the other. Like the iconic doughnut and coffee cup, they may not be geometrically equivalent in the sense of Euclid, but they are **topologically equivalent** in the sense of topology since with a little time and effort one can pull, stretch, kneed one into the other (see Figure 5.50).

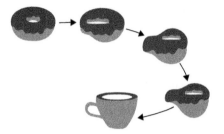

Figure 5.50 The iconic doughnut and coffee cup equivalence.

We have often heard the story of the poor topologist who cannot distinguish a doughnut from a simple coffee cup, but the story is false and the joke is on everyone else. The topologist *can* tell the difference, it is just that the topologist does not care. Topology relaxes the rigid rules of geometry and seeks out new properties of objects remain invariant under more general transformations. The topologist discards the old geometric tools of ruler and protractor, replacing them with the new tool of *continuity* (see Figure 5.51).

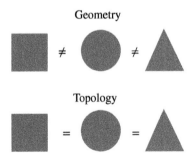

Figure 5.51 Topologically equivalent objects.

We go about our daily lives not giving much thought about many fundamental difference between objects. Look around you and think about some qualitative difference between objects. Figure 5.52 shows a few typical household objects with some fundamental differences between them that interest topologists.

The previous discussion lead us to the formal definition of a homeomorphism and homeomorphic (or topologically equivalent) objects.

Definition   A mapping $f: A \rightarrow B$ between sets is called a **homeomorphism** if it is a one-to-one correspondence between A and B that is continuous and has a continuous inverse. The sets A to B are then called **homeomorphic**, or **topologically equivalent**, which is denoted by $A \approx B$. Intuitively, a homeomorphism is a continuous stretching and bending of an object into a new shape.

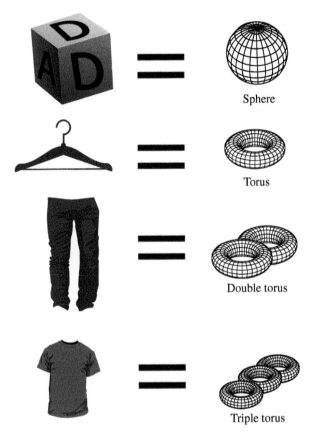

Figure 5.52 Household objects with topological similarities.

5.3.2 Topological Equivalent Objects

Example 1   Topological Equivalent Letters

The letters of the alphabet can be treated as graphs like the ones studied in the Section 5.1 that can be stretched and bent. However, no matter how much stretching and bending of the letters, there are two properties that never change (see Table 5.2).

One property that remains invariant is the property of having a given number of "holes" (or faces in the language of graphs) like in the letter, like "Q" will always have one hole in it, while B always has two, and M will never contain a hole. Another property of the letters that never changes is the number of lines connecting certain points on the letter, or vertices in the language of graphs. For example the letter X contains a point (or vertex) where the letter meets itself

Table 5.2 Topologically homeomorphic letters.

Topological invariants	Homeomorphic letters
no holes, no tails	{C, G, I, J, L, M, N, S, U, V, W, Z}
no holes, three tails	{E, F, T, Y}
no holes, four tails	{X}
one hole, no tails	{D,O}
one hole, one tail	{P,Q}
one hole, two tails	{A, R}
two holes, no tail	{B}
no holes, four tails	{H, K}

four times (a degree four vertex), and no matter how much stretching and bending, the letter X it will always have exactly one point that connects to four lines. Table 5.2 groups the alphabet into equivalence classes according to holes and 'tails' of the letters adjacent to a vertex.

> **Important Note** Topology is mostly the study of position without any regard to distance and angle. It is concerned with connections between objects, much like we saw when we studied graphs in the previous sections. Although topology is studied for its own sake, like many areas of pure mathematics, many of its ideas apply to the real world. Topological chemistry is an active area of research helping to understand molecular structures.

Although many objects can be seen to be topologically equivalent visually, an analytic solution is preference if possible.

Example 2   Open Real Intervals

There exists a homeomorphism between a bounded and unbounded set. Suppose

$$f(x) = \frac{1}{x}$$

This function is a homeomorphism between the bounded set (0, 1) and (1, ∞). It is interesting to think that we are able to "stretch" the bounded set (0, 1) to the set (1, ∞) of infinite length.

5.3.3   Homeomorphisms as Equivalence Relations

In Section 3.3, we saw that an equivalence relation partitions a set into equivalence classes, where the members of each equivalence class share a common property. We now show the relation of being homeomorphic between two sets is an equivalence relation, thus enabling one to partition surfaces, curves, and other objects into distinct equivalence classes whose members share a common property.

Theorem 1   The relation of two objects being topologically equivalent is an equivalence relation.

Proof
We must show the relation is reflexive, symmetric, and transitive.

Reflexive: A set is topologically equivalent with itself since the identity map $f(x) = x$ satisfies the conditions of a homeomorphism.

Symmetric: If sets A and B are topologically equivalent, this implies there exists a one-to-one correspondence f from A to B that is continuous with a continuous inverse, which in turn implies that f^{-1} exists and is also a one-to-one correspondence from B to A with a continuous inverse. Hence, the relation of two sets being homeomorphic is symmetric.

Transitive: If A and B are topologically equivalent, then there exists a continuous, one-to-one correspondence from $g : A \rightarrow B$ with a continuous inverse g^{-1}. If B is topologically equivalent to C, then there exists a continuous, one-to-one correspondence between $f : B \rightarrow C$ that has a inverse f^{-1}. These two facts imply that the composition

$$(f \circ g)(x) = f[g(x)]$$

also satisfies the conditions of a homeomorphism and hence, the relation of two sets being homeomorphic is transitive. ∎

5.3.4   Topological Invariants

A major problem in geometric topology is to determine if two given objects are topologically equivalent. To show two objects are topologically equivalent it suffices to find just one homeomorphism between them. However, to prove that two objects are not homeomorphic, there is no special function that identifies nonhomeomorphic objects. To show that two objects are not homeomorphic, the strategy is to look at properties, called **topological invariants**, that are shared by homeomorphic objects. Thus, if two objects do not share a topological invariant, they are not homeomorphic.

One topological invariant of a surface is the property of being able to continuously shrink any simple closed curve on the surface to a single point on the surface. This is true for a sphere, but not true for a torus so we know the sphere is not topologically equivalent to a torus. (A string tied around the inside circle of a torus cannot be shrunk to a point.) (See Figure 5.53).

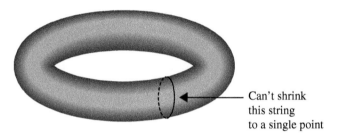

Can't shrink
this string
to a single point

Figure 5.53 Topological test.

Other topological properties are the property of being connected (i.e. not the union of two disjoint pieces), the cardinality of a set, and the number of sides of a surface.

Table 5.3 lists a few topological properties for solid figures and surfaces. The reader may not be familiar with many of them, but should appreciate the fact that topologists have categorized dozens of such properties.

Table 5.3 A few topological properties.

Topological properties of sets in the plane
Path-wise connected
The number of path-wise components
The number of cut points of similar types
Closed and bounded
Connectedness
Euler characteristic
Number of sides of a surface
A curve being simply connected

5.3.5   Euler Characteristic for Graphs, Polyhedra, and Surfaces

We saw in Section 5.1 that a connected, planar graph with v vertices, e edges, and f faces satisfied the relation

$$v - e + f = 2$$

called the Euler characteristic. It is not difficult to visualize that if the graph was drawn on a rubber sheet and no much you stretch or bend the rubber sheet, this number will always be two.

Although the Euler characteristic is valid for planar connected graphs in the plane, its importance is how it relates to polyhedra surfaces. A **polyhedron** is a solid whose surface consists of faces like squares, triangles, pentagons, and so on. Figure 5.44 shows the five famous polyhedra from Euclid's Elements, called the **Platonic Solids**. These solids, the tetrahedron, cube, octahedron, dodecahedron, and icosahedron, along with their planar graph representations, are drawn in Figure 5.54.

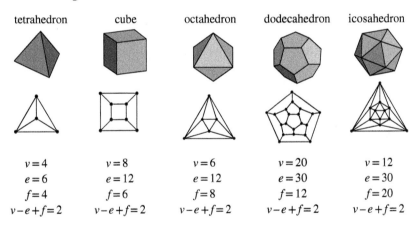

tetrahedron	cube	octahedron	dodecahedron	icosahedron
$v = 4$	$v = 8$	$v = 6$	$v = 20$	$v = 12$
$e = 6$	$e = 12$	$e = 12$	$e = 30$	$e = 30$
$f = 4$	$f = 6$	$f = 8$	$f = 12$	$f = 20$
$v - e + f = 2$	$v - e + f = 2$	$v - e + f = 2$	$v - e + f = 2$	$v - e + f = 2$

Figure 5.54 Platonic solids and their planar graph representations.

Theorem 2   Euler Characteristic for Polyhedra

Every simply connected, convex, polyhedra with v vertices, and e edges, and f faces, satisfies the equation

$$v - e + f = 2$$

known as the Euler characteristic.

Proof

Euler never actually proved his famous equation and many mathematicians at the time tried and failed. It was finally proven in 1811 by 20-year-old Augustin-Louis

Cauchy. Cauchy's idea was to think of a polyhedron as hollow objects and their surfaces made of rubber. He then cut off one of its faces and stretched out what remained onto a flat surface as a planar graph. We illustrate in Figure 5.55 how Cauchy did this starting with a cube.

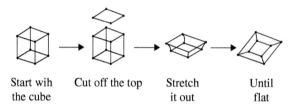

Start wih   Cut off the top   Stretch   Until
the cube   it out   flat

Figure 5.55 Converting a cube to its planar graph.

Observe that after completing the maneuver of cutting off the top of the cube, the topless cube and the resulting planar graph both have 8 vertices, 12 edges, and 5 faces. The five faces of the cube are the result of removing the top face before bending. If we agree to associate this top face with an "outside" face of the planar graph as we did in Section 5.2, then both the cube with its top face and a planar graph with its "outside' face both have six faces. Hence, the cube and its planar representation both have the same number of vertices, edges, and faces. But, as we saw in Section 5.1, the Euler characteristic of a connected, planar graph is 2, and hence the same holds for a cube. Although this proof is not a general proof for arbitrary convex polyhedra, the general proof follows along the same lines. ∎

> **Historical Note**   The word "topology" was first used in a 1847 paper by Johann Listing called *Vorstudien zur Topologie*. Although that paper was not particularly important, he published a more noteworthy paper in 1861 in which he described the one-sided Möbius strip four years before Möbius.

5.3.6   The Euler Characteristic of a Surface

The role of Euler's Characteristic in topology is that it helps determine if surfaces are topologically equivalent. If two surfaces do *not* have the same Euler characteristic, they are *not* topologically equivalent. For example the Euler characteristic of a sphere is two, while the Euler characteristic of a torus (doughnut) is zero, thus the sphere is not topologically equivalent to the torus. It is not hard to imagine that you cannot stretch and bend a sphere into the shape of a torus, or vice-versa, no matter how hard you try.

Mathematicians find the Euler characteristic of a surface by a process of covering the surface with a net of polygons, often triangles consisting of vertices, edges, and faces. To explain how this is done, we demonstrate how a sphere can be "triangulated" with a series of vertices, edges, and faces in the form of a spherical tetrahedron.[1] This surface tetrahedron suggests an inscribed tetrahedron as shown in Figure 5.56, and since the inscribed tetrahedron has Euler characteristic of two, and does not change as the tetrahedron is inflated to the shape of the sphere, we define the Euler characteristic of the sphere as that of the tetrahedron, which is two.

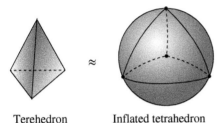

Terehedron   Inflated tetrahedron

Figure 5.56 Finding the Euler characteristic of a sphere.

It is not necessary to use triangles for the net of polygons of the sphere. For example a net of squares leads to a topologically equivalent inscribed cube, which also having Euler characteristic of two and is unchanged as the cube is inflated to the shape of a sphere, thus defining the Euler characteristic of the sphere again as two (see Figure 5.57).

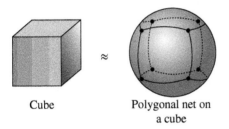

Cube   Polygonal net on
a cube

Figure 5.57 Finding the Euler characteristic of a sphere.

To find the Euler characteristic of the torus, we again construct a net of polygons that will inflate to that of the torus (see Figure 5.58).

We now compute the Euler characteristic by carefully counting the vertices, edges, and faces of the (nonconvex) polygon in Figure 5.48. We cannot really see every side of the polygon in the drawing, but we can still imagine what is on the other side. Counting vertices, edges, and faces, we find the numbers given in Table 5.4:

---

1 In the spherical tetrahedron, the faces are not flat and the edges are not straight.

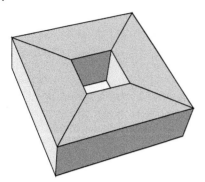

Figure 5.58 Polygon inscribed in a torus.

Table 5.4 Euler characteristic of a torus.

	Top	Bottom	Vertical	Total
Vertices	8	8	0	16
Edges	12	12	8	32
Faces	4	4	8	16

Hence, we have the Euler characteristic of

$$v - e + f = 16 - 32 + 16 = 0$$

Other types of surfaces like multihole tori containing multiple holes are not topologically equivalent to either the sphere or single-hole torus. It turns out that each hole in a sphere decreases its Euler characteristic by two, thus a double-hole torus has Euler characteristic of –2, a triple-hole torus an Euler characteristic of –4, and so on.

Problems

1. **Experimenting with the Möbius Band**
   Take a piece of paper about a foot long and an inch wide. Bring the ends of the paper together and give the ends of the paper a half twist then tape the ends together. You have now created a Möbius strip drawn in Figure 5.59.[2] You are now ready to carry out the following experiments.

---

2 Möbius strips are not that uncommon in the real world. Often, in industrial settings, conveyor belts are given a half twist so each side of the belt wears evenly.

Figure 5.59 Möbius strip.

a) Take a red pencil and color one of the edges of the Möbius strip. Start anywhere on the strip and continue until you arrive back at the starting point. How many edges are there? Surprise, only one edge.

b) Now let us color the sides of the strip. Draw a line down the middle of the strip until you arrive back at the starting point. How many sides are there to the strip? Double surprise.

c) Things just get more curious with the Möbius strip. Now take a pair of scissors and cut the band lengthwise down the middle. What do you think will happen?

d) Now create a second Möbius strip but this time instead of cutting the band down the middle, cut it about one-third the way from one of the edges. The results now are even more surprising.

2. **The Doughnut and Coffee Cup**
The doughnut and the coffee cup are two objects everyone knows are homeomorphic. Can you find other objects that are not homeomorphic to the doughnut or coffee cup? What are some topological properties of these objects that are different from those of the doughnut or coffee cut?

3. **Euler Characteristic Experiment**
Start with a planar graph consisting of only one point and compute the Euler characteristic. Then start adding vertices and edges in any way you please. Note the Euler characteristic will always be 2.

4. **Euler Characteristic for Planar Polygons**
Show that the Euler characteristic $v - e + f$ of each of the polygons in the plane, square, polygon, hexagon, and octagon is one, although if you count the unbounded region outside the polygon as a face, then the Euler Characteristic is 2.

5. **Euler's Formula**
   Carry out the steps that Cauchy used to convert a polyhedron to a planar graph for a tetrahedron and an octahedron.

6. **Truncated Cube**
   A **truncated solid** is a polyhedra with its vertices clipped off drawn in Figure 5.60. Find the number of vertices, edges, and faces of the truncated cube and show that it has an Euler characteristic of two.

Figure 5.60 Truncated cube.

7. **Nonconvex Polyhedra**
   Euler's characteristic does not always have the value of two for nonconvex polyhedra. Show that the nonconvex polyhedra in Figure 5.61 has an Euler characteristic of 3.

Figure 5.61 Nonconvex polyhedra.

8. **Toroidal Polyhedra**
   The (nonconvex) polyhedron drawn in Figure 5.62 called a toroidal polyhedra, which is topologically equivalent to the torus. Show that the Euler

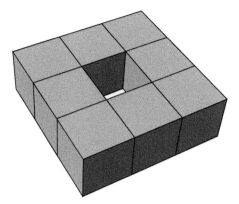

Figure 5.62 Toroidal polyhedra.

characteristic of this polyhedron is zero, thus defining the Euler characteristic for the torus.

9. **Double Torus**

   The (nonconvex) polyhedra shown in Figure 5.63 is topologically equivalent to the double-holed torus. Show that the Euler characteristic of this polyhedron is minus two, thus defining the Euler characteristic for the double-holed torus.

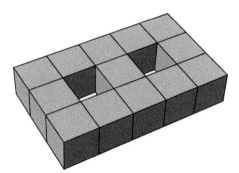

Figure 5.63 Double-holed torus.

10. **Triple-Hole Torus**

    For each additional hole in the sphere, the Euler characteristic decreases by 2. Hence, the sphere has Euler characteristic 2, the single-hole torus has Euler characteristic 0, the double-hole torus −2, and the following triple-hole torus −4. Show that the Euler characteristic of the triple-hole torus drawn in Figure 5.64 is −4 Count carefully.

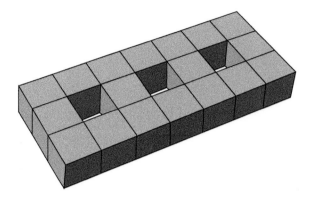

Figure 5.64 Triple-hole torus.

11. **Internet Research**

There is a wealth of information related to topics introduced in this section just waiting for curious minds. Try aiming your favorite search engine toward *Euler's characteristic of different surfaces, homeomorphisms, different topological properties,* and *visualizing topological equivalence.*

5.4

Point-Set Topology on the Real Line

Purpose of Section   To introduce some topological concepts of the real number system, such as open and closed sets, interior, boundary, and exterior points of a set, and limit points. These act as the foundation of many concepts in analysis.

5.4.1   Introduction

There are several areas of topology, such as graph theory and network topology, geometric topology, algebraic topology, differential topology, combinatorial topology, and point-set (or general) topology. Some areas of topology are sufficiently diverse that practicing researchers in one area may have only a passing knowledge of other areas. While graph theory is interested in connections between objects, and geometric topology was interested in shapes of objects, point-set topology is about "closeness." By closeness, we mean closeness of numbers, points, functions, matrices, operators, and other mathematical objects. The interest in closeness lies in the fact it gives rise to limits, infinite series, convergence, continuity, and mathematical concepts associated with calculus, differential equations, real and complex analysis, as well as areas of science and engineering.

Point-set topology can be studied in any dimension, even infinite, but in this section, we restrict our attention to the one-dimensional real line, mostly because it is easier to visualize than in two and three dimensions, and a whole lot simpler than dimensions larger than three.[1] The basic concepts of point-set topology begin with open intervals and the concept of neighborhoods.

---

1 Someone once said, the way to visualize higher dimensions like 4, 5, 6, was to just close your eyes, sit back, and visualize three dimensions.

Advanced Mathematics: A Transitional Reference, First Edition. Stanley J. Farlow.
© 2020 John Wiley & Sons, Inc. Published 2020 by John Wiley & Sons, Inc.
Companion website: www.wiley.com/go/farlow/advanced-mathematics

Definition Let $a \in \mathbb{R}$ and $\delta > 0$. A δ-**neighborhood** of a is the open interval $N_\delta(a) = (a - \delta, a + \delta)$ of radius δ centered at a. Alternate ways of writing this are

$$N_\delta(a) = \{x \in \mathbb{R} : a - \delta < x < a + \delta\} = \{x \in \mathbb{R} : |x - a| < \delta\}.$$

See Figure 5.65.

Figure 5.65 δ-neighborhood.

Important Note Point set topology depends strongly on the ideas of set theory introduced by Georg Cantor in the late 1800s.

This brings us to the unifying concept of this section, and in much of point-set topology, the open set.

Definition A subset of real numbers $A \subseteq \mathbb{R}$ is an **open set** if for every $a \in A$ there exists a $\delta > 0$ such that $N_\delta(a) \subseteq A$. That is

$$A \subseteq \mathbb{R} \text{ open} \Leftrightarrow (\forall a \in A)(\exists \delta > 0)(N_\delta(a) \subseteq A)$$

See Figure 5.66

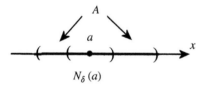

Figure 5.66 Open set.

In the definition of an open set, when we say "there exists a δ greater than zero," we normally are thinking of a small δ, not a large one. In plain language, a set is open if you can "wiggle" any point in the set around, and if you do not wiggle it around too much it will still be in the set. Still another way of thinking about open sets is that every point in an open set is surrounded by other points in the set, i.e. every point in the set is insulated from the outside.

Note Because one often thinks of real numbers as points on the number line, we often refer to real numbers as "points."

Example 1 Open Sets
The following subsets of the real numbers are open sets:

a) Open intervals $(a, b) = \{x \in \mathbb{R} : a < x < b\}$ are open sets.
b) Unbounded intervals of the following form are open sets.

$$(a, \infty) = \{x \in \mathbb{R} : x > a\}$$
$$(-\infty, b) = \{x \in \mathbb{R} : x < b\}$$

What does it mean for a set *not* to be open? To answer that question, we negate[2] the definition of an open set and find

$$A \subseteq \mathbb{R} \text{ open} \Leftrightarrow (\forall a \in A)(\exists \delta > 0)(N_\delta(a) \subseteq A)$$
$$A \subseteq \mathbb{R} \text{ not open} \Leftrightarrow (\exists a \in A)(\forall \delta > 0)(N_\delta(a) \nsubseteq A)$$

In other words, a set is not open if there exists at least one point $a \in A$ right on the boundary of A and not insulated from the outside by other points of A. The interval $(a, b]$ is not open since the point $x = b$ cannot be "wiggled" *any* amount without finding itself outside $(a, b]$.

The most important properties of open sets relate to their union and intersection.

Theorem 1 Main Theorem of Open Sets

- The union of *any* number of open sets, finite or infinite (countable or uncountable infinite), is an open set.[3]
- The intersection of any *finite* number of open sets is an open set.

Proof
Union of Open Sets Is Open
We show that if $\{A_\alpha\}_{\alpha \in \Delta}$ is an arbitrary family of open sets, then

$$\bigcup_{\alpha \in \Delta} A_\alpha$$

---

2 You can see the benefit of the predicate logic notion that allows one to negate sentences very easily.
3 All sets in this section are subsets of the real numbers unless otherwise specified.

is an open set. To show this let

$$a \in \bigcup_{\alpha \in \Delta} A_\alpha$$

Hence, a belongs to some neighborhood $N_\delta(a)$ in some member A_β of the family. Hence, we have

$$(\exists \delta > 0) \left(a \in N_\delta(a) \subseteq A_\beta \subseteq \bigcup_{\alpha \in \Delta} A_\alpha \right)$$

which implies that the union of open sets is open. ∎

Finite Intersection of Open Sets Is Open
The intersection of any finite number of open sets is open.

Proof
Let $\{A_k\}_{k=1}^n$, be a finite family of open sets. To show that the intersection

$$\bigcap_{k=1}^n A_k$$

is open. To show this, let

$$a \in \bigcap_{k=1}^n A_k$$

Hence, a belongs to some neighborhood $N_{\delta_k}(a) \subseteq A_k$ for each $k = 1, 2, \ldots, n$. In other words,

$$(\forall a \in A_k)(\exists \delta_k > 0)(a \in N_{\delta_k}(a) \subseteq A_k)$$

If we pick[4] $\delta = \min \{\delta_k : k = 1, 2, \ldots, n\} > 0$, we have

$$a \in N_\delta(a) \subseteq \bigcap_{k=1}^n A_k$$

which proves the result. ∎

---

4 It is necessary that the number of open sets be *finite*, else the values of δ_k might *not* have a minimum value.

Important Note Infinite intersections of open sets may not be open. See Figure 5.67.

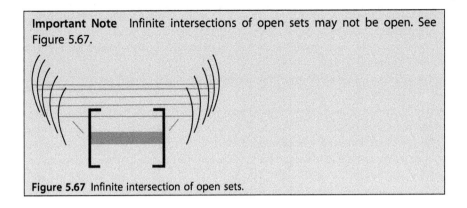

Figure 5.67 Infinite intersection of open sets.

Example 2 Unions and Intersections of Open Sets

a) $\displaystyle\bigcup_{n=1}^{\infty}\left(0, 2-\frac{1}{n}\right) = (0,1) \cup \left(0, \frac{3}{2}\right) \cup \left(0, \frac{5}{3}\right) \cup \cdots = (0,2)$ (open)

b) $\displaystyle\bigcap_{n=1}^{N}\left(0, 1+\frac{1}{n}\right) = (0,2) \cap \left(0, \frac{3}{2}\right) \cap \cdots \left(0, 1+\frac{1}{N}\right) = \left(0, 1+\frac{1}{N}\right)$ (open)

c) $\displaystyle\bigcap_{n=1}^{\infty}\left(0, 1+\frac{1}{n}\right) = (0,2) \cap \left(0, \frac{3}{2}\right) \cap \left(0, \frac{4}{3}\right) \cap \cdots = (0,1]$ (not open)

So what do open sets look like in general? Basically, they are nothing more than unions of open intervals, although the union can be infinite, countable, and even uncountable. We state without proof the general characterization of open sets.

Theorem 2 Characterization of Open Sets
A set $A \subseteq \mathbb{R}$ of real numbers is open if and only if A is an open interval or a finite or countable union of disjoint open intervals.

Closed Sets
The concept of open sets leads us to what might be called the opposite of an open set, a *closed* set.

Definition A set $A \subseteq \mathbb{R}$ is **closed** if and only if its complement $\bar{A}$ is open.

Example 3 Closed Sets
a) An interval like $A = [a, b]$, which is often the domain for functions in calculus is closed since its complement $\bar{A} = (-\infty, a) \cup (b, \infty)$ is open. For example, the intervals $[0, 1]$, $[-2, 3]$ are closed sets.
b) The unbounded intervals $A = [a, \infty)$, $B = (-\infty, b]$ are also closed since their compliments $\bar{A} = (-\infty, a)$, $\bar{B} = (b, \infty)$ are open. For example, $[0, \infty)$, $(-\infty, 0]$ are closed sets.

c) Any singleton set $\{a\}$ consisting of a single real number is a closed set since its complement $(-\infty, a) \cup (a, \infty)$ is open. In fact, any finite set $\{a_1, a_2, \dots, a_n\}$ is closed since its complement is the union of open intervals, which by Theorem 1 is open.

d) The natural numbers $\mathbb{N}$ and the integers $\mathbb{Z}$ are both closed sets.

Keep in mind not all sets are open or closed; the sets $A = (0, 1]$ and $B = [-3, 2)$ are neither open nor closed.

The set of real numbers $\mathbb{R}$ is open from the definition of an open set, and hence its complement, the empty set $\emptyset$ is closed. But the empty set is also open vacuously by definition since there is no point $a \in \emptyset$ to check the condition $a \in N_\delta(a) \subseteq \emptyset$. But if the empty set is open, that means $\mathbb{R}$ is closed. This means $\mathbb{R}$ and $\emptyset$ are *both* open and closed sets. In fact, they are the *only* sets of real numbers that are both open and closed. All other sets are either open, closed, or neither.

We have seen that the union of an arbitrary number of open sets is open and the intersection of a finite number of open sets is open. We now ask if there are corresponding properties for closed sets. The following theorem answers this question which shows the "dual" nature of these properties.

Theorem 3   Main Theorem of Closed Sets
- The intersection of an arbitrary collection of closed sets is closed.
- The union of a *finite* number of closed sets is closed.

Proof
The proof is based on DeMorgan's laws whose proof is left for the reader. See Problem 13. ∎

Important Note   The infinite union of closed sets may not be closed. See Figure 5.68.

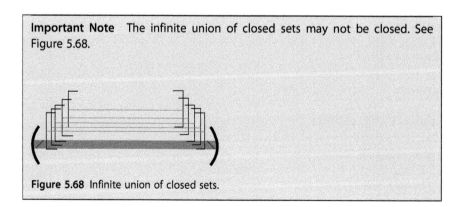

Figure 5.68 Infinite union of closed sets.

Example 4   Unions and Intersections of Closed Sets

The following examples illustrate that the intersection of an arbitrary number of closed sets is closed, but that the union of closed sets may not be closed, unless it is the union of a finite number of closed sets.

a) $\displaystyle\bigcap_{n=1}^{\infty}\left[0,1+\frac{1}{n}\right] = [0,2]\cap\left[0,\frac{3}{2}\right]\cap\left[0,\frac{4}{3}\right]\cap\cdots = [0,1]$ (closed)

b) $\displaystyle\bigcup_{n=1}^{N}\left[0,2-\frac{1}{n}\right] = [0,1]\cup\left[0,\frac{3}{2}\right]\cup\left[0,\frac{5}{3}\right]\cup\cdots = \left[0,1-\frac{1}{N}\right]$ (closed)

c) $\displaystyle\bigcup_{n=1}^{\infty}\left[0,2-\frac{1}{n}\right] = [0,1]\cup\left[0,\frac{3}{2}\right]\cup\left[0,\frac{5}{3}\right]\cup\cdots = [0,2)$ (not closed)

5.4.2   Interior, Exterior, and Boundary of a Set

Three important concepts of topology are the concepts of interior, exterior, and boundary of a set.[5]

Interior Point of a Set   We define a point $a \in \mathbb{R}$ to be an **interior point** of a set $A \subseteq \mathbb{R}$ if and only if there exists a $\delta > 0$ such that $a \in N_\delta(a) \subseteq A$. We denote the interior points of a set A by Int (A) called the **interior** of the set. In the language of predicate logic:

$$a \in \text{Int}(A) \Leftrightarrow (\exists \delta > 0)(N_\delta(a) \subseteq A)$$

An interior point of A always belongs to A. The interior Int (A) of A is always an open set. See Figure 5.69.

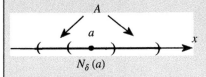

$N_\delta(a)$

Figure 5.69 Interior point.

---

5 We are studying basic topology of the real numbers, which allows us to talk about closeness, convergence, and so on. In general, a **topology** on a set is a family of subsets, called **open sets**, which are closed under unions and finite intersections.

Boundary Point of a Set   A point $a \in \mathbb{R}$ is a **boundary point** of A if and only if for any $\delta > 0$ the δ-neighborhood of a intersects both A and the complement of A. The boundary points of A are denoted by Bdy(A). In the language of predicate logic, we have

$$a \in \mathrm{Bdy}(A) \Leftrightarrow (\forall \delta > 0)[(N_\delta(a) \cap A \neq \emptyset) \wedge (N_\delta(a) \cap \bar{A} \neq \emptyset)].$$

A boundary point of a set A may or may not belong to A. The set of boundary points Bdy(A) of a set is always a closed set. See Figure 5.70.

Boundary points of A

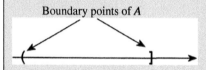

Figure 5.70 Boundary points.

Exterior Point   A point $a \in \mathbb{R}$ is an **exterior point** of a set A if and only if there exists a $\delta > 0$ such that $a \in N_\delta(a) \subseteq \bar{A}$. We denote the set of exterior points, called the **exterior** of A, by Ext (A). In the language of predicate logic, we have

$$a \in \mathrm{Ext}(A) \Leftrightarrow (\exists \delta > 0)(a \in N_\delta(a) \subseteq \bar{A})$$

An exterior point of A will never belong to A. The set of exterior points Ext (A) of a set is always an open set.

Important Note   Intuitively, interior points of a set are points not "right on the edge" of the set. Boundary points are points "right on the edge" of the set, and exterior points the set cannot get "close" to.

5.4.3   Interiors, Boundaries, and Exteriors of Common Sets

Table 5.5 shows the interiors, boundaries, and exteriors of some common sets. Note that the interior and exterior of a set are always open. Also, note that the exterior is the complement of the union of the interior and boundary, and is always closed. Note too that the union of the interior, boundary, and exterior of a set is everything. In other words, the real numbers $\mathbb{R}$.

Table 5.5 Interiors, boundaries, and exteriors of sets.

$A \subseteq \mathbb{R}$	$Int(A) \subseteq A$	$Bdy(A)$	$Ext(A)$
$\varnothing$	$\varnothing$	$\varnothing$	$\mathbb{R}$
(a, b)	(a, b)	$\{a, b\}$	$(-\infty, a) \cup (b, \infty)$
$[a, b]$	(a, b)	$\{a, b\}$	$(-\infty, a) \cup (b, \infty)$
$(a, b]$	(a, b)	$\{a, b\}$	$(-\infty, a) \cup (b, \infty)$
$\{a\}$	$\varnothing$	$\{a\}$	$(-\infty, a) \cup (a, \infty)$
$\{a, b, c\}$	$\varnothing$	$\{a, b, c\}$	$\mathbb{R} - \{a, b, c\}$
$\left\{1, \frac{1}{2}, \frac{1}{3}, \ldots\right\}$	$\varnothing$	$\left\{1, \frac{1}{2}, \frac{1}{3}, \frac{1}{4}, \ldots, 0\right\}$	$\mathbb{R} - \left\{1, \frac{1}{2}, \frac{1}{3}, \ldots, 0\right\}$
$(0, 1) \cup \{2\}$	$(0, 1)$	$\{0, 1, 2\}$	$(-\infty, 0) \cup (1, 2) \cup (2, \infty)$
$(-1, 0) \cup (0, 1)$	$(-1, 0) \cup (0.1)$	$\{-1, 0, 1\}$	$(-\infty, -1) \cup (1, \infty)$
$\mathbb{N}$	$\varnothing$	$\mathbb{N}$	$\mathbb{R} - \mathbb{N}$
$\mathbb{Z}$	$\varnothing$	$\mathbb{Z}$	$\mathbb{R} - \mathbb{Z}$
$\mathbb{Q}$	$\varnothing$	$\mathbb{R}$	$\varnothing$
$\mathbb{R}$	$\mathbb{R}$	$\varnothing$	$\varnothing$

5.4.4 Limit Points

The concept of a limit is fundamental in calculus and analysis. The reader will recall that the two fundamental operations of the calculus, the derivative and integral, are both limits. We can thank the French mathematician Augustin-Louis Cauchy (1789–1857) and German mathematician Karl Weierstrass (1815–1897) for providing rigorous definition of the limit, the so-called epsilon, delta (ε, δ) definition, which allows mathematicians to reason with precision the ideas of calculus and avoid the imprecise reasoning of the past.

Definition A number a is a **limit point** of a set $A \subseteq \mathbb{R}$ if and only if for every $\delta > 0$ the δ-neighborhood of a contains a member of A other than a itself. A limit point of a set may or may not belong to the set. We denote the set of limit points of a set A by Limits (A) (see Figure 5.71).

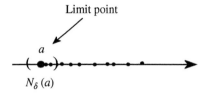

Figure 5.71 Limit point.

> **Important Note** Intuitively, a limit point of a set is a point that can be "approached" by points in the set. In other words, a set likes to "snuggle up" to its limit points.

5.4.5 Closed Sets Contain Their Limit Points

We have seen that a good way to determine if a set is closed is to show that its complement is open. Although this indirect procedure is useful, there is a direct way to determine if a set is closed. The following theorem makes this precise.

Theorem 4 A subset $A \subseteq \mathbb{R}$ is closed if and only if it contains its limit points.

Proof
($\Rightarrow$) Suppose A is closed and let x be a limit point of A. The goal is to show that $x \in A$. Assume the contrary that $x \notin A$ which means $x \in \bar{A}$. But $\bar{A}$ is open and so there exists a neighborhood of x lying completely in $\bar{A}$ which means x is not a limit point of A and contradicts the fact that x was assumed to be a limit point of A. Hence, we have that $x \in A$ meaning A contains its limit points.

($\Leftarrow$) If A contains its limit points, then any $x \in \bar{A}$ in the compliment is not a limit point of A, which means there exists an open neighborhood of x lying in $\bar{A}$, which means $\bar{A}$ is open and hence A is closed. ∎

Table 5.6 gives the limit points of some common sets of real numbers.

5.4.6 Topological Spaces

We saw in Chapter 4 how *relations* (ordering, equivalence, function, ...) can be defined on a set so that the elements of the set can be compared, ordered, and classified. We now introduce a different structure on a set, a topology, which places a general "distance" structure on a set, allowing one to make precise

Table 5.6 Limit points of some common sets.

$A \subseteq \mathbb{R}$	Limit points(A)	Limit points(A) $\subseteq A$?
$\emptyset$	$\emptyset$	Closed
(a, b)	$[a, b]$	Open
$[a, b]$	$[a, b]$	Closed
$(a, b]$	$[a, b]$	Neither open or closed
$\{a\}$	$\emptyset$	Closed
$\{a, b, c\}$	$\emptyset$	Closed
$\left\{1, \frac{1}{2}, \frac{1}{3}, \frac{1}{4}, \ldots\right\}$	$\{0\}$	Neither open or closed
$(0, 1) \cup \{2\}$	$[0, 1]$	Neither open or closed
$[0, 1] \cup \{2\}$	$[0, 1]$	Closed
$\mathbb{N}$	$\emptyset$	Closed
$\mathbb{Z}$	$\emptyset$	Closed
$\mathbb{Q}$	$\mathbb{R}$	Neither open or closed
$\mathbb{R}$	$\mathbb{R}$	Closed

concepts like closeness, limits, and all those concepts used in calculus, like continuous functions, derivatives, and integrals.

We now show how open sets are the foundation of a topological structure on a set, in our case the real numbers $\mathbb{R}$.

Topological Space   A **topological space** (X, J) is a set X together with a collection J of *subsets* of X, the sets in the collection J called **open sets** that satisfy:

1) The union of any collection of sets in J is again a member of J.
2) The intersection of any finite collection of sets in J is again a member of J.
3) Both the empty set and the entire set X belong to J (i.e. are open sets).

Example 5   Usual Topology on $\mathbb{R}$

The most famous topological space (as far as we are concerned) is the pair $(\mathbb{R}, J)$ of real numbers along with the collection J of open subsets of $\mathbb{R}$. We have already seen how open subsets of real numbers satisfy conditions 1), 2), 3) and thus $(\mathbb{R}, J)$

is a topological space, called the topological space of the real numbers with the *usual* topology, the word "usual" meaning open sets with neighborhoods, etc., as discussed in this section. We have seen how open sets give rise to concepts like closeness, interiors, boundaries, exteriors, limits, and so on. It is this *family of open subsets* of $\mathbb{R}$ that "does the job" allowing us to dig deep into the "metrical" structure of the real numbers.[6]

5.4.7 Calculus Without Topology Is No Calculus

Recall that functions $f: \mathbb{R} \to \mathbb{R}$ are the basis of many areas of mathematics, like calculus, which the reader no doubt has some familiarity. If there is no topological structure on $\mathbb{R}$, there are no limits, continuity, differentiable and integral functions, and so on. When a student of calculus thinks about the real numbers, the student most likely is thinking about the topological structure of the real numbers and the distance between points, and so on.[7]

Topology allows one to make precise many of the important concepts of mathematics, one of the most important being continuous functions. We now give the topological definition of continuity.

Definition   A function $f: X \to Y$ is **continuous** if for *any* open set $U \subseteq Y$ its inverse image $f^{-1}(U) \subseteq X$ is an open set in X.

Example 6   Continuity and Open Sets
Show that the function $f: \mathbb{R} \to \mathbb{R}$ defined by $f: x \to x^3$ is continuous.

Proof
We show that the inverse image of any open set in the range of f (i.e. real numbers) is an open set in the domain of f (also the real numbers). To do this let $U \subseteq \mathbb{R}$ be an arbitrary open set in the range of f. The goal is to show $f^{-1}(U) \subseteq \mathbb{R}$ is an open set in the domain of f. Taking an arbitrary $y = x^3 \in U \subseteq \mathbb{R}$, and since U is assumed open, there exists an $\varepsilon > 0$ such that

$$\left(x^3 - \varepsilon, x^3 + \varepsilon\right) \subseteq U$$

---

6 It is an easy matter to extend open sets to higher dimensions, hence topological spaces in the plane, 3-dimensions, and so on.
7 $\mathbb{R}$ is also an *algebraic field* where we can add, subtract, multiply and divide, as well as a *complete ordered field* where numbers are ordered, but now we are thinking about the topological structure of the real numbers.

Students of calculus, who have studied the (ε, δ) method for showing continuity of functions, know that $x \in \mathbb{R}$, the preimage of $f(x) = x^3$, has a δ neighborhood $(x - \delta, x + \delta)$ that maps completely inside $(x^3 - \varepsilon, x^3 + \varepsilon)$. In other words

$$f(x - \delta, x + \delta) \subseteq (x^3 - \varepsilon, x^3 + \varepsilon) \subseteq U$$

or

$$(x - \delta, x + \delta) \subseteq f^{-1}(U)$$

In other words, the inverse image $f^{-1}(U)$ of an arbitrary open set in the range of $f(x) = x^3$ is an open set in its domain. Hence $f: x \to x^3$ is continuous. ∎

Problems

1. Tell if the following sets subsets of $\mathbb{R}$ are open, closed, both, or neither.
   a) $(-1, 0) \cup (0, 1)$
   b) $[0, \infty)$
   c) $(0, \infty)$
   d) $\mathbb{N}$
   e) $\mathbb{Z}$
   f) $\mathbb{Q}$
   g) $A = \{0, 1, 2, \ldots, 100\}$
   h) $\{x : |x - 1| < 3\}$
   i) $\varnothing$
   j) $\left\{ 1, \dfrac{1}{2}, \dfrac{1}{3}, \dfrac{1}{4}, \ldots \right\}$
   k) $\left\{ 1, \dfrac{1}{2}, \dfrac{1}{3}, \dfrac{1}{4}, \ldots \right\} \cup \{0\}$
   l) $\{x : x^2 > 0\}$
   m) $\displaystyle\bigcup_{n=1}^{\infty} \left(\dfrac{1}{n}, 3 - \dfrac{1}{n} \right)$
   n) $\displaystyle\bigcap_{n=1}^{\infty} \left(-\dfrac{1}{n}, \dfrac{1}{n} \right)$
   o) $\displaystyle\bigcup_{n=2}^{\infty} \left[\dfrac{1}{n}, 1 - \dfrac{1}{n} \right]$
   p) $\displaystyle\bigcap_{k=1}^{\infty} \left[0, \dfrac{1}{k} \right]$

2. **Interiors, Boundaries, and Exteriors**
   Fill the blanks in Table 5.7.

Table 5.7 Interiors, boundaries, and exteriors.

	A	Int(A)	Bdy(A)	Ext(A)
a)	$\mathbb{Z}$			
b)	$\{\sin x : 0 < x < 2\pi\}$			
c)	$(0, \infty)$			
d)	$(0, 1) \cup \{2\}$			
e)	$\{1, 2\}$			
f)	$\{\sin x : 0 \le x \le \pi\}$			
g)	$(-1, 0) \cup (0, 1)$			
h)	$\left\{\dfrac{1}{n} : n \in \mathbb{N}\right\} \cup \{0\}$			

3. **True or False**
   Answer true or false about the following sets of real numbers.
   a) A nonempty set can be both open and closed.
   b) A point can lie both in the interior and on the boundary.
   c) Finite sets are always closed.
   d) Infinite sets are always open.
   e) The boundary of a set is always finite.

4. **Mystery Sets**
   Find two sets, which are not equal, but have the same interior, boundary, and exterior.

5. **Finding Examples**
   Find the following sets of real numbers.
   a) Set with two boundary points in the set, one not in the set.
   b) Set with four boundary points in the set, three not in the set.
   c) A set with three boundary points, none in the set.
   d) A set with three boundary points, all in the set.

6. **Finite Sets Closed**
   Show that the finite set $A = \{1, 2\}$ is closed by finding its complement and showing the complement is an open set.

7. **Limit Points**
   If they exist, find the limit points of the following sets.
   a) $\mathbb{N}$
   b) $\mathbb{Q}$
   c) $\mathbb{R}$

d) $(2, 4) \cup (4, 5)$

e) $\{(-1)^n : n \in \mathbb{N}\}$

f) $\varnothing$

g) $\mathbb{Q} \cap (0, 1)$

h) $\left\{ \dfrac{m}{2^n} : m,n \in \mathbb{N} \right\}$

i) $\left\{ m + \dfrac{1}{n} : m,n \in \mathbb{N} \right\}$

8. Closed Sets

A set is closed if and only if it contains its limit points. Find the limit points of the following sets and determine if the sets are closed.

a) $\mathbb{Z}$ **Ans:** Limits $(\mathbb{Z})$ = $\varnothing$, hence, $\mathbb{Z}$ is closed.

b) $\mathbb{Q}$

c) $\mathbb{R}$

d) $(2, 4) \cup (4, 5)$

e) $\{(-1)^n : n \in \mathbb{N}\}$

f) $\varnothing$

g) $\mathbb{Q} \cap (0, 1)$

h) $\left\{ \dfrac{m}{2^n} : m,n \in \mathbb{N} \right\}$

i) $\left\{ m + \dfrac{1}{n} : m,n \in \mathbb{N} \right\}$

9. Examples

Give examples of the following.

a) A bounded set with no limit points.

b) An unbounded set with one limit point.

c) A set with two limit points.

d) An unbounded set whose limit points have cardinality $\aleph_0$.

e) An unbounded with one limit point.

f) An open set with no limit points.

10. Sets and Limits

Find examples of a set A of real numbers with the following properties:

a) A set that is equal to its limit points.

b) A set that is a subset of its limit points.

c) A set that contains all its limit points.

d) A set that is not a subset of its limit points and its limit points are not a subset of the set.

11. Intersections and Unions of Closed Sets

Show the following properties for collections of closed sets.

a) The intersection of any family (finite or infinite) of closed sets is closed.

b) The union of a *finite* number of closed sets is closed.
Hint: Use the properties of open sets and DeMorgan's laws.

12. **Cantor Set**

Let $I = [0, 1]$. Remove the open middle third

$$\left(\frac{1}{3}, \frac{2}{3}\right)$$

and call A_1 the set that remains; that is

$$A_1 = \left[0, \frac{1}{3}\right] \cup \left[\frac{2}{3}, 1\right].$$

Now remove the open third intervals from each of these two parts of A_1, and call the remaining part A_2. Thus

$$A_2 = \left[0, \frac{1}{9}\right] \cup \left[\frac{2}{9}, \frac{1}{3}\right] \cup \left[\frac{2}{3}, \frac{7}{9}\right] \cup \left[\frac{8}{9}, 1\right]$$

Continuing in this manner, remove the open middle third of each segment in A_k and call the remaining set A_{k+1}. Note that we will get

$$A_1 \supset A_2 \supset A_3 \supset \cdots A_k \supset \cdots$$

Continue this process indefinitely, always removing the middle third of existing segments (see Figure 5.72). The limiting set of this infinite process is called the Cantor set, and is defined as

$$C = \bigcap_{k=1}^{\infty} A_k.$$

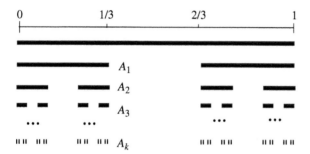

Figure 5.72 Cantor set.

a) Are there any points left in the Cantor set?
b) Show the Cantor set is closed.[8]

13. **Intersection of Open Sets**
Find an example of a family of open sets whose intersection is not open.

14. **Union of Closed Sets**
Find an example of a family of closed sets whose union is not closed.

15. **Topologies on $\{a, b, c\}$**
A topology on a set X is a family of subsets of X that is closed under unions and finite intersections. Show whether the families $T_1, T_2 \subseteq P(X)$ are or are not topologies on $X = \{a, b, c\}$.
a) $T_1 = \{\emptyset, \{a\}, \{a, b\}, \{a, b, c\}\}$
b) $T_2 = \{\emptyset, \{a\}, \{c\}, \{a, b\}, \{a, b, c\}\}$

16. **Continuous Image of an Open Set**
We saw that for a continuous function, the *inverse* image of an open set is open in their respective sets, but that does not mean the image of an open set is open under a continuous mapping. Give an example of an open set in $U \subseteq \mathbb{R}$ and a continuous function $f: \mathbb{R} \to \mathbb{R}$ whose image $f(U) \subseteq \mathbb{R}$ is not open.

17. **Internet Research**
There is a wealth of information related to topics introduced in this section just waiting for curious minds. Try aiming your favorite search engine toward *history of point-set topology, what is point-set topology, types of topology, interior,* and *boundary and exterior of a set.*

---

8 The Cantor set has a variety of interesting mathematical properties. It has no interior, every point in the Cantor set is a limit point. The Cantor set is uncountable, but at the same time has total "length" (measure) of zero.

Chapter 6

Algebra

6.1

Symmetries and Algebraic Systems

Purpose of Section   To introduce the idea of a symmetry of an object in the plane which will act as an introduction to the study of the algebraic group.

6.1.1   Abstraction and Abstract Algebra

The ability to think abstractly is a unique feature of human thought, a capacity not shared by "lower forms" of living creatures.[1] The power to capture the essence of what we experience is so engrained in our mental processes that we never give it a second thought. If the human mind did not have the capability to abstract commonalities in daily living, we would be living in a different world. Imagine lacking the capacity to grasp the "essence" of what makes up a chair. We would be forced to call every chair by a different name in order to communicate to others what we are referring to. The statement "the chair in the living room" would have no meaning unless we knew exactly what chair was being mentioned. Parents point to a picture of a dog in a picture book and tell their one-year-old infant, "dog," and it is a proud moment for the parent when the child sees a strange dog in the yard and says, "dog!"

The concept of number is a crowning achievement of human's ability to abstract the essence of size of sets. It is not necessary to talk about "three people," "three days," "three dogs," and so on. We have abstracted among those things the commonality of *threeness,* so there is no need to say "three goats plus five goats is eight goats," or "three cats plus five cats is eight cats," we simply say three plus five equals eight.

The current chapter is a glimpse into some ideas of what is called *abstract algebra.* Before defining what we mean by an abstract algebra, you should realize

---

1 At least we humans think so.

Advanced Mathematics: A Transitional Reference, First Edition. Stanley J. Farlow.
© 2020 John Wiley & Sons, Inc. Published 2020 by John Wiley & Sons, Inc.
Companion website: www.wiley.com/go/farlow/advanced-mathematics

you have already studied some abstract algebras whether you know it or not. The integers are an example of an abstract algebra, although you probably have never called them abstract or even an algebra. The integers are a set of objects, supplied with binary operations of addition, subtraction, multiplication, and a collection of rules the operations must obey. Abstract algebra *abstracts* the essence of the integers and other mathematical structures, and says, "let's not study just this or that, let's study *all* things which have certain properties of interest." Not all that different from when the infant first says "dog," realizing there are more dogs than just the one in the picture book. Abstract mathematics allows one to think about attributes and relationships, and not focus on specific objects that possess those attributes and relationships.

The benefits of abstraction are many; it uncovers relationships between different areas of mathematics by allowing one to "rise up" above the nuances of a particular area of study and see things from a broader viewpoint, like seeing the forest and not simply the trees, like the saying goes. A disadvantage of abstraction, if there is one, is that abstract concepts are more difficult to grasp and require more "mathematical maturity" before they can be appreciated. It also might be argued that by seeing things from afar, we are unable to get into the nitty-gritty of a discipline. In summary, abstract algebra studies general mathematical structures with given properties, important structures being *groups, rings,* and *fields.*

However, before we start a formal discussion of algebraic groups in the next section, we motivate their study with the introduction of symmetries.

> **Important Note** One hundred years ago when the ideas of abstract algebraic systems were starting to percolate into popular textbooks, the subject was called "modern algebra." However, over the years that term has become out of date, and it is now simply called "algebra," not to be confused with the basic algebra studied in middle and high school.

6.1.2   Symmetries

We are all familiar with symmetrical objects, which we generally think of as objects of beauty, and although you may not be prepared to give a mathematical definition of symmetry, you know one when you see it. Most people would say a square is more symmetrical than a rectangle, and a hexagon more symmetrical than a square, and a circle is the most symmetrical object of all.

Regular patterns and symmetries are known to all cultures and societies. Although we generally think of symmetry in terms of geometric objects, we can also include physical objects as well, like a molecule, the crystalline structure of a mineral, a plant, an animal, the solar system, or even the universe. The concept of symmetry embodies processes like chemical reactions, the scattering of elementary particles, a musical score, the evolution of the solar system, and even mathematical equations. In physics, symmetry has to do with the invariance (i.e. unchanging) of natural laws under space and time transformations. A physical law that has space/time symmetries establishes that the law is independent of translating, rotating, or reflecting the coordinates of the system. The symmetries of a physical system are fundamental to how the system acts and behaves. The equation

$$x^2 + y^2 + z^2 + 3xyz + 1 = 0$$

is symmetric in the three variables x, y, z, since after interchanging any two the equation is unchanged.

Symmetry also plays an important role in calculus. The graph of a real valued function f of a real variable that satisfies $f(x) = f(-x)$ is symmetric about the y-axis, and when $f(x) = -f(-x)$ the graph of the function is symmetric through the origin and unchanged when the graph is rotated $180°$ about the origin.

6.1.3 Symmetries in Two Dimensions

For a single (bounded) figure in two dimensions, there are two types of symmetries.[2] There is symmetry across a line in which one side of the object is the mirror image of its other half. This bilateral symmetry, or the symmetry of left and right, and is common in the structure of many animals, especially humans. This type of symmetry is called **line symmetry** (or **reflective** or **mirror** symmetry). Figure 6.1 shows an isosceles triangle with a line symmetry through its vertical median.

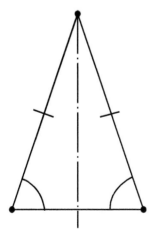

Figure 6.1 Line symmetry.

---

2 We do not include translation symmetries here since we are considering only bounded geometric objects.

We measure the extent to which an object is "symmetric" by counting the number of its symmetries. The parallelogram, rectangle, and square in Figure 6.2 have zero, two, and four lines of symmetry. You can envision yourself other objects that have various lines of symmetry. Chemists are well aware of lines of symmetry of molecules since they relate to how chemical compounds behave.

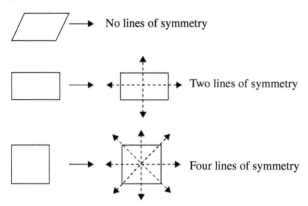

Figure 6.2 Degrees of symmetry.

A second type of symmetry is **rotational** (or **radial**) **symmetry**. An object has a rotational symmetry if the object appears exactly the same when rotated a certain number of times around a central point. The triangle in Figure 6.1 has no rotational symmetry, whereas the three figures in Figure 6.3 have various levels of rotational symmetry. The letter "Z" has no line symmetries, but repeats itself when rotated 0 and 180°,[3] so we say it has rotational symmetry of degree two. The object the middle, again has no line symmetries, but repeats itself when rotated 0, 120, and 240°, so it has rotational symmetries of degree three. Finally, we have the most symmetric planar object of all, the circle which has both an infinite number of rotational and line symmetries.

Figure 6.3 Rotational symmetries.

---

3 It is a convention to call a 0° rotation a rotational symmetry. Hence, all objects have at least one rotational symmetry.

Some objects have both reflective and rotational symmetries as illustrated by the regular polygons[4] in Figure 6.4 which have the same number of reflective and rotational symmetries. The equilateral triangle has three rotational symmetries (rotations of 0°, 120°, and 240° about a center point) and three reflective symmetries through median lines passing through the vertices. A regular polygon with n vertices has n rotation symmetries (each rotation 360/n degrees) and n lines of symmetry.

Equal number of rotation and line symmetries		
Figure	Rotation symmetries	Line symmetries
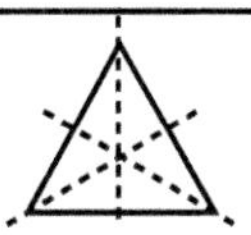	Three rotations 0°, 120°, 240°	Three line symmetries
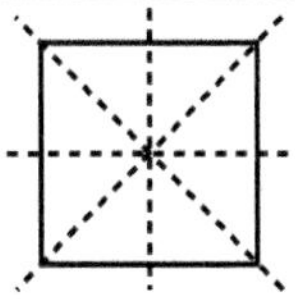	Four rotations 0°, 90°, 180°, 270°	Four line symmetries
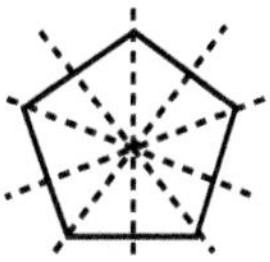	Five rotations 0°, 72°, 144°, 216°, 288°	Five line symmetries
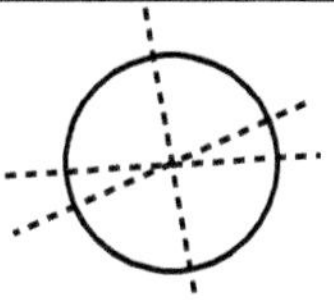	Infinite number of rotation symmetries	Infinite number of line symmetries

Figure 6.4 Figures having both rotational and reflective symmetries.

6.1.4 Symmetry Transformations

Although you can think of symmetries as a property of an object, there is another interpretation of symmetries that is more beneficial for our purposes. A symmetry is a function or mapping or transformation.

---

4 Recall that a regular polygon is a polygon whose sides have the same length and whose angles are the same.

Definition   A one-to-one correspondence (bijection) of an object onto itself is called a **symmetry** if it preserves the shape of the object and the image of the object is indistinguishable from the original object. Visually, the image of an object under a symmetry transformation looks exactly like the original object. See Figure 6.5. The extent of symmetry of an object is measured by the number of transformations that map the object to itself.

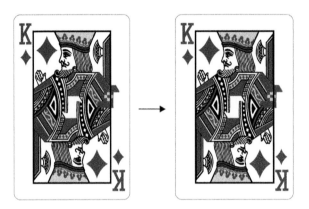

Figure 6.5 Rotation symmetry of 180°.

Important Note   There are three types of symmetries in the plane, rotations, and reflections, and translations. However, if an object is bounded (i.e. inside some circle with finite radius), then a translation is not a symmetry since it alters the location of the object. For our purposes, we are only interested in line and rotation symmetries of bounded objects.

6.1.5   Symmetries of a Rectangle

Figure 6.6 shows a rectangle where the length and width are different and the corners are labeled *A, B, C, D,*

A	B
D	C

Figure 6.6 Four symmetries of a rectangle.

The rectangle has two rotational symmetries of 0°, 180° and two line (or flip) symmetries, where the lines of symmetry are the horizontal and vertical midlines. These four symmetries[5] are illustrated in Figure 6.7.

Motion	Symbol	First position		Final position	
No motion	e	A B D C		A B D C	
Rotate 180°	R_{180}	A B D C		C D B A	
Flip over horizontal median	H	A B D C		D C A B	
Flip over vertical median	V	A B D C		B A C D	

Figure 6.7 Four symmetries of a rectangle.

So what do these symmetries have to do with algebraic structures, which is the focus of this chapter? Since a symmetry is a transformation that maps the points of an object back onto itself, we can define the product of two symmetries as the composition of two symmetries and the result will be a new symmetry. If each symmetry leaves the object unchanged so does the composition of two symmetries. Hence, the composition of symmetries defines a *product* of two symmetries, just like the product of two numbers $2 \times 3 = 6$ getting a new number.

If we perform a 180° rotation,[6] denoted by R_{180}, followed by H, a flip through the horizontal midline, we denote this composition or product by $R_{180} H$ reading left to right. The net result (product) of these two symmetry operations is illustrated in Figure 6.8, and is the same as performing the single symmetry V, a flip through the vertical midline. Hence, we write the product $R_{180} H = V$. It is

---

5 Note that the two rotation symmetries keep the orientation the same (letters ABCD go around clockwise), while the reflection change the orientation where ABCD go around counterclockwise. Also note that to perform the reflections, one must move the two-dimensional rectangle into three dimensions to perform the operation.

6 It is our convention that all rotations are done counterclockwise.

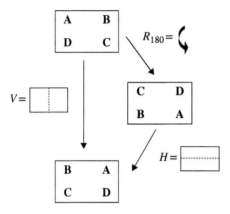

Figure 6.8 Product of symmetries $R_{180}H = V$.

important to note that symmetry operations are carried out from "left-to-right" in the product notation.

Note that the "do nothing" symmetry e (rotation of zero degrees), called the **identity symmetry**, is analogous to the number one in the multiplication of integers. Note that some operations like e, R_{180}, V, H return the figure to its original position after operating two times. For that reason, we say these symmetries are equal to their inverse, which we illustrate by the equations in Table 6.1.

Table 6.1 Typical symmetry products.

Symmetry	Symmetry products
e	$ee = e \Rightarrow e = e^{-1}$
H	$HH = e \Rightarrow H = H^{-1}$
V	$VV = e \Rightarrow V = V^{-1}$
R_{180}	$R_{180}R_{180} = e \Rightarrow R_{180} = R_{180}^{-1}$

Since the composition of symmetries, which we call products, yields a new symmetry, we can construct multiplication tables for symmetries, much like we did in grade school for multiplication tables of numbers. The multiplication table for symmetries of a rectangle is shown in Figure 6.9. The product of two symmetries lies at the intersection of the row and column symmetries, where the row symmetry is carried out first. For example, the intersection of the row labeled R_{180} and column labeled H is V, which means $R_{180}H = V$. The borders of the cells containing the identity symmetry e are darkened as an aid in reading the table.

	e	R_{180}	H	V
e	e	R_{180}	H	V
R_{180}	R_{180}	e	V	H
H	H	V	e	R_{180}
V	V	H	R_{180}	e

Figure 6.9 Multiplication table for symmetries of a rectangle.

6.1.6 Observations

1) Every row and column of the multiplication table contains one and exactly one of the four symmetries. It is a Latin square.
2) The main diagonal contains the identity symmetry e, which means every symmetry is its own inverse.
3) The table is symmetric about the main diagonal which means the multiplication of symmetries is **commutative**. In other words, $AB = BA$, just like multiplication of numbers. We call this a **commutative** algebraic system.
4) The four symmetries e, R_{180}, V, H along with their product as defined by the table, forms what is called an **algebraic group**. Observe how this system is analogous to the integers with the operation of addition, with some similarities and some differences.

> **Important Note** Symmetries in physics are different from the symmetries of geometric objects introduced in this section. In physics, a (continuous) symmetry of a physical system refers to a feature of a physical system that is unchanged under some transformations. The importance of symmetries in physics became evident after the 1918 paper by German mathematician Emmy Noether, who proved than symmetries of physical systems correspond to conservation laws. Research at the most advanced level of theoretical physics is interested in the symmetries of physical systems.

6.1.7 Symmetries of an Equilateral Triangle

We now examine the equilateral triangle drawn in Figure 6.10. It is "more symmetric" than the triangle drawn in Figure 6.1 that had one line symmetry. The equilateral triangle has three rotational symmetries where the triangle is rotated 0°, 120°, 240° about its center, and three line symmetries where the triangle is reflected through lines passing through vertices as drawn as dotted line segments.

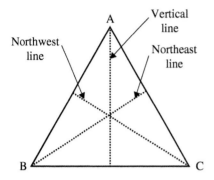

Figure 6.10 Six symmetries of an equilateral triangle.

We denote these symmetry mappings by

$$e, R_{120}, R_{240}, F_v, F_{nw}, F_{ne}$$

where

- $e = R_0$ is the identity (rotation by $0°$) symmetry
- R_{120} = counterclockwise rotation of $120°$
- R_{240} = counterclockwise rotation of $240°$
- F_v = flip through the vertical median
- F_{nw} = flip through the northwest median
- F_{ne} = flip through the northeast median

These symmetries are illustrated in Figure 6.11.

As we did for the rectangle, we can construct a multiplication table for the symmetries called a Cayley table shown in Figure 6.12.

Again, we have drawn darker around the identity symmetry $e = R_0$ as an aid in understanding and interpreting its results. We have also shaded the "northeast" and "southwest" blocks in the table as an aid in reading the table.

Important Note Some crystals in chemistry exhibit rotational symmetry but no mirror symmetry. Such shapes are called **enantiomorphic**.

Important Note Some objects have no symmetry (other than the identity map), such as a left-hand glove or the letters G, Q, J. Take a look around you. Most objects have no symmetry at all.

Example 1 Commutative Operations
Are the symmetry operations for the equilateral triangle commutative? In other words, does the order of the symmetries that are performed make a difference to the outcome?

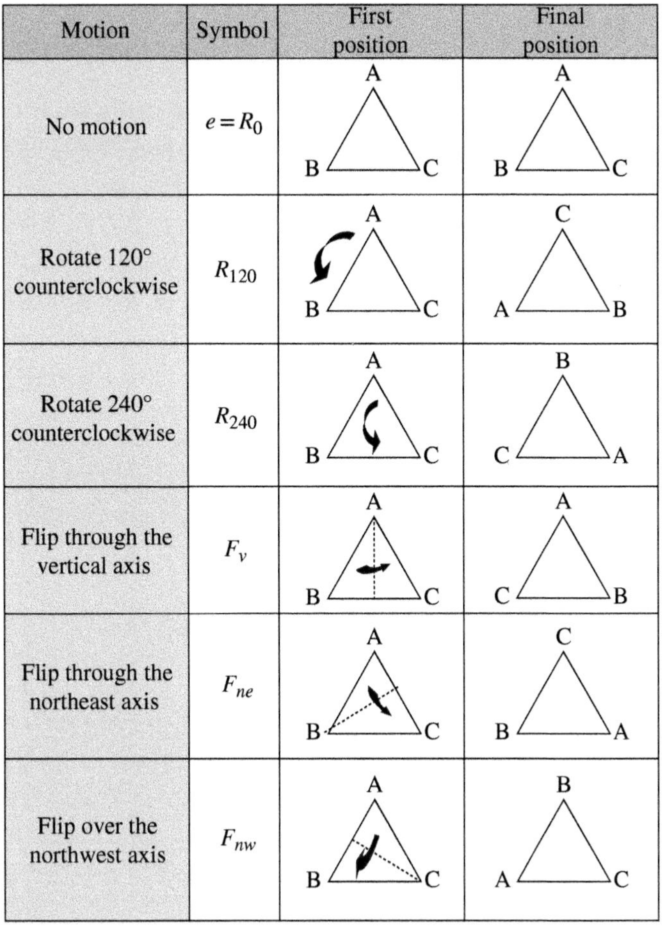

Motion	Symbol	First position	Final position
No motion	$e = R_0$	A / B—C	A / B—C
Rotate 120° counterclockwise	R_{120}	A / B—C	C / A—B
Rotate 240° counterclockwise	R_{240}	A / B—C	B / C—A
Flip through the vertical axis	F_v	A / B—C	A / C—B
Flip through the northeast axis	F_{ne}	A / B—C	C / B—A
Flip over the northwest axis	F_{nw}	A / B—C	B / A—C

Figure 6.11 Six symmetries of an equilateral triangle.

	$e = R_0$	R_{120}	R_{240}	F_v	F_{ne}	F_{nw}
$e = R_0$	R_0	R_{120}	R_{240}	F_v	F_{ne}	F_{nw}
R_{120}	R_{120}	R_{240}	R_0	F_{ne}	F_{nw}	F_v
R_{240}	R_{240}	R_0	R_{120}	F_{nw}	F_v	F_{ne}
F_v	F_v	F_{nw}	F_{ne}	R_0	R_{240}	R_{120}
F_{ne}	F_{ne}	F_v	F_{nw}	R_{120}	R_0	R_{240}
F_{nw}	F_{nw}	F_{ne}	F_v	R_{240}	R_{120}	R_0

Figure 6.12 Cayley table for symmetries of an equilateral triangle.

Solution

We can determine if symmetries are commutative by looking at the products in the multiplication table in Figure 6.12. If the table is symmetric around its diagonal elements, the symmetries are commutative. In this case, the table is not symmetric for all symmetries, so the symmetry operations are not commutative. Note that $F_{nw} F_{ne} \neq F_{ne} F_{nw}$, although $R_{120} R_{240} = R_{240} R_{120} = R_0$.

Example 2 Inverse Symmetries

What is the inverse of each symmetry of the equilateral triangle?

Solution

Note that

$$R_0^2 = F_v^2 = F_{ne}^2 = F_{nw}^2 = R_0 \quad (R_0 = e)$$

which means R_0, F_v, F_{ne}, F_{nv} are their own inverses, which is denoted by

- $R_0^{-1} = R_0$
- $F_v^{-1} = F_v$
- $F_{ne}^{-1} = F_{ne}$
- $F_{nv}^{-1} = F_{nv}$
- $R_{120}^{-1} = R_{240}$
- $R_{240}^{-1} = R_{120}$

The fact that there is exactly one identity symmetry $e = R_0$ in every row and column means that each symmetry has exactly one inverse.

Pure Mathematics   The story is told how Abraham Lincoln, failing to convince his cabinet how their reasoning was faulty, asked them, "How many legs does a cow have?" When they said four, he then continued, "Well then, if a cow's tail was a leg, how many legs does it have?" When they said five, obviously, Lincoln said, "That's where you are wrong. Just calling a tail a leg doesn't make it a leg." This story may be true in the real world, but in the world of *pure mathematics* it is wrong. If you call a cow's tail a leg, then it *is* a leg. In pure mathematics, we care not what things are, only the rules or axioms that govern them.

6.1.8   Rotation Symmetries of Polyhedra

In addition to symmetries in the plane, there are symmetries in higher dimensions that play an important role in many areas of science. For instance in crystallography, which shows how atoms and molecules can be arranged within crystals, chemists are interested in the **symmetry axes** of various polyhedra.

Since polyhedra have vertices (v), edges (e), and faces (f), the symmetry axes can be one of six types vv, ee, ff, ve, vf, ef. A vv symmetry means the axis of symmetry passes through two vertices (vv), whereas a vf symmetry means the axis of symmetry passes through a vertex and an opposite face, and so on.

6.1.9 Rotation Symmetries of a Cube

Table 6.2 shows the 24 rotational symmetries of the cube of the form ff, ee, vv, meaning the axis or rotation always passes through opposite faces, edges, and vertices.

Table 6.2 Rotational symmetries of a cube.

Rotation symmetries of a cube	Symmetry angles	Total symmetries
3 ff symmetry axes	90°, 180°, 270°	9
4 vv symmetry axes	120°, 240°	8
6 ee symmetry axes	180°	6
Identity map	0°	1

These 24 symmetries are visualized in Figure 6.13. The reader can try to visualize these symmetries or obtain a child's block to simulate them.

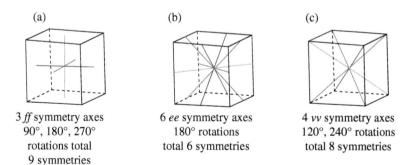

(a)

3 ff symmetry axes
90°, 180°, 270°
rotations total
9 symmetries

(b)

6 ee symmetry axes
180° rotations
total 6 symmetries

(c)

4 vv symmetry axes
120°, 240° rotations
total 8 symmetries

Figure 6.13 Rotational symmetries of a cube.

Problems

1. **Finding Symmetries**
    Determine the symmetries of the letters of the alphabet. The more the symmetries, the more symmetric the letter. Which letter is the most/least symmetric?

    ABCDEFGHIJKLMNOPQRSTUVWXYZ

2. **Drawing Symmetries**
    Draw a figure that has the following symmetries. We do not allow the identity symmetry to be a symmetry in these classifications.
    a) 0 rotational and 0 line symmetries
    b) 1 rotational and 0 line symmetries
    c) 0 rotational and 1 line symmetry
    d) 2 rotational and 0 line symmetries
    e) 1 rotational and 0 line symmetries
    f) 3 rotational and 0 line symmetries

3. **Symmetries of a Parallelogram**
    Describe the symmetries of a parallelogram that is neither a rhombus nor a rectangle.

4. **Symmetries of an Ellipse**
    Describe the symmetries of an ellipse

5. **Representation of D_2 with Matrices**
    Show that the matrices

    $$e = \begin{bmatrix} 1 & 0 \\ 0 & 1 \end{bmatrix}, \quad A = \begin{bmatrix} -1 & 0 \\ 0 & -1 \end{bmatrix}, \quad B = \begin{bmatrix} 0 & 1 \\ 1 & 0 \end{bmatrix}, \quad C = \begin{bmatrix} 0 & -1 \\ -1 & 0 \end{bmatrix}$$

    with operation of matrix multiplication obey the multiplication table in Figure 6.14. This is called the dihedral group D_2.

O	e	A	B	C
e	e	A	B	C
A	A	e	C	B
B	B	C	e	A
C	C	B	A	e

    Figure 6.14 Dihedral multiplication table.

6. **Symmetries of a Square**
   The following matrices define the six symmetries of a square. Find the multiplication table for these symmetries.

$$e = \begin{bmatrix} 1 & 0 \\ 0 & 1 \end{bmatrix}, \quad R_{90} = \begin{bmatrix} 0 & -1 \\ 1 & 0 \end{bmatrix}, \quad R_{180} = \begin{bmatrix} -1 & 0 \\ 0 & -1 \end{bmatrix}, \quad R_{270} = \begin{bmatrix} 0 & 1 \\ -1 & 0 \end{bmatrix}$$

$$V = \begin{bmatrix} -1 & 0 \\ 0 & 1 \end{bmatrix}, \quad H = \begin{bmatrix} 1 & 0 \\ 0 & -1 \end{bmatrix}, \quad F_{ne} = \begin{bmatrix} 0 & 1 \\ 1 & 0 \end{bmatrix}, \quad F_{nw} = \begin{bmatrix} 0 & -1 \\ -1 & 0 \end{bmatrix},$$

7. **Symmetry Groups**
   Find the symmetries of the following letters and make a multiplication table for the symmetries of each letter.
   a) S
   b) T
   c) J

8. **Symmetries of a Tetrahedron**
   Can you find the seven axes of symmetries of the regular tetrahedron with four identical triangular sides as illustrated in Figure 6.15.

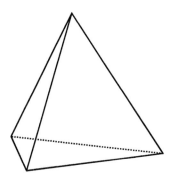

Figure 6.15 Tetrahedron.

9. **Cayley Table for D_3**
   Note that the Cayley table for the symmetries of an equilateral triangle bunched together into four distinct blocks, two blocks consisting of rotations, and two consisting of blocks of flips. From this table, tell if the following statements are true or false.
   a) rotation followed by a rotation is the rotation, i.e. $RR = R$
   b) rotation followed by a flip is a rotation, i.e. $RF = R$

c) rotation followed by a rotation is a flip, i.e. $RR = F$
d) rotation followed by a flip is a flip, i.e. $RF = F$
e) flip followed by a flip is a rotation, i.e. $FF = R$
f) flip followed by a rotation is a rotation, i.e. $FR = R$
g) flip followed by a flip is a flip, i.e. $FF = F$
h) flip followed by a rotation is a flip, i.e. $FR = F$

10. **Symmetries of Solutions of Differential Equations**
The solutions of the differential equation $dy/dx = y$ are the functions of the form $y = ce^x$, where c is an arbitrary constant. Show that the transformation $x' = x + h, y' = y$, where h is an arbitrary real number that maps the set of solutions back into the set of solutions, and hence is a symmetry transformation of the solutions of the differential equation.

11. **Internet Research**
There is a wealth of information related to topics introduced in this section just waiting for curious minds. Try aiming your favorite search engine toward *lines of symmetry* (*look under the image tab*), *rotational symmetry, symmetry in theoretical physics,* and *history of symmetry.*

6.2

Introduction to the Algebraic Group

> **Purpose of Section**   To introduce the mathematical structure of an algebraic group and to illustrate group concepts, we introduce two important groups: the cyclic and dihedral groups.

6.2.1   Basics of a Group

The idea of a symmetry is fundamental in science and its in-depth study requires the mathematical machinery of the algebraic group. Group theory is one of the most useful mathematical tools for mathematicians and scientists alike, including particle physicists, where group theory is the language for understanding the structure of particle physics.

> **Historical Note**   The word *group* was first used by the 20-year-old French genius Evariste Galois in 1830, who wrote his seminal paper on the unsolvability of the fifth order polynomial equation on the night before he was killed in a duel. Although many of the greatest mathematicians at the time did not appreciate Galois's brilliance, a letter Galois wrote eventually ended up in the possession of the French mathematician Joseph Liouville, who published Galois's results in 1846. Over the next 50 years, mathematicians gradually came to understand the genius of Galois' ideas and his work gradually developed into the theory of groups.

The most fundamental component of a group is its binary operation.

6.2.2   Binary Operations and the Group

A **binary operation** on a set A is a rule, which assigns to each pair of elements of A a unique element of A. Thus, a binary operation is a function $f: A \times A \to A$. Two common binary operations familiar to the reader are $+$, $\times$ which assign

Advanced Mathematics: A Transitional Reference, First Edition. Stanley J. Farlow.
© 2020 John Wiley & Sons, Inc. Published 2020 by John Wiley & Sons, Inc.
Companion website: www.wiley.com/go/farlow/advanced-mathematics

the sum $a + b \in \mathbb{R}$ and product $a \times b \in \mathbb{R}$ to a pair $(a, b) \in \mathbb{R} \times \mathbb{R}$ of real numbers. We now give a formal definition of a group.

Definition   An **algebraic group** G is a set of elements with a **binary operation,** often denoted by "$*$," that satisfies the following properties:

- **Closure:** The operation $*$ is a closed operation. That is, if a, $b \in G$, then $a * b \in G.$,
- **Associative:** The operation $*$ is **associative**. That is, for all a, b, $c \in G$, we have $(a * b) * c = a * (b * c)$.
- **Identity:** G has a unique **identity**[1] e. That is, there exists a unique $e \in G$ satisfying $a * e = e * a = a$ for all $a \in G$.
- **Inverse:** Every element $a \in G$ has a unique **inverse.**[2] That is, for every $a \in G$, there exists a unique $a^{-1} \in G$ that satisfies $a * a^{-1} = a^{-1} * a = e$.

We often denote a group G with operation $*$ by $\{G, *\}$.

Often, it happens that $a * b = b * a$ for all a, $b \in G$. When this happens, the group is called a **commutative** (or **Abelian**) group. We also denote the group operation $a * b$ simply as ab, or maybe by $\oplus$ if the group operation is addition or resembles addition. A group is called **finite** if it contains a finite number of elements and the number of elements in the group is called the **order** of the group. If the order of a group G is n, we denote this by writing $|G| = n$. If the group is not of finite order, we say it is of **infinite order**.

In Plain English

- **Associative:** The associative property

$$(a*b)*c = a*(b*c)$$

says that when three elements a, b, $c \in G$ (keeping them in the same order) are combined, the result is the same regardless of which two elements are combined first. Although most groups have an associative binary relation, there are important examples in mathematics where binary operations are not associative, including the cross product of vectors in vector analysis, and the difference between two numbers.[3] There is also the hypercomplex number system of octonians, where octonian multiplication is not associative.

---

1 It is not necessary to state that the identity is unique since it can be *proven* that there is only one identity.

2 Again, it can be proven that the inverse is unique so it is not necessary to assume uniqueness.

3 $(a - b) - c \neq a - (b - c)$.

- **Identity:** The identity element of a group is the element $e \in G$ that leaves every element $a \in G$ unchanged when combined with e. In the group of the integers $\mathbb{Z}$ with the binary operation + (addition), the identity is 0 since $a + 0 = 0 + a = a$ for every integer a.
- **Inverse:** The inverse a^{-1} of an element a depends not only on a, but on the identity e. For example the inverse of an integer a with group operation addition + and identity 0, is $-a$ since $a + (-a) = (-a) + a = 0$.

Table 6.3 shows some binary operations of different sets.

Table 6.3 Properties of binary operations.

	Properties of binary operations			
Operation	Associative	Commutative	Identity	Inverse
$\cup$ on $P(A)$	Yes	Yes	Yes	No
$\cap$ on $P(A)$	Yes	Yes	Yes	No
gcd on $\mathbb{N}$	Yes	Yes	No	No
+ on $\mathbb{R}$	Yes	Yes	Yes	Yes
$-$ on $\mathbb{R}$	No	No	No	No
$\times$ on $\mathbb{Q}$	Yes	Yes	Yes	Yes
min on $\mathbb{R}$	Yes	Yes	No	No

Example 1   Group Test

Tell which of the following sets and binary operations define a group.

a) $\{\mathbb{Z}, +\}$
b) $\{\mathbb{Z}, (m + n)/2\}$
c) $\{\mathbb{Z}, -\}$

Solution

a) We leave it to the reader to show $\{\mathbb{Z}, +\}$ is a group.
b) Taking the average of two integers does not always result in an integer. Hence, the group operation is not closed in $\mathbb{Z}$. No need to check the other properties.
c) The integers $\mathbb{Z}$ with the difference operation is not a group since subtraction is not associative, which can be seen from

$$m - (n - p) \neq (m - n) - p.$$

6.2.3 Cayley Table

The binary operation of a finite group can sometimes be illustrated by means of a Cayley table as drawn in Figure 6.16, which shows the products $g_i g_j$ of two members g_i and g_j of the group. It is much like the addition or multiplication tables the reader studied as a child. A Cayley table is an example of a *Latin square*, meaning that every element of the group occurs once and exactly once in every row and column. We examine the Cayley table to learn about the inner workings of a group.

$*$	$g_1 = e$	g_2	g_3	$\cdots$	g_j	$\cdots$
$g_1 = e$	e	g_2	g_3	$\cdots$	g_j	$\cdots$
g_2	g_2	g_2^2	$g_2 g_3$	$\cdots$	$g_2 g_j$	$\cdots$
$\cdots$	$\cdots$	$\cdots$	$\cdots$	$\cdots$	$\cdots$	$\cdots$
g_i	$g_i g_1$	$g_i g_2$	$g_i g_3$	$\cdots$	$g_i g_j$	$\cdots$
$\cdots$	$\cdots$	$\cdots$	$\cdots$	$\cdots$	$\cdots$	$\cdots$

Figure 6.16 Cayley table for a group.

Example 2 Order 2 and Order 3
For the two groups of order 2 and 3 illustrated in Figure 6.17 by their Cayley tables, answer the following questions. These are the only groups of order 2 and 3.

a) Are the groups are commutative?
b) Find the inverse of each element in each group.
c) Show both groups are associative.

(a)

$*$	e	a
e	e	a
a	a	e

(b)

$*$	e	a	b
e	e	a	b
a	a	b	e
b	b	e	a

Figure 6.17 (a) Order 2. (b) Order 3.

Solution
We leave this fun for the reader.

The graphs demonstrated in Figure 2 are the only graphs of order 2 and 3, respectively. For order 4 however, there are two possible groups, one is the Klein four-group and the other the cyclic group of order 4. The group in Example 3 is the cyclic group of order 4. The Klein four-group will come later.

Example 3 Order 4

The set $G = \{a, b, c, d\}$ and binary operation $*$ define the group of order 4 illustrated in Figure 6.18.

a) What is the identity element of the group.
b) Find the inverse of each element.
c) Is the group commutative?
d) Does the associative property $a * (b * c) = (a * b) * c$ hold?

*	a	b	c	d
a	a	b	c	d
b	b	c	d	a
c	c	d	a	b
d	d	a	b	c

Figure 6.18 Order 4 group.

Solution

a) Identity is a since $ab = ba = b$, $ac = ca = c$, $ad = da = d$.
b) $a^{-1} = a$, $b^{-1} = d$, $c^{-1} = c$, $d^{-1} = b$
c) The Cayley table is symmetric so the group is commutative.
d) Yes, $a * (b * c) = a * d = d$ and $(a * b) * c = b * c = d$.

In general, there is no quick way to verify the associative property like there is for the commutative property. One must check *all* possible arrangements to verify the property. On the other hand, if one instance where the associative property fails, then the binary operation $*$ is not associative.

Example 3 illustrates one of the two groups of order 4. We now present the other group of order 4, the Klein four-group.

Example 4 Klein Four-Group

Show that the set of four members $G = \{e, a, b, c\}$ described by the following multiplication table in Figure 6.19 forms a group. This group is called the **Klein**[4] **four-group** and is the symmetry group of a rectangle studied in Section 6.1.

*	e	a	b	c
e	e	a	b	c
a	a	e	c	b
b	b	c	e	a
c	c	b	a	e

Figure 6.19 Klein four-group.

---

4 Felix Klein (1849–1925) was a German geometer of the nineteenth century.

Solution

We verify the following criterion for a group.

Closure: The binary operation is closed since all the members of the multiplication table are members of G.

Identity: The element e is the identity since multiplying an element by e yields the same element.

Inverse: To find the inverse r^{-1} of an element r follow along the row labeled "r" until you get to the group identity e, then the inverse r^{-1} is the column label above e. In the Klein four-group, each element e, a, b, c is its own inverse.

Associativity: The hardest requirement to verify for a group is the associative property, which requires we check $(r * s) * t = r * (s * t)$, where r, s, t can be any of the elements e, a, b, c. Unfortunately, what this means is we must check $4^3 = 48$ equations. The computations that can be simplified by observing the group is commutative. Other shortcuts can be used (as well as computer algebra systems) to shorten the list of elements you must check. For this group, we observe that the group operation $*$ is simply the composition of functions and we can resort to the fact that composition of functions is associative.

Familiar Groups You are familiar with more groups that you probably realize. Table 6.4 shows just a few algebraic groups you might have seen in earlier studies.

Table 6.4 Common groups.

Group	Elements	Operation	Identity	Inverse
$\mathbb{Z}$	$n \in \mathbb{Z}$	Addition	0	$-n$
$\mathbb{Q}+$	$m/n \ m,n > 0$	Multiplication	1	n/m
$\mathbb{Z}_n$	$k \in \{0, 1, 2, \dots, n-1\}$	Addition mod n	0	$n - k$
$\mathbb{R} - \{0\}$	x nonzero real number	Multiplication	1	$1/x$
$\mathbb{R}^2$	$(a, b) \in \mathbb{R}^2$	Vector addition	$(0, 0)$	$(-a, -b)$

> **Important Note** All objects have a symmetry group consisting of the symmetry transformations that leave the shape of the object unchanged. The more elements in the group, the more "symmetric" the object. Chemists are well aware of the symmetry groups of molecules. Interested readers can find the symmetry groups of different molecules online. Water has three members in its symmetry group in addition to the identity symmetry, one rotation and two mirror symmetries.

6.2.4   Cyclic Groups: Modular Arithmetic

The most common and most simple of all groups are the **cyclic groups**, which are well-known to every child who has learned to keep time.

Definition   A **finite cyclic group** $(\mathbb{Z}_n, *)$ of order n is a group that contains an element $g \in Z_n$ called the **generator** of the group, such that

$$\mathbb{Z}_n = \left\{ e, g, g^2, g^3, ..., g^{n-1} \right\}.$$

where the "powers" of g are simply repeated multiplications[5] of g; that is $g^2 = g * g$, $g^3 = g^2 * g$, ... If the group operation is addition, then we would write the group as

$$\mathbb{Z}_n = (e, g, 2g, 3g, ..., (n-1)g)$$

An alternate notation for the finite cyclic group with generator g is $<g>$.

For example, the three rotational symmetries $\{e, R_{120}, R_{240}\}$ of an equilateral triangle form a cyclic group $\mathbb{Z}_3$ with generator $g = R_{120}$ since $R_{120}^2 = R_{240}, R_{120}^3 = e$.

Cyclic groups also describe modular (or clock) arithmetic, which is the arithmetic we perform when keeping time on 12-hour clock. The 12-hour clock leads us to the cyclic group Z_{12} with elements

$$Z_{12} = \{0, 1, 2, 3, 4, 5, 6, 7, 8, 9, 10, 11\}$$

and group operation

$$a \oplus b = (a + b) \bmod 12$$

which is basically the arithmetic you do when keeping time. The "mod 12" simply refers to computing $a \oplus b$ by computing the ordinary sum $(a + b)$, then taking its remainder after dividing by 12. We denote the group operation by $\oplus$ to remind us that the operation is addition, only reduced modulo 12. In clock arithmetic, the equation $2 = (9 + 5) \bmod 12$, which can be interpreted as meaning five hours after 9 p.m. is 2 a.m.

The 12 hours of the clock, 0 through 11, along with the binary operation of addition modulo 12 is an Abelian group of order 12 called the **cyclic group of order 12**, denoted by $(\mathbb{Z}_{12}, \oplus)$. The Cayley table for this group is shown in Figure 6.20.

> **Important Note**   The evolution of group theory has resulted in three main areas of application: (i) the theory of algebraic equations, (ii) number theory, and (iii) geometry. Early researchers in group theory were Joseph-Louis Lagrange (1736–1813), Niels Abel (1802–1829), and Evariste Galois (1811–1832).

---

5 We use the word "multiplication" here, but keep in mind the group operation can mean any binary operation, even addition.

⊕	0	1	2	3	4	5	6	7	8	9	10	11
0	0	1	2	3	4	5	6	7	8	9	10	11
1	1	2	3	4	5	6	7	8	9	10	11	0
2	2	3	4	5	6	7	8	9	10	11	0	1
3	3	4	5	6	7	8	9	10	11	0	1	2
4	4	5	6	7	8	9	10	11	0	1	2	3
5	5	6	7	8	9	10	11	0	1	2	3	4
6	6	7	8	9	10	11	0	1	2	3	4	5
7	7	8	9	10	11	0	1	2	3	4	5	6
8	8	9	10	11	0	1	2	3	4	5	6	7
9	9	10	11	0	1	2	3	4	5	6	7	8
10	10	11	0	1	2	3	4	5	6	7	8	9
11	11	0	1	2	3	4	5	6	7	8	9	10

Figure 6.20 Cayley table for the cyclic group of 12 elements.

Figure 6.21 shows various clocks that give rise to different cyclic groups.

Cyclic groups $\mathbb{Z}_n$		
	Z_2 Cyclic group order 2	<table><tr><td>⊕</td><td>0</td><td>6</td></tr><tr><td>0</td><td>0</td><td>6</td></tr><tr><td>6</td><td>6</td><td>0</td></tr></table>
	Z_3 Cyclic group order 3	<table><tr><td>⊕</td><td>0</td><td>4</td><td>8</td></tr><tr><td>0</td><td>0</td><td>4</td><td>8</td></tr><tr><td>4</td><td>4</td><td>8</td><td>0</td></tr><tr><td>8</td><td>8</td><td>0</td><td>4</td></tr></table>
	Z_4 Cyclic group order 6	<table><tr><td>⊕</td><td>0</td><td>3</td><td>6</td><td>9</td></tr><tr><td>0</td><td>0</td><td>3</td><td>6</td><td>9</td></tr><tr><td>3</td><td>3</td><td>6</td><td>9</td><td>0</td></tr><tr><td>6</td><td>6</td><td>9</td><td>0</td><td>3</td></tr><tr><td>9</td><td>9</td><td>0</td><td>3</td><td>6</td></tr></table>
	$\mathbb{Z}_6$ Cyclic group order 6	<table><tr><td>⊕</td><td>0</td><td>2</td><td>4</td><td>6</td><td>8</td><td>10</td></tr><tr><td>0</td><td>0</td><td>2</td><td>4</td><td>6</td><td>8</td><td>10</td></tr><tr><td>2</td><td>2</td><td>4</td><td>6</td><td>8</td><td>10</td><td>0</td></tr><tr><td>4</td><td>4</td><td>6</td><td>8</td><td>10</td><td>0</td><td>2</td></tr><tr><td>6</td><td>6</td><td>8</td><td>10</td><td>0</td><td>2</td><td>4</td></tr><tr><td>8</td><td>8</td><td>10</td><td>0</td><td>2</td><td>4</td><td>6</td></tr><tr><td>10</td><td>10</td><td>0</td><td>2</td><td>4</td><td>6</td><td>8</td></tr></table>

Figure 6.21 Finite cyclic groups.

Example 5 Relatively Prime Group

Two integers are called **relatively prime** if they have no common factors other than 1, or equivalently, if their greatest common divisor is 1. For example, 4 and 15 are relatively prime, but 4 and 14 are not. The positive integers less than 10 that are relatively prime with 10 are 1, 3, 7, and 9. We call this set U (10) = {1, 3, 7, 9}, and along with the binary operation of multiplication modulo 10, it is a group with the following Cayley table drawn in Figure 6.22. What is interesting is that this is the same group as the numbers 1, 2, 3, and 4 that are relative prime to 5 with multiplication modulo 5. Also, since there are only two groups of order 4, this group must be either the cyclic group of order 4 or the Klein four-group. It may not be obvious which group is simply looking at the Cayley table because the members of the group are not arranged in the Cayley table in the same order. However, the group is commutative so that what it indicates is the cyclic group of order 4.

⊗	1	3	7	9
1	1	3	7	9
3	3	9	1	7
7	7	1	9	3
9	9	7	3	1

Figure 6.22 Cayley table for U(10).

Note that the group is Abelian and that $3^{-1} = 7$, $7^{-1} = 3$ since $3 \times 7 = 1$.

6.2.5 Isomorphic Groups: Groups that are the Same

Sometimes groups appear different when looking at their Cayley tables, but after relabeling their elements, one discovers they are the same group. For example, consider two groups illustrated in Figure 6.23.

⊕	0	1	2	3
0	0	1	2	3
1	1	2	3	0
2	2	3	0	1
3	3	0	1	2

×	1	i	−1	−i
1	1	i	−1	−i
i	i	−1	−i	1
−1	−1	−i	1	i
−i	−i	1	i	−1

Figure 6.23 Isomorphic groups.

The group at the left is the cyclic group $\mathbb{Z}_4$ = {0, 1, 2, 3} and the group at the right consists of the four numbers G = {1, i, − 1, − i} which lie on the unit circle in the complex plane, where the group operation is multiplication (see Figure 6.24).

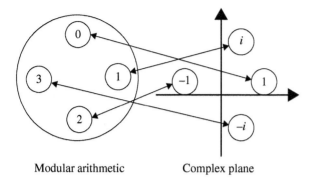

Modular arithmetic   Complex plane

Figure 6.24 Isomorphism.

Looking carefully at the two Cayley tables in Table 6.4, you will see that the numbers 0, 1, 2, 3 in the table for $\mathbb{Z}_4$ are arranged in the same order as the numbers 1, i, -1, $-i$ in the table for G. If we make the correspondences

$$0 \leftrightarrow 1, \quad 1 \leftrightarrow i, \quad 2 \leftrightarrow -1, \quad 3 \leftrightarrow -i, \quad \oplus \leftrightarrow \times$$

we see that the groups $\mathbb{Z}_4$ and G are the same, they simply use different symbols. When two groups are the same, but only different in their symbols used in their description, the groups are called **isomorphic**. In this example, the described one-to-one correspondence is called an **isomorphism** between the groups. This motivates the following formal definition of an isomorphism.

Definition   Let $\{G_1, *\}$ and $\{G_2, \oplus\}$ be two groups with respective group operations $*$ and $\oplus$. An **isomorphism** $T: G_1 \rightarrow G_2$ from G_1 to G_2 is a one-to-one and onto mapping from G_1 onto G_2 that preserves group operations. That is

$$T(a*b) = T(a) \oplus T(b)$$

for all a, $b \in G_1$. When there exists an isomorphism from one group to another group, the groups are called **isomorphic** (i.e. the same from an abstract point of view). See Figure 6.25.

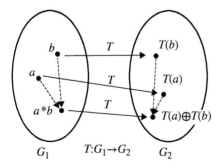

Figure 6.25 Isomorphism.

Roughly speaking, for isomorphic groups, the mapping and the group operations can be carried out in either order. You can operate then map $T(a * b)$, or you can map and then operate, $T(a) \oplus T(b)$, the results are the same.

Example 6   Relatively Prime Group
The group $U(10)$ described in Example 5 of positive integers less than 10 relatively prime to 10, with group operation of multiplication modulo 10, is described by the following Cayley table in Figure 6.26

$\otimes$	1	3	7	9
1	1	3	7	9
3	3	9	1	7
7	7	1	9	3
9	9	7	3	1

Figure 6.26 Relative prime group.

and is isomorphic to the cyclic group $\mathbb{Z}_4 = \{0, 1, 2, 3\}$. We can see this by making the correspondence

$$1 \leftrightarrow 0, \quad 3 \leftrightarrow 1, \quad 9 \leftrightarrow 2, \quad 7 \leftrightarrow 3$$

from which we arrive at the familiar Cayley table for $\mathbb{Z}_4$ drawn in Figure 6.27.

$\otimes$	0	1	2	3
0	0	1	2	3
1	1	2	3	0
2	2	3	0	1
3	3	0	1	2

Figure 6.27 Cyclic group of order 4.

Example 7   An Isomorphism You Know
Let $G_1 = (0, \infty)$ be the positive real numbers with group operation of multiplication, and $G_2 = \mathbb{R}$ the real numbers with group operation of addition. The bijection $T : (0, \infty) \to \mathbb{R}$ defined by

$$T(x) = \log x$$

is an isomorphism since it satisfies

$$T(xy) = \log(xy) = \log x + \log y = T(x) + T(y).$$

The inverse function $T^{-1}(x) = e^x$ from $\mathbb{R}$ to $(0, \infty)$ is also an isomorphism since

$$T(x + y) = e^{x+y} = e^x e^y = T(x)T(y).$$

6.2.6 Dihedral Groups: Symmetries of Regular Polygons

In Section 6.1, we saw how the symmetries of a figure form an arithmetic system where one can "multiply" symmetries much like we multiply numbers. The set of symmetries of an object along with the arithmetic operation of composition of symmetries forms a group of symmetries for the figure. Every figure, no matter how "nonsymmetric," has *one* symmetric group, the trivial symmetry, however, the more symmetric a figure, the larger its symmetry group. A rectangle has four symmetries, whereas the more "symmetrical" square has eight symmetries. Can you find them?

A polygon is called **regular** if all its sides have the same length and all its interior angles are equal. An equilateral triangle is a regular 3-gon, a square is a regular 4-gon, a pentagon a regular 5-gon, and so on. The symmetry group of a regular n-gon has n rotational symmetries and n flip symmetries for a total of $2n$ symmetries. This group is called the **dihedral group** of the n-gon and denoted by D_n Can you find the 10 symmetries of the dihedral group D_5 of the pentagon drawn in Figure 6.28?

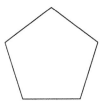

Figure 6.28 Find the symmetric group D_5.

Important Note A cyclic group of order n defines n symmetry rotations of an object about a point. A dihedral group of order $2n$ defines n symmetry rotations of an object about a point, plus n symmetry reflections of the object through a line.

Historical Note At the International Congress of Mathematicians in 1900, the German mathematician David Hilbert posed 23 problems for mathematicians to solve in the next century. Hilbert's 18th problem asked whether crystallographic groups in n dimensions were always finite. The problem was solved in 1910 by German mathematician L. Bieberbach, who proved they are finite in every dimension, however, finding the *number* of these groups is another matter. In $\mathbb{R}^3$, there are 230 symmetry groups; in $\mathbb{R}^4$, there are 4783.

6.2.7 Multiplying Groups

There are only two (nonisomorphic) groups of order 6, the dihedral group D_3 and the cyclic group $\mathbb{Z}_6$. Sometimes groups can be factored into smaller groups somewhat like the number six that can be factored as $6 = 3 \times 2$. The cyclic group $\mathbb{Z}_6$ can be factored as the Cartesian product

$$\mathbb{Z}_6 = \mathbb{Z}_3 \times \mathbb{Z}_2 = \{(m,n) : m \in \mathbb{Z}_3, n \in \mathbb{Z}_2\}$$

of the two smaller cyclic groups $\mathbb{Z}_3$ and $\mathbb{Z}_2$ where if we define the group operation on the Cartesian product as

$$(m_1, n_1) \oplus (m_2, n_2) = ((m_1 + m_2)\bmod 3, (n_1 + n_2)\bmod 2)$$

we arrive at the Cayley table in Figure 6.29.

$\oplus$	(0,0)	(1,1)	(2,0)	(0,1)	(1,0)	(2,1)
(0,0)	(0,0)	(1,1)	(2,0)	(0,1)	(1,0)	(2,1)
(1,1)	(1,1)	(2,0)	(0,1)	(1,0)	(2,1)	(0,0)
(2,0)	(2,0)	(0,1)	(1,0)	(2,1)	(0,0)	(1,1)
(0,1)	(0,1)	(1,0)	(2,1)	(0,0)	(1,1)	(2,0)
(1,0)	(1,0)	(2,1)	(0,0)	(1,1)	(2,0)	(0,1)
(2,1)	(2,1)	(0,0)	(1,1)	(2,0)	(0,1)	(0,0)

Figure 6.29 $\mathbb{Z}_3 \times \mathbb{Z}_2$.

Also, if we make the following identification

$$0 \leftrightarrow (0,0), \quad 1 \leftrightarrow (1,1), \quad 2 \leftrightarrow (2,0), \quad 3 \leftrightarrow (0,1), \quad 4 \leftrightarrow (1,0), \quad 5 \leftrightarrow (2,1)$$

the multiplication table in Figure 6.29 is the multiplication table for the cyclic group $\mathbb{Z}_6$, as shown in Figure 6.30.

$\oplus$	0	1	2	3	4	5
0	0	1	2	3	4	5
1	1	2	3	4	5	0
2	2	3	4	5	0	1
3	3	4	5	0	1	2
4	4	5	0	1	2	3
5	5	0	1	2	3	4

Figure 6.30 Multiplication table for $\mathbb{Z}_6$.

> **Important Note** Groups occur everywhere in nature, science, and mathematics, often as transformations of a set that preserves interesting properties. Chemists are interested in symmetries of molecules and the number and types of symmetries. Physicists study space–time symmetry transformations of the laws of physics, realizing that symmetries are associated with conservation principles.

Problems

1. **Groups?**
   Do the following sets with given binary operations to form a group? If they do, find the identity element and the inverse of each element. If it does not form a group, explain why not.
   a) even integers, addition
   b) {−1, 1}, multiplication
   c) nonzero complex numbers, multiplication
   d) nonzero rational numbers, multiplication
   e) positive rational numbers, multiplication
   f) complex numbers 1, − 1, i, − i, multiplication
   g) positive irrational numbers together with 1, multiplication
   h) integers, subtraction.

2. **Odd Integers Under Addition**
   Give reasons why the odd integers under addition do not form a group?

3. **Finish the Group**
   Complete the following Cayley table in Figure 6.31 for a group of order 3.

*	e	a	b
e	e	a	b
a	a		
b	b		

Figure 6.31 Order 3 Cayley table.

4. **Finish the Group**
   Complete the following Cayley table in Figure 6.32 for a group of order 4 without looking at the Cayley tables of the Klein four-group or the cyclic group of order 4 in the text.

*	e	a	b	c
e	e	a	b	c
a	a			
b	b			
c	c			

Figure 6.32 Klein four-group.

5. **Verification of a Group**
   Do the nonzero integers with the operation of multiplication form a group?

6. **Group You are Well Familiar**
   Show $\{\mathbb{Z}, +\}$ is a group.

7. **Property of a Group**
   Verify that for all elements a, b in a group, the following identity holds.

   $$(ab)^{-1} = b^{-1}a^{-1}.$$

   Hint: Show $(ab)(b^{-1}a^{-1}) = e$.

8. **Heisenberg Group**
   A group that plays an important role in quantum mechanics is the Heisenberg group which consists of all matrices of the form

   $$\begin{pmatrix} 1 & x & y \\ 0 & 1 & z \\ 0 & 0 & 1 \end{pmatrix}$$

   where x, y, z are real numbers, and the group operation is matrix multiplication. Show that this is a group.

9. **Direct Product of Groups**
   Define a group $(\mathbb{Z}_2 \times \mathbb{Z}_2, \oplus)$ consisting of the Cartesian product

   $$\mathbb{Z}_2 \times \mathbb{Z}_2 = \{(a,b) : a,b \in \mathbb{Z}_2\}$$

   and binary operation

   $$(a,b) \oplus (c,d) = ((a+c)\bmod 2, (b+d)\bmod 2)$$

   Find the Cayley table for this group.

10. **Modulo 5 Multiplication**
    Create the multiplication table for the integers 1, 2, 3, and 4, where multiplication is defined as mod(5) arithmetic.

11. **Modulo 4 Multiplication**
    Create the multiplication table for the integers 1, 2, and 3 for modular
    arithmetic mod(4) and show that this does not define a group. In other
    that the numbers 1, 2, ... , $n - 1$ forms a group under mod(n) multiplica-
    tion, it must be true that n is a prime number.

12. **Relative Prime Group $U(10)$**
    For each positive integer n, the set of positive integers 1, 2, ..., n that are
    relatively prime[6] with n is denoted by $U(n)$. For example, $U(10) = \{1, 3, 7,$
    $9\}$. The set $U(n)$ is a group under multiplication modulo n.
    a) Draw the Cayley table for $U(10)$.
    b) Is the group Abelian?
    c) What is the inverse of each element?

13. **Relative Prime Group $U(8)$**
    For each positive integer n, the set of positive integers 1, 2, ..., n that are
    relatively prime with n is denoted by $U(n)$. For example $U(8) = \{1, 3, 5, 7\}$.
    The set $U(n)$ is a group under multiplication modulo n.
    a) Draw the Cayley table for $U(8)$.
    b) Is the group Abelian?
    c) What is the inverse of each element?

14. **Isomorphic Groups**
    Show that the following Group C and Group D in Figure 6.33 are isomor-
    phic by interchanging the third and fourth columns of C, and then the
    third and fourth rows to get the table for D.

(a)

$\oplus$	0	1	2	3
0	0	1	2	3
1	1	2	3	0
2	2	3	0	1
3	3	0	1	2

(b)

$\otimes$	1	2	3	4
1	1	2	3	4
2	2	4	1	3
3	3	1	4	2
4	4	3	2	1

Figure 6.33 (a) Group C. (b) Group D.

___
6 Two numbers are relatively prime if and only if the greatest common divisor of both numbers
is one.

15. Infinite Group

Show that the set of all rational numbers x in the interval $[0, 1)$ form an infinite group if the group operation is defined as

$$x + y = \begin{cases} x+y & \text{if } 0 \le x+y < 1 \\ x+y-1 & \text{if } x+y \ge 1 \end{cases}$$

16. Groups and Latin Squares

The Cayley table for a group forms what is called a **Latin square**. That is, every element of the group occurs exactly once in every row and exactly once in every column. The converse is not true, however, since there are Latin squares that do not form groups. Find a Latin square for the numbers $\{0, 1, 2\}$ that does not form a group.

17. Modular Fun

Compute the following sums and products in the given cyclic group $\mathbb{Z}_n = \{0, 1, 2, \ldots, n-1\}$.

a) $\mathbb{Z}_4 : 1 + 7$ and 3×7
b) $\mathbb{Z}_5 : 9 + 7$ and 5×7
c) $\mathbb{Z}_6 : 10 + 7$ and 2×7
d) $\mathbb{Z}_9 : 11 + 20$ and 3×7
e) $\mathbb{Z}_{10} : 100 + 7$ and 30×10
f) $\mathbb{Z}_{11} : 11 + 11$ and 11×7
g) $\mathbb{Z}_{11} : 12 + 7$ and 10×10
h) $\mathbb{Z}_{15} : 14 + 1$ and 3×6

18. Group $a + b\sqrt{2}$

If a, b are rational numbers, not both zero, then show the set $a + b\sqrt{2}$ forms a group under multiplication

$$\left(a_1 + b_1\sqrt{2} \right)\left(a_2 + b_2\sqrt{2} \right) = (a_1 a_2 + 2b_1 b_2) + (a_1 b_2 + a_2 b_1)\sqrt{2}$$

19. Verification of a Group

Prove that the nonzero real numbers $\mathbb{R}^* = \mathbb{R} - \{0\}$ with binary operation $*$ is a group, where

$$a * b = \frac{1}{2}ab$$

20. Homomorphisms

The most important functions between two groups are homomorphisms, those that "preserve" the group operations. More precisely, a function $f : G \to H$ between two groups is a homomorphism when

$f(xy) = f(x)f(y)$ for all x and y in G

where xy denotes the group operation in G and $f(x)f(y)$ denotes the group operation in H. The following identities describe homomorphisms between two groups, where group operations are understood. What are the groups G, H, and what is the homomorphism between the groups?

a) $c(x + y) = cx + cy$

b) $|xy| = |x||y|$

c) $(xy)^2 = x^2 y^2$

d) $\log_a (xy) = \log_a x + \log_a y$

e) $\sqrt{xy} = \sqrt{x}\sqrt{y}$

21. **Internet Research**

There is a wealth of information related to topics introduced in this section just waiting for curious minds. Try aiming your favorite search engine toward *applications of group theory, history of group theory,* and *examples of groups of different sizes.*

6.3

Permutation Groups

Purpose of Section   To introduce the idea of a permutation of a set and how a composition of permutations can be interpreted as a binary operation yielding a new permutation. This binary operation gives rise to an important group of permutations called the **symmetric group S_n.**

6.3.1   Permutations and Their Products

In Section 2.3, we introduced the concept of a permutation (or arrangement) of a set of objects. We now return to the subject, but now our focus is different. Instead of thinking of a permutation as an arrangement of objects (which it is, of course), we think of a permutation as a one-to-one correspondence (bijection) from a set onto itself. For example, a permutation of elements of the set $A = \{1, 2, 3, \dots, n\}$ can be thought of a one-to-one mapping of the set onto itself, represented by

$$P = \begin{pmatrix} 1 & 2 & \cdots & k & \cdots & n \\ 1^P & 2^P & \cdots & k^P & \cdots & n^P \end{pmatrix}$$

which gives the image k^P of each element $k \in A$ in the first row as the element directly below it in the second row.

For example, a typical permutation of the four elements $A = \{1, 2, 3, 4\}$ is

$$P = \begin{pmatrix} 1 & 2 & 3 & 4 \\ 2 & 3 & 4 & 1 \end{pmatrix}$$

A good way to think about this permutation is to think of a tomato, strawberry, lemon, and apple, arranged from left to right in positions we call 1, 2, 3, and 4. If we apply the permutation P mapping, we get the new arrangement shown in Figure 6.34.

Advanced Mathematics: A Transitional Reference, First Edition. Stanley J. Farlow.
© 2020 John Wiley & Sons, Inc. Published 2020 by John Wiley & Sons, Inc.
Companion website: www.wiley.com/go/farlow/advanced-mathematics

Figure 6.34 Permutation mapping.

The tomato that was originally in position 1 has moved to position 2, the strawberry that was in position 2 has moved to position 3, and so on. The important thing to know is that although the individual items have moved, the positions 1, 2, 3, and 4 *remain the same*.

Another way to think of a permutation is with a directed graph, as drawn in Figure 6.35. Here, we see the movement of the fruits as everything is shifted to the right, and the apple at the end goes to the front of the line. Again, think of the fruits as moving, but the positions are fixed.

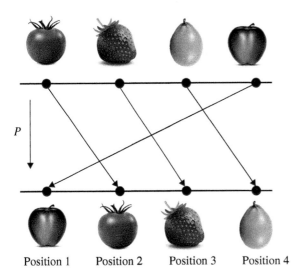

Figure 6.35 Visualization of a permutation.

6.3.1.1 Product of Permutations

Starting with the permutation:

$$P = \begin{pmatrix} 1 & 2 & 3 & 4 \\ 2 & 3 & 4 & 1 \end{pmatrix}$$

suppose we follow this permutation by a second permutation Q. In other words, the composition of the permutation P is followed by a second permutation Q, which gives rise to a "reshuffling of a reshuffling." This leads us to the definition of the product of two permutations.

Definition The composition of permutations P and Q is the **product** of P and Q, and denoted[1] by PQ.

The product of the permutations P and Q is shown in Figure 6.36.

$$PQ = \overset{P}{\begin{pmatrix} 1 & 2 & 3 & 4 \\ \downarrow & & & \\ 2 & 3 & 4 & 1 \end{pmatrix}} \overset{Q}{\begin{pmatrix} 1 & 2 & 3 & 4 \\ 2 & 1 & 4 & 3 \end{pmatrix}} = \overset{PQ}{\begin{pmatrix} 1 & 2 & 3 & 4 \\ 1 & 4 & 3 & 2 \end{pmatrix}}$$

Figure 6.36 Permutation product.

> **History** The idea of a permutation or arrangement of things has received attention in various cultures throughout history. In the *Chinese Book of Changes*, attention is given to "arrangements" of the mystic trigrams. The Greek historian Plutarch writes that the philosopher Xenocrates (350 BCE) computed the number of possible syllables as 1 000 000 000 000, which hints at taking permutations of syllables.

Figure 6.37 illustrates the movement of the four fruits under the action of this product.

1 In compositions of functions $(f \circ g)(x) = f[g(x)]$ we evaluate from "right to left," evaluating the function g first and f second, but here, in the case of permutation functions, we evaluate from "left to right."

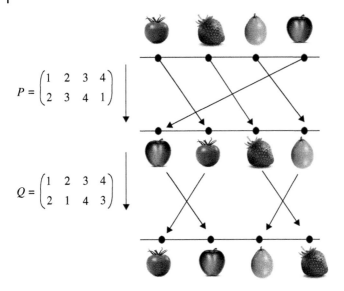

$$P = \begin{pmatrix} 1 & 2 & 3 & 4 \\ 2 & 3 & 4 & 1 \end{pmatrix}$$

$$Q = \begin{pmatrix} 1 & 2 & 3 & 4 \\ 2 & 1 & 4 & 3 \end{pmatrix}$$

Figure 6.37 Product (composition) of two permutations.

A second visualization of this permutation product is shown in Figure 6.38. This time we use Greek symbols.

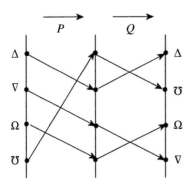

Figure 6.38 Another representation of the product of permutations.

Example 1   Multiplying Permutations
Find the product PQ of

$$P = \begin{pmatrix} 1 & 2 & 3 & 4 \\ 4 & 3 & 2 & 1 \end{pmatrix}, \quad Q = \begin{pmatrix} 1 & 2 & 3 & 4 \\ 2 & 1 & 4 & 3 \end{pmatrix}$$

Solution

We illustrate the product in Figure 6.39. Note that

$$PQ(1) = 3, \quad PQ(2) = 4, \quad PQ(3) = 1, \quad PQ(4) = 2$$

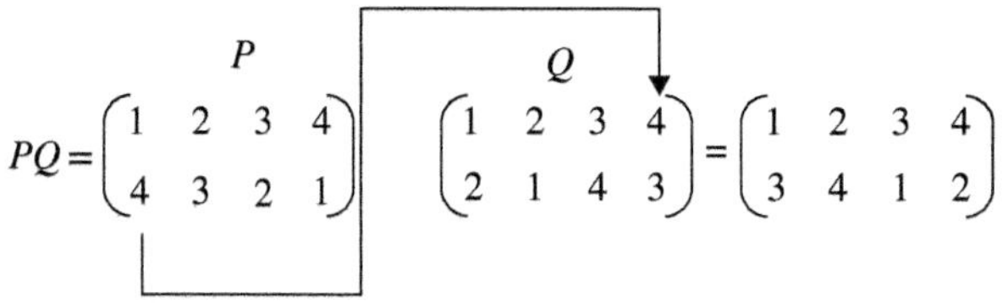

$$PQ = \begin{pmatrix} 1 & 2 & 3 & 4 \\ 4 & 3 & 2 & 1 \end{pmatrix} \begin{pmatrix} 1 & 2 & 3 & 4 \\ 2 & 1 & 4 & 3 \end{pmatrix} = \begin{pmatrix} 1 & 2 & 3 & 4 \\ 3 & 4 & 1 & 2 \end{pmatrix}$$

Figure 6.39 Product of permutations.

The graph illustration of the product is shown in Figure 6.40.

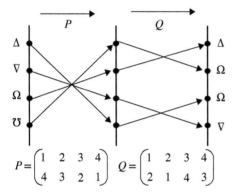

$$P = \begin{pmatrix} 1 & 2 & 3 & 4 \\ 4 & 3 & 2 & 1 \end{pmatrix} \quad Q = \begin{pmatrix} 1 & 2 & 3 & 4 \\ 2 & 1 & 4 & 3 \end{pmatrix}$$

Figure 6.40 Composition of two permutations.

If we carry out the permutations P, Q in Example 1 in the opposite order, we find

$$QP = \begin{pmatrix} 1 & 2 & 3 & 4 \\ 2 & 1 & 4 & 3 \end{pmatrix} \begin{pmatrix} 1 & 2 & 3 & 4 \\ 4 & 3 & 2 & 1 \end{pmatrix} = \begin{pmatrix} 1 & 2 & 3 & 4 \\ 3 & 4 & 1 & 2 \end{pmatrix} = PQ$$

which leads us to believe that it makes no difference in the order the permutations are performed. However, this is not true in general as the following example shows.

$$PQ = \begin{pmatrix} 1 & 2 & 3 \\ 2 & 1 & 3 \end{pmatrix} \begin{pmatrix} 1 & 2 & 3 \\ 3 & 2 & 1 \end{pmatrix} = \begin{pmatrix} 1 & 2 & 3 \\ 2 & 3 & 1 \end{pmatrix}$$

$$QP = \begin{pmatrix} 1 & 2 & 3 \\ 3 & 2 & 1 \end{pmatrix} \begin{pmatrix} 1 & 2 & 3 \\ 2 & 1 & 3 \end{pmatrix} = \begin{pmatrix} 1 & 2 & 3 \\ 3 & 1 & 2 \end{pmatrix}$$

6.3.2   Inverses of Permutations

If a permutation P maps k to k^P, then its **inverse** P^{-1} maps k^P back onto k. In other words, the inverse of a permutation can be found by simply interchanging the top and bottom rows of the permutation P. For convenience in reading however, we reorder the top row 1, 2, n from left to right. For example

$$Q = \begin{pmatrix} 1 & 2 & 3 & 4 \\ 4 & 3 & 1 & 2 \end{pmatrix} \Rightarrow Q^{-1} = \begin{pmatrix} 1 & 2 & 3 & 4 \\ 3 & 4 & 2 & 1 \end{pmatrix}$$

$$P = \begin{pmatrix} 1 & 2 & 3 & 4 \\ 2 & 3 & 4 & 1 \end{pmatrix} \Rightarrow P^{-1} = \begin{pmatrix} 1 & 2 & 3 & 4 \\ 4 & 1 & 2 & 3 \end{pmatrix}$$

The reader can verify that

$$PP^{-1} = QQ^{-1} = \begin{pmatrix} 1 & 2 & 3 & 4 \\ 1 & 2 & 3 & 4 \end{pmatrix}.$$

6.3.3   Cycle Notation for Permutations

A more streamlined way to display permutations is by the use of the **Cauchy cycle** (or **cyclic**) notation. To illustrate how this works, consider the permutation[2]

$$P = \begin{pmatrix} 1 & 2 & 3 & 4 & 5 & 6 \\ 3 & 2 & 5 & 6 & 1 & 4 \end{pmatrix}$$

To write this permutation in cyclic notation, we start at the upper left-hand corner with 1 and write (1 and then follow it with its image $1^P = 3$, that is (13. Next, note that P maps 3–5, so we write (135. Then P maps 5 back to the original 1, so we have our first cycle (135). We then continue on with 2 (next unused element in the first row) and observe that P maps 2 to itself, so we have a 1-cycle (2). Finally, we see that P maps 4–6, so we write (46 and since 6 maps back to 4 we have our final cycle, the 2-cycle (46). Hence, P is written in what is called the **product** of three cycles: a 3-cycle, a 1-cycle, and a 2-cycle,

$$P = (135)(2)(46) = (135)(46)$$

where we dropped the 1-cycle (2) to streamline the notation.

---

[2] Sometimes, only the bottom row of the permutation is given since the first row is ambiguous. Hence, the permutation listed here could be expressed as {325614}.

Example 2   Cycle Notation

Each of the following permutations is displayed in both function and cycle notation. Make sure you can go "both ways" in these equations.

a) $\begin{pmatrix} 1 & 2 & 3 & 4 & 5 & 6 \\ 2 & 4 & 3 & 5 & 6 & 1 \end{pmatrix} = (12456)(3) = (12456)$

b) $\begin{pmatrix} 1 & 2 & 3 & 4 \\ 4 & 3 & 2 & 1 \end{pmatrix} = (14)(23)$

c) $\begin{pmatrix} 1 & 2 & 3 & 4 & 5 \\ 1 & 2 & 4 & 5 & 3 \end{pmatrix} = (1)(2)(345) = (345)$

d) $\begin{pmatrix} 1 & 2 & 3 \\ 3 & 1 & 2 \end{pmatrix} = (132)$

e) $\begin{pmatrix} 1 & 2 & 3 \\ 1 & 2 & 3 \end{pmatrix} = (1)(2)(3) = (\ )$

Note the identity permutation in Example 2^e) is sometimes written as ( ).

Important Note   The cycle notation was introduced by the French mathematician Cauchy in 1815. The notation has the advantage that many properties of permutations can be seen from a glance.

6.3.4   Products of Permutations in Cycle Notation

When computing the product of permutations expressed in cycle notation, one reads from left to right and thinks of each cycle as a function and the entire product as the composition of functions.[3] The process is best explained with an example. Consider the product of four cycles

$$(145)(23)(24)(51) = (1234)(5) = (1234)$$

Starting with 1 in the left-most cycle, we see it maps to 4, whereupon we move to the second cycle that does not contain 4, so we move to the third cycle that maps 4–2, whereupon we move to the last cycle that does not contain a 2, so we conclude

---

3 The "left-to-right" convention for multiplying permutations is **not** universal. Another convention is to write the order the permutations are carried out from right to left and so products are computed from "right-to-left."

that the product 1 maps to 2, whereupon we start the product permutation as (12...). We then carry out the same process starting at the leftmost permutation to find the image of 2, whereupon the first cycle does not contain 2 so we move to the second cycle and see that 2 maps to 3, and since none of the remaining cycles contain 3, we conclude the product maps 2 maps to 3, so we write the product as (123...). Continuing this process, we obtain the final product of (1234).

Example 3   Permutations in Cycle Form
For the set

$$A = \{1,2,3,4,5\}$$

we have the following products.

a) $(12357)(2476) = (147)(2356) = \begin{pmatrix} 1 & 2 & 3 & 4 & 5 & 6 & 7 \\ 4 & 3 & 5 & 7 & 6 & 2 & 1 \end{pmatrix}$

b) $(1234)(1432) = (1)(2)(3)(4) = (\ ) = \begin{pmatrix} 1 & 2 & 3 & 4 \\ 1 & 2 & 3 & 4 \end{pmatrix}$

c) $(1342)^{-1} = (1243) = \begin{pmatrix} 1 & 2 & 3 & 4 \\ 2 & 4 & 1 & 3 \end{pmatrix}$

d) $(14)^{-1} = (14)$ since $(14)(14) = (1)(4) = (\ )$

e) $(125)(34) = (34)(125)$ (cycles commute)

Note that in the inverse permutation, the orientation goes in the opposite direction (i.e., counterclockwise versus clockwise). Although the commutative law does not hold in general for permutations, there are cases where permutations do commute. For example, if two cycles share no common element, then the order can be switched as in the case $(123)(45) = (45)(123)$. However, $(13)(12) \neq (12)(13)$.

Example 4   Permutation Group
The following permutations are written as the product of transpositions.

$$(1234...n) = (12)(13)(14)...(1n)$$

$$(4321) = (43)(42)(41)$$

$$(15324) = (15)(13)(12)(14)$$

Example 5   Products of Cycles
Find the product $(1532)(35)(14)$ of three permutations written in product form.

Ans: The easiest way is to make a directed graph shown in Figure 6.44.

Viewing the results of Figure 6.41, we find the product to be

$$(1532)(35)(14) = \begin{pmatrix} 1 & 2 & 3 & 4 & 5 \\ 3 & 4 & 2 & 1 & 5 \end{pmatrix}$$

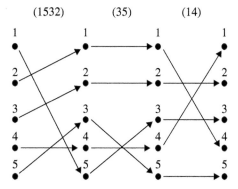

Figure 6.41 Product of cycles.

Using the directed graph in Figure 6.41 as a guide, see if you are able to carry out the multiplication (1532)(35)(14) of these cycle permutations. Think of someone giving three commands one after the other to a group of five people standing in a row, then ask yourself where will each of the people end up. Think of following each person moving from position to position at each command. We see the person starting in location "1" moves to position "5" on the first command, then to position "3" on the second command, and stays at position "3"on the third command.

6.3.5 Transpositions

A permutation that interchanges just two elements of a set and leaves all others unchanged is called a **transposition** (or **2-cycle**). For example

$$\begin{pmatrix} 1 & 2 & 3 & 4 \\ 1 & 4 & 3 & 2 \end{pmatrix} = (24)$$

$$\begin{pmatrix} 1 & 2 & 3 & 4 & 5 \\ 1 & 3 & 2 & 4 & 5 \end{pmatrix} = (23)$$

are transpositions. What may not be obvious is that any permutation can be written as the product of transpositions. In other words, any permutation of elements can be carried out by repeated interchanges of just two elements. For example Figure 6.42 shows four girls lined up from left to right waiting to

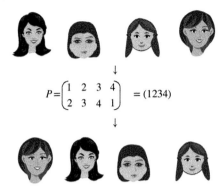

$$P = \begin{pmatrix} 1 & 2 & 3 & 4 \\ 2 & 3 & 4 & 1 \end{pmatrix} = (1234)$$

Figure 6.42 Rotation permutation.

get their picture taken. The photographer asks the three on the left to move one place to their right, and the end girl to move to the left position, which is a result of the following permutation.

The question then arises, is it possible to carry out this maneuver by repeated interchanges of members, two at a time? The answer is yes, and the equation is

$$(1234) = (12)(13)(14)$$

To see how this works, watch how they move in Figure 6.43.

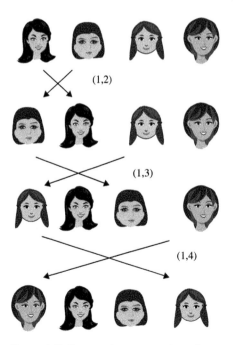

Figure 6.43 Permutation as a product of transpositions.

Example 6   Factoring Permutations as Transpositions
Observe that the following permutation in Figure 6.44 can be expressed as a
series of transpositions.

$$\begin{pmatrix} 1 & 2 & 3 & 4 & 5 & 6 \\ 4 & 3 & 6 & 1 & 5 & 2 \end{pmatrix} = (34)(45)(23)(12)(56)(23)(45)(34)(23)$$

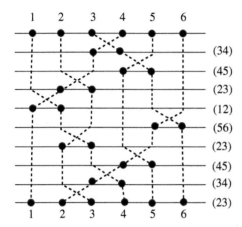

Figure 6.44 Permutation as the product of transpositions.

6.3.6 Symmetric Group S_n

We now show that the set of all permutations of a set forms a group.

Theorem 1   The set of all permutations of the set $A = \{1, 2, 3, \dots, n\}$ with n
members whose group operation is the composition of permutations is a group.
The group is called the **symmetric group** S_n on n elements, and the order of the
group is $|S_n| = n!$.

Proof
To show that the symmetric group is a group, we need to verify the group CIIA
axioms of closure, identity, inverse, and associativity.

   Closure: Each permutation is a one-to-one mapping from $A = \{1, 2, \dots, n\}$
onto itself, so repeated permutations PQ is also a one-to-one mapping of $\{1,
2, \dots, n\}$ onto itself.

   Identity: The permutation that assigns every member to itself serves as the
identity of the group.

   Unique Inverse: Permutations are one-to-one correspondences between
members of the set and itself and so it has a unique inverse.

   Associative Law: The group operation is associative since it is the composition
of functions that are always associative. Hence, the axioms of a group are satisfied. ∎

6.3.7 Symmetric Group S_3

In Section 6.2, we constructed the group of rotational and reflective symmetries of an equilateral triangle called the dihedral group D_3. What we did not realize at the time was that this dihedral group of symmetries is the same as the symmetric group S_3 of all permutations of the three vertices {1, 2, 3} of a triangle. Figure 6.45 illustrates this relationship.

Group of permutations of {1,2,3}	Group of symmetries of an equilateral triangle	Interpretation
$P_1 = \begin{pmatrix} 1 & 2 & 3 \\ 1 & 2 & 3 \end{pmatrix}$ (1)(2)(3)	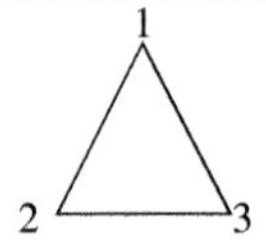	Do nothing
$P_2 = \begin{pmatrix} 1 & 2 & 3 \\ 2 & 3 & 1 \end{pmatrix}$ (123)	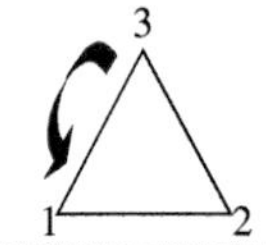	Counterclockwise rotation of 120°
$P_3 = \begin{pmatrix} 1 & 2 & 3 \\ 3 & 1 & 2 \end{pmatrix}$ (132)	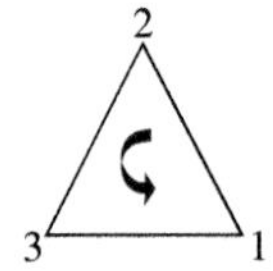	Counterclockwise rotation of 240°
$P_4 = \begin{pmatrix} 1 & 2 & 3 \\ 1 & 3 & 2 \end{pmatrix}$ (23)	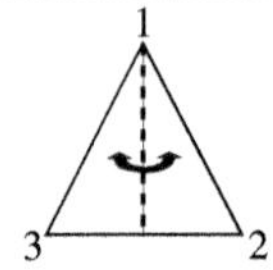	Flip through vertex 1
$P_5 = \begin{pmatrix} 1 & 2 & 3 \\ 3 & 2 & 1 \end{pmatrix}$ (13)	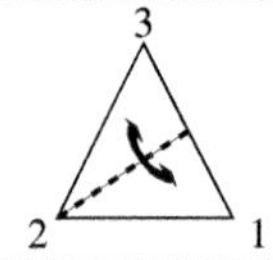	Flip through vertex 2
$P_6 = \begin{pmatrix} 1 & 2 & 3 \\ 2 & 1 & 3 \end{pmatrix}$ (12)	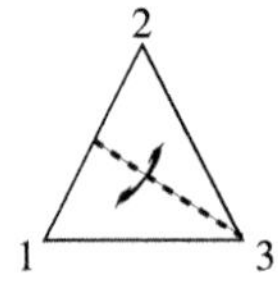	Flip through vertex 3

Figure 6.45 Abstract equivalence of S_3 and D_3.

The Cayley table for the group symmetric group S_3 is shown in Figure 6.46.

PQ		Q					
		$e=()$	(123)	(132)	(12)	(13)	(23)
P	$e=()$	e	(123)	(132)	(12)	(13)	(23)
	(123)	(123)	(132)	e	(23)	(12)	(13)
	(132)	(132)	e	(123)	(13)	(23)	(12)
	(12)	(12)	(13)	(23)	e	(123)	(132)
	(13)	(13)	(23)	(12)	(132)	e	(123)
	(23)	(23)	(12)	(13)	(123)	(132)	e

Figure 6.46 Symmetric group S_3.

6.3.8 Alternating Group

From the Cayley table of S_3 drawn in Figure 6.46, there is an obvious subgroup located at the upper left of the table in the unshaded region called the alternating group A_3 with members

$$A_3 = \{e, (123), (132)\}.$$

The property that characterizes this subgroup may not be obvious at first glance, but when its members are expressed in terms of fundamental transpositions, they all have an even number as seen by

$()$ 0 transpositions (even number)

$(123) = (13)(32)$ 2 transpositions (even number) ,

$(132) = (12)(23)$ 2 transpositions (even number)

whereas the other three permutations of S_3, namely (12), (13), (23) have an odd number of transpositions (namely one). It is a general property of the symmetric group S_n that half the $n!$ permutations have an even number of transpositions, called the even permutations, and half the permutations have an odd number of transpositions, called the odd permutations. The even permutations (which always contain the identity permutation) form an alternating subgroup A_n of order $n!/2$. Many of these alternating groups have interesting properties. The alternating group A_4 of the 12 even permutations of S_4 are the symmetries of the regular tetrahedron.

Problems

1. **Finding Permutations**
    Given the permutations

    $$P = \begin{pmatrix} 1 & 2 & 3 & 4 \\ 2 & 4 & 3 & 1 \end{pmatrix}, \quad Q = \begin{pmatrix} 1 & 2 & 3 & 4 \\ 4 & 2 & 1 & 3 \end{pmatrix}$$

    find:
    a) PQ
    b) P^{-1}
    c) QP^{-1}
    d) P^2
    e) $(PQ)^{-1}$

2. **Permutation Identity**
    For permutations

    $$P = \begin{pmatrix} 1 & 2 & 3 & 4 \\ 4 & 3 & 2 & 1 \end{pmatrix}, \quad Q = \begin{pmatrix} 1 & 2 & 3 & 4 \\ 2 & 1 & 4 & 3 \end{pmatrix}$$

    prove or disprove $(PQ)^{-1} = Q^{-1}P^{-1}$.

3. **Cycle Notation**
    Fill in the blanks in the permutation

    $$P = \begin{pmatrix} 1 & 2 & 3 & 4 & 5 \\ - & - & - & - & - \end{pmatrix}$$

    represented by the following cyclic products.
    a) $(13)(24)$
    b) $(123)(45)$
    c) (1432)
    d) $(1)(2)(35)(4)$
    e) $(135)(42)$

4. **Composition of Permutations**
    Given the permutations:

    $$P = \begin{pmatrix} 1 & 2 & 3 & 4 & 5 \\ 3 & 2 & 4 & 5 & 1 \end{pmatrix}, \quad Q = \begin{pmatrix} 1 & 2 & 3 & 4 & 5 \\ 2 & 4 & 1 & 3 & 5 \end{pmatrix}, \quad R = \begin{pmatrix} 1 & 2 & 3 & 4 & 5 \\ 2 & 5 & 3 & 1 & 4 \end{pmatrix}$$

a) Show that $PQ \neq QP$
b) Verify $(PQ)R = P(QR)$
c) Verify $(PQ)^{-1} = Q^{-1}P^{-1}$

5. **Cycles as the Product of Two-Cycles**
   A two-cycle is an exchange of two elements of a set, such as the permutation (23) of interchanging 2 and 3, leaving the other elements of the set unchanged. Every permutation of a finite set can be written (not uniquely) as the product of two-cycles. Write the permutation (12345) as the product or composition of two-cycles. Take five different objects and put them in a row and verify that your answer is correct by shuffling them in both ways.

6. **Symmetric Group S_2**
   Given the set $A = \{1, 2\}$.
   a) Construct the Cayley table for the group of permutations on A.
   b) What is the order of this group?
   c) Is the group Abelian?
   d) What is the inverse of each element of the group?

7. **Transpositions**
   Verify the following products.
   a) $(1234\ldots n) = (12)(13)(14)\cdots(1n)$
   b) $(214) = (21)(24) = (24)(41)$
   c) $(4321) = (43)(42)(41)$
   d) $(15324) = (15)(13)(12)(14)$

8. **Do Transpositions Commute?**
   Do transpositions commute in general? For the set $\{1, 2, 3\}$, is it true that $(12)(13) = (13)(12)$?

9. **Decomposition into Transitions Is Not Unique**
   Show that the decomposition of the permutation (12345) can be written in any of the three forms:

$$
\begin{aligned}
(12345) &= (12)(13)(14)(15) \\
&= (15)(25)(35)(45) \\
&= (23)(24)(25)(21)
\end{aligned}
$$

Cartesian (or Direct) Product of Groups

It is possible to piece together smaller groups to form larger groups. If H and G are groups, their Cartesian product[4] is

$$H \times G = \{(h,g) : h \in H, g \in G\},$$

where the group operation $*$ in $H \times G$ is

$$(h,g)*(h',g') = (hh',gg'),$$

where hh' is the group operation in group H, and gg' is the group operation in group G. The following problems illustrate some Cartesian products of groups.

10. **Cartesian Product $\mathbb{Z}_2 \times \mathbb{Z}_2$**
    Consider the cyclic groups $\mathbb{Z}_2 = \{0, 1\}$, where the group operation is addition mod 2. Find the Cartesian product $\mathbb{Z}_2 \times \mathbb{Z}_2$ and construct its multiplication table. Show the table is the same as the multiplication table for the Klein four group of symmetries of a rectangle. In other words, the Klein four group is **isomorphic** to $\mathbb{Z}_2 \times \mathbb{Z}_2$.

11. **Cartesian Product Group $\mathbb{Z}_2 \times \mathbb{Z}_3$**
    Find the elements of the Cartesian product $\mathbb{Z}_2 \times \mathbb{Z}_3$. What is the order of the group? What is the Cayley table for the group? Hint: Keep in mind that the product $(a, b)(c, d) = (e, f)$, where $e = (a + c) \bmod 2, f = (b + d) \bmod 3$.

12. **Permutation Matrices**
    The permutation

    $$P = \begin{pmatrix} 1 & 2 & 3 \\ 1 & 3 & 2 \end{pmatrix}$$

    can be carried out with a 3×3 matrix.
    a) Find the matrix?
    b) Find the six matrices that represent the six permutations of S_3.

13. **Internet Research**
    There is a wealth of information related to topics introduced in this section just waiting for curious minds. Try aiming your favorite search engine toward *permutations in cycle notation, symmetric group, permutation group,* and *puzzles with permutations.*

---

4 The Cartesian product is often written $H \times G$. The Cartesian product can be extended to the product on any number of groups, like $G_1 \times G_2 \times \cdots \times G_n$.

6.4

Subgroups

Groups Inside a Group

> **Purpose of Section**   To introduce the concept of a **subgroup** and find the
> subgroups of various cyclic groups. We also introduce the idea of cosets of a
> subset and the concept of a quotient group.

6.4.1   Introduction

Recall the six symmetries of an equilateral triangle: the identity map, three flips
about the midlines, and two (counterclockwise) rotations of 120° and 240° illustrated in Figure 6.47.

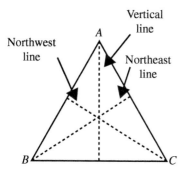

Figure 6.47 Symmetries of an equilateral triangle.

Although we have seen that these symmetries, along with the operation of compositions forms an algebraic group, what is also true is that the group is made up
of several smaller groups, In the given case of the six symmetries of an equilateral
triangle, the subset of three rotational symmetries $\{e, R_{120}, R_{240}\}$ whose multiplication table is shown in Figure 6.48, can easily be verified to form a group.

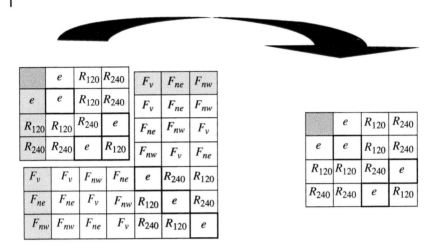

Figure 6.48 Subgroup of rotations of symmetries of an equilateral triangle.

The above discussion motivates the following definition of "groups within groups," or subgroups.

Definition   Let $\{G, *\}$ be a group with operation $*$. If a subset $H \subseteq G$ itself forms a group with the same operation $*$, then H is called a **subgroup** of G.

6.4.1.1   At Least Two Subgroups

Although all groups have two subgroups, the group itself and the trivial group consist of only the identity $\{e\}$, and we are mainly interested in the other subgroups, called **proper subgroups**, although we often refer to them simply as subgroups.

Example 1   Subgroups of Symmetries
Find the subgroups of the dihedral group D_3 of symmetries of an equilateral triangle.

Solution
The Cayley table for the dihedral group D_3 of symmetries of an equilateral triangle and its subgroups are displayed in Figure 6.49. There are four subgroups[1]

---

1 Of course, $(G, *)$ is a subgroup of *itself*, but when we say subgroups, we mean *proper* subgroups when $H \neq G$.

(a)

*	e	R_{120}	R_{240}	F_v	F_{ne}	F_{nw}
e	e	R_{120}	R_{240}	F_v	F_{ne}	F_{nw}
R_{120}	R_{120}	R_{240}	e	F_{ne}	F_{nw}	F_v
R_{240}	R_{240}	e	R_{120}	F_{nw}	F_v	F_{ne}
F_v	F_v	F_{nw}	F_{ne}	e	R_{240}	R_{120}
F_{ne}	F_{ne}	F_v	F_{nw}	R_{120}	e	R_{240}
F_{nw}	F_{nw}	F_{ne}	F_v	R_{240}	R_{120}	e

(b)

*	e	R_{120}	R_{240}	F_v	F_{ne}	F_{nw}
e	e	R_{120}	R_{240}	F_v	F_{ne}	F_{nw}
R_{120}	R_{120}	R_{240}	e	F_{ne}	F_{nw}	F_v
R_{240}	R_{240}	e	R_{120}	F_{nw}	F_v	F_{ne}
F_v	F_v	F_{nw}	F_{ne}	e	R_{240}	R_{120}
F_{ne}	F_{ne}	F_v	F_{nw}	R_{120}	e	R_{240}
F_{nw}	F_{nw}	F_{ne}	F_v	R_{240}	R_{120}	e

(c)

*	e	R_{120}	R_{240}	F_v	F_{ne}	F_{nw}
e	e	R_{120}	R_{240}	F_v	F_{ne}	F_{nw}
R_{120}	R_{120}	R_{240}	e	F_{ne}	F_{nw}	F_v
R_{240}	R_{240}	e	R_{120}	F_{nw}	F_v	F_{ne}
F_v	F_v	F_{nw}	F_{ne}	e	R_{240}	R_{120}
F_{ne}	F_{ne}	F_v	F_{nw}	R_{120}	e	R_{240}
F_{nw}	F_{nw}	F_{ne}	F_v	R_{240}	R_{120}	e

(d)

*	e	R_{120}	R_{240}	F_v	F_{ne}	F_{nw}
e	e	R_{120}	R_{240}	F_v	F_{ne}	F_{nw}
R_{120}	R_{120}	R_{240}	e	F_{ne}	F_{nw}	F_v
R_{240}	R_{240}	e	R_{120}	F_{nw}	F_v	F_{ne}
F_v	F_v	F_{nw}	F_{ne}	e	R_{240}	R_{120}
F_{ne}	F_{ne}	F_v	F_{nw}	R_{120}	e	R_{240}
F_{nw}	F_{nw}	F_{ne}	F_v	R_{240}	R_{120}	e

Figure 6.49 (a) $H_1 = \{e, F_v\}$ flip around vertical axis. (b) $H_2 = \{e, F_{nw}\}$ flip around the northwest axis. (c) $H_3 = \{e, F_{ne}\}$ flip around the northeast axis. (d) Three rotations of a rectangle.

of D_3; the rotational subgroup $\{e, R_{120}, R_{240}\}$ of order 3 and three "flip" subgroups $\{e, F_v\}$, $\{e, F_{ne}\}$, $\{e, F_{nw}\}$, each of order 2.

We let the reader verify each of these subgroups satisfy the required conditions to be a group.

6.4.2 Subgroups of the Klein Four-Group

Recall that the group of symmetries of a rectangle forms the Klein four-group with elements

$$G = \{e, R_{180}, H, V\}.$$

Figure 6.50 shows the Cayley table of these symmetries and three subgroups of order 2.

*	e	R_{180}	H	V
e	e	R_{180}	H	V
R_{180}	R_{180}	e	V	H
H	H	V	e	R_{180}
V	V	H	R_{180}	e

Group of symmetries of a rectangle

*	e	R_{180}	H	V
e	e	R_{180}	H	V
R_{180}	R_{180}	e	V	H
H	H	V	e	R_{180}
V	V	H	R_{180}	e

Subgroup of horizontal flips

*	e	R_{180}	H	V
e	e	R_{180}	H	V
R_{180}	R_{180}	e	V	H
H	H	V	e	R_{180}
V	V	H	R_{180}	e

Subgroup of vertical flips

*	e	R_{180}	H	V
e	e	R_{180}	H	V
R_{180}	R_{180}	e	V	H
H	H	V	e	R_{180}
V	V	H	R_{180}	e

Subgroup of rotational symmetries

Figure 6.50 Symmetry group of a rectangle and three subgroups.

Note that the order of a subgroup divides the order of the group. This fundamental property of subgroups is called Lagrange's Theorem after the French/Italian mathematician Joseph-Louis Lagrange (1736–1813).

6.4.3 Test of Subgroups

Although a subset H of a group G is a group only if it satisfies the requirements for a group, it is only necessary to verify that the group operation $*$ is closed in H and that every element of H has an inverse in H. There is no need to show the associative property or the existence of an identity since the identity in G is also an identity in H. This result is summarized in the following theorem.

Theorem 1 Conditions for Being a Subgroup
If $\{G, *\}$ is a group and H a (nonempty) subset of G, then H with operation $*$ is a **subgroup** of $\{G, *\}$, provided the following conditions are satisfied:

i) The operation $*$ is closed in H. That is,

$$\forall x, y \in H \Rightarrow x*y \in H$$

ii) Every element h in H has an **inverse** $h^{-1} \in H$ such that $h * h^{-1} = h^{-1} * h = e \in G$.
That is,

$$(\forall h \in H)(\exists h^{-1} \in H)(h*h^{-1} = h^{-1}*h = e).$$

where "e" is the identity element in G.

Proof
We assume properties (i) and (ii) and show that $(H, *)$ satisfies the conditions of closure, identity, inverse, and associativity.
Closure in H: This is the assumed property (i).
Identity in H: If $h \in H$, by condition (ii) there exists a $h^{-1} \in H$ that satisfies $h * h^{-1} = e \in G$. But the closure assumption (i) says $e = h * h^{-1} \in H$.
Inverse in H: This is the assumed property (ii).
Associative condition: The associative law.

$$(a*b)*c = a*(b*c)$$

holds for all a, b, $c \in H$ since it holds for all elements of G. ∎

Example 2 Test of Subgroup
Let $G = \mathbb{Z} = \{0, \pm 1, \pm 2, ...\}$ be the group of integers with addition $+$ as the group operation. Show that the set of even integers $2\mathbb{Z} = \{0, \pm 2, \pm 4, ...\}$ is a subgroup of G.

Solution
We verify the two conditions for a subset of a group to be a subgroup.

Closure of Addition: If $m = 2k_1$ and $n = 2k_2$ are even integers, so is their sum as can be seen from the equation

$$m + n = 2(k_1 + k_2) \in 2\mathbb{Z}.$$

Inverse in Subset: Every even integer $2k \in 2\mathbb{Z}$ has an inverse, namely $-2k \in 2\mathbb{Z}$, as can be seen from

$$2k + (-2k) = [2 + (-2)]k = 0 \in 2\mathbb{Z}. ∎$$

Example 3   Group of Infinite Order
The points (a, b) in the Cartesian plane with group operation

$$(a,b) \oplus (c,d) = (a+c, b+d)$$

form a group. Show that $H = \{(x, 0) : x \in \mathbb{R}\}$ is a subgroup.

Solution
We verify the two conditions that ensure a subset of a group to be a subgroup.

Closure: The x-axis is a subset of the plane, and the operation $\oplus$ is closed in H since

$$\left.\begin{array}{l}(x_1,0) \in H \\ (x_2,0) \in H\end{array}\right\} \Rightarrow (x_1,0) \oplus (x_2,0) = (x_1 + x_2, 0) \in H$$

Inverse: Since $(x, 0) + (-x, 0) = (0, 0)$ every element $(x, 0) \in H$ has the unique inverse $(-x, 0)$. ■

Example 4   Subgroups of the Octic Group
Figure 6.51 shows the eight symmetries of the dihedral group

$$D_4 = \{e, R_{90}, R_{180}, R_{270}, F_V, F_H, F_{nw}, F_{ne}\}$$

which represents the eight symmetries of a square, which is also called the **octic** group.

a) Does the group commute? Hint: Compare $R_{270}F_{ne}$ and $F_{ne}R_{270}$.
b) There are ten subgroups of the octic group. Find them.

Solution
a) The reader can check that

$$R_{270}F_{ne} \neq F_{ne}R_{270}.$$

Hence, the octic group is not commutative.

b) The eight proper subgroups of the octic group D_4 are[2]

$$\boxed{\begin{array}{l}\{e, V\}, \{e, H\}, \{e, R_{180}\}, \{e, F_{nw}\}, \{e, F_{ne}\}, \\ \qquad\qquad \{e, R_{180}, V, H\}, \{e, R_{90}, R_{180}, R_{270}\}, \{e, R_{180}, F_{nw}, F_{ne}\}\end{array}}$$

---

2 The octic group, which is a subgroup of itself, is not included here.

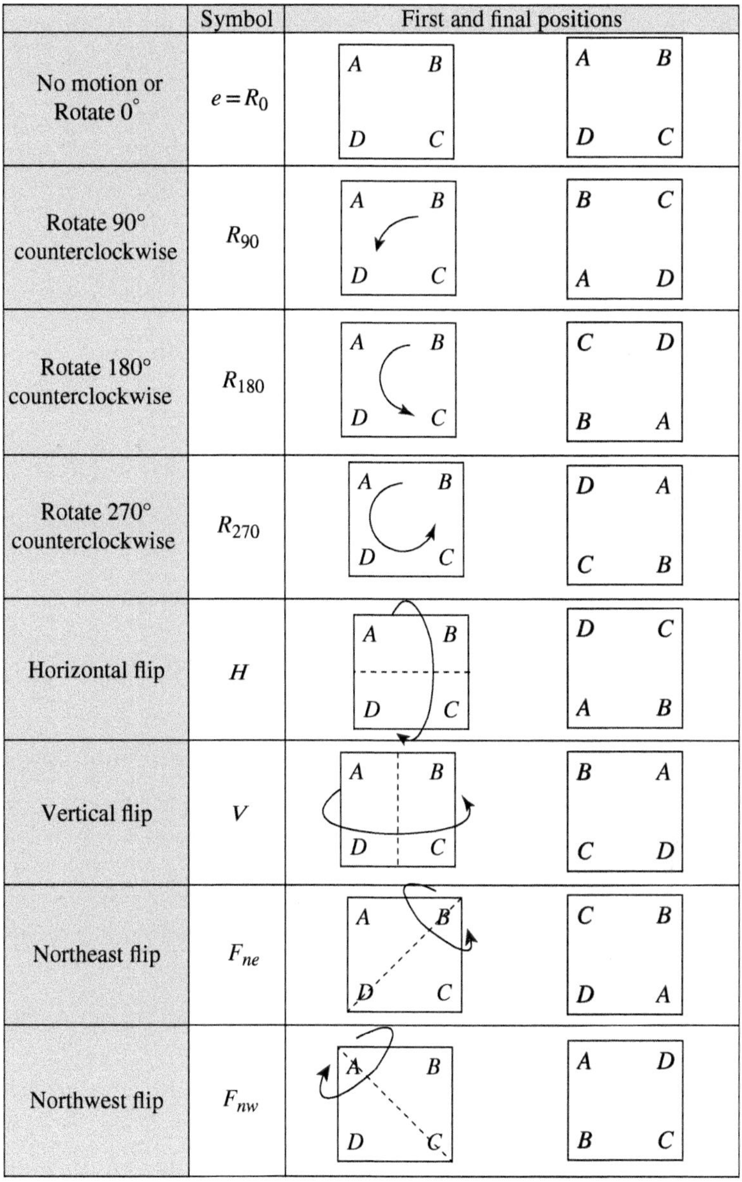

	Symbol	First and final positions	
No motion or Rotate 0°	$e = R_0$	A B D C	A B D C
Rotate 90° counterclockwise	R_{90}	A B D C	B C A D
Rotate 180° counterclockwise	R_{180}	A B D C	C D B A
Rotate 270° counterclockwise	R_{270}	A B D C	D A C B
Horizontal flip	H	A B D C	D C A B
Vertical flip	V	A B D C	B A C D
Northeast flip	F_{ne}	A B D C	C B D A
Northwest flip	F_{nw}	A B D C	A D B C

Figure 6.51 Eight symmetries of a square.

The subgroups of the symmetry group of a square form a partially ordered set, ordered by set inclusion, illustrated in Figure 6.52. The octic group D_4 is itself a subgroup of the group S_4 of 24 permutations of four elements.

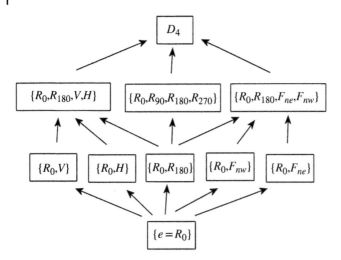

Figure 6.52 Hasse diagram for the subgroups of the octic group D_4.

6.4.4 Subgroups of Cyclic Groups

We have seen that the cyclic group Z_n is generated by an element in the group. That is, there exists a $g \in Z_n$ such that

$$\langle g \rangle \equiv \{e, g, g^2, g^3, \ldots, g^{n-1}\} = Z_n$$

To find the subgroups of Z_n, we begin with an arbitrary element $h \in Z_n$ and compute the set $\langle h \rangle$ generated by h. The elements generated by h may or may not be all of Z_n, but they will always create a subgroup of Z_n. We then pick a second member $h' \in Z_n$ which is not in the first generated set $\langle h \rangle$ and compute $\langle h' \rangle$. This will yield another subset of Z_n. Continuing this process will eventually yield all subgroups of Z_n.

Let us apply this technique to find all the subsets of the cyclic group

$$Z_{12} = \{0, 1, 2, 3, 4, 5, 6, 7, 8, 9, 10, 11\},$$

where the group operation is addition modulo 12. Starting with 1, we generate powers of $g = 1$, remembering that powers of one in this group are really adding 1. Hence, we have

$$\langle 1 \rangle = \{0, 1, 2, 3, 4, 5, 6, 7, 8, 9, 10, 11\} \subseteq Z_{12}$$

which generates the entire group Z_{12}. We now select the element $g = 2$, which generates the subgroup

$$\langle 2 \rangle = \{0, 2, 4, 6, 8, 10\} \subseteq Z_{12}.$$

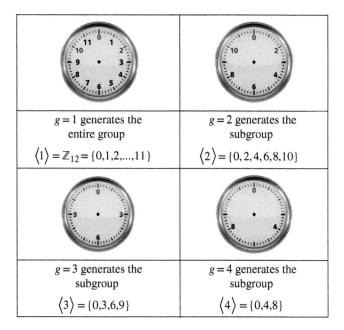

Figure 6.53 Four typical subgroups generated by elements of the group.

Figure 6.53 shows the subgroups generated by $g = 1, 2, 3, 4$. Do you see why $\langle 5 \rangle = \mathbb{Z}_{12}$ and $\langle 6 \rangle = \{0, 6\}$.

You might notice that the numbers $1, 5, 7, 11$ are relative prime to 12 to generate the entire group $\mathbb{Z}_{12}$. Do you also see a pattern between the orders of those subgroups generated by numbers *not* relatively prime with 12? That is, the orders of the subgroups $\langle 2 \rangle$, $\langle 3 \rangle$, $\langle 4 \rangle$, $\langle 6 \rangle$, $\langle 8 \rangle$, $\langle 9 \rangle$, $\langle 10 \rangle$?

Table 6.5 shows the subgroups generated by each element of the group and the order of the subgroup generated by the generator.

6.4.5 Cosets and the Quotient Group

If G is a group with group operation "+" and subgroup H, then for any $g \in G$ the "g-translation" of H

$$g + H = \{g + h : h \in H\}$$

is called the left coset[3] of H with respect to g. Similarly, the right cosets are defined by

$$H + g = \{h + g : h \in H\}$$

---

3 We could express the group operation multiplicatively as gH, rather than $g + H$. It is strictly a matter of preference.

Table 6.5 Generators of subsets of $\mathbb{Z}_{12}$.

Generator	Order of the generator
$\langle 1 \rangle = \mathbb{Z}_{12}$	$12(1^{12} = 0)$
$\langle 2 \rangle = \{0, 2, 4, 6, 8, 10\}$	$6(2^6 = 0)$
$\langle 3 \rangle = \{0, 3, 6, 9\}$	$4(3^4 = 0)$
$\langle 4 \rangle = \{0, 4, 8\}$	$3(4^3 = 0)$
$\langle 5 \rangle = \mathbb{Z}_{12}$	$12(5^{12} = 0)$
$\langle 6 \rangle = \{0, 6\}$	$2(6^2 = 0)$
$\langle 7 \rangle = \mathbb{Z}_{12}$	$12(7^{12} = 0)$
$\langle 8 \rangle = \{0, 4, 8\}$	$3(8^3 = 0)$
$\langle 9 \rangle = \{0, 3, 6, 9\}$	$4(9^4 = 0)$
$\langle 10 \rangle = \{0, 2, 4, 6, 8, 10\}$	$6(10^6 = 0)$
$\langle 11 \rangle = \mathbb{Z}_{12}$	$12(11^{12} = 0)$

If the left and right coset coincide,[4] as they do for commutative groups or for certain subgroups called normal subgroups, we simply refer to them as cosets without specifying left or right. You can think of the cosets of a subgroup as what you get when you "shift" the subgroup to the left or right with members $g \in G$, that is "translated" subgroups. The subgroup H itself is a coset when the "shifting factor" $g \in G$ is the identity element in G. By selecting various $g \in G$, one can create a partition of the group G into equivalence classes of cosets. This is useful since subdividing a group into disjoint parts allows one to identify parts of the group that are similar from some point of view.

For example, the (commutative) group $G = \mathbb{Z}_6$ of integers mod 6 has a subgroup $H = \{0, 3\}$ of two elements 0 and 3. Translating the subgroup with the six elements $g \in \mathbb{Z}_6$, we arrive at three cosets of the form

$$g + H = \{g + \{0,3\}, g \in \mathbb{Z}_6\}$$

which are

$$0 + H = 3 + H = \{0,3\}$$
$$1 + H = 4 + H = \{1,4\}$$
$$2 + H = 5 + H = \{2,5\}$$

Note that the subgroup $H = \{0, 3\}$ itself is a coset when $g = 0 \in \mathbb{Z}_6$.

---

4 The left and right cosets coincide if G is commutative or if the subgroup H satisfies $g + H = H + g$ for all $g \in G$. Subgroups H with this property are called **normal** subgroups.

The question now arises, is it possible to do "arithmetic" with the cosets, and the answer if yes, when the left and right cosets coincide. Without verification, it is not difficult to show that the binary operation

$$(g_1 + H) \oplus (g_2 + H) = (g_1 + g_2) + H \quad \text{additively}$$

$$(g_1 H) \otimes (g_2 H) = g_1 g_2 H \qquad\qquad \text{multiplicatively}$$

acting on the cosets forms a group. This group is called the **quotient** (or **factor**) **group** of G modulo H, and are written

$$G/H = \{g + H : g \in G\}$$

pronounced $G \bmod H$.

The following example illustrates how an infinite (commutative) group can be "factored" into just three equivalence classes, each having a common property.

Example 5   Cosets in $\mathbb{Z}$

Let $G = \mathbb{Z}$ be the group of integers with group operation addition, and $H = 3\mathbb{Z} = \{0, \pm 3, \pm 6, \ldots\}$ the subgroup of multiples of 3. By selecting different "shifting values" $g \in G = \mathbb{Z}$, we can shift H to obtain its cosets, which are

$$0 + 3\mathbb{Z} = \{\ldots, -6, -3, 0, 3, 6, \ldots\} = \text{equivalence class of } 0 \bmod 3$$

$$1 + 3\mathbb{Z} = \{\ldots, -5, -2, 1, 4, 7, \ldots\} = \text{equivalence class of } 1 \bmod 3$$

$$2 + 3\mathbb{Z} = \{\ldots, -4, -1, 2, 5, 8, \ldots\} = \text{equivalence class of } 2 \bmod 3$$

Although the cosets of $3\mathbb{Z}$ are formed by allowing the shifting value $g \in \mathbb{Z}$ to roam over all integers $\mathbb{Z}$, only three cosets are created. Note that they partition $\mathbb{Z}$ into three disjoint subsets.

The set $3\mathbb{Z}$ and its two cosets

$$\mathbb{Z}/3\mathbb{Z} = \{3\mathbb{Z}, 1 + 3\mathbb{Z}, 2 + 3\mathbb{Z}\}$$

with binary operation $\oplus$ defined by

$$(a + 3\mathbb{Z}) \oplus (b + 3\mathbb{Z}) = (a + b) + 3\mathbb{Z}$$

where $a, b \in \mathbb{Z}$ defines the quotient group $\mathbb{Z}/3\mathbb{Z}$ whose Cayley table is shown in Figure 6.54.[5]

---

5 If the group G is not a commutative, then one must require the subgroup H to be a **normal subgroup**, meaning that for all $g \in G$ one has $g + H = H + g$, which allows one to uniquely define the cosets of H.

	$0+3\mathbb{Z}$	$1+3\mathbb{Z}$	$2+3\mathbb{Z}$
$0+3\mathbb{Z}$	$0+3\mathbb{Z}$	$1+3\mathbb{Z}$	$2+3\mathbb{Z}$
$1+3\mathbb{Z}$	$1+3\mathbb{Z}$	$2+3\mathbb{Z}$	$0+3\mathbb{Z}$
$2+3\mathbb{Z}$	$2+3\mathbb{Z}$	$0+3\mathbb{Z}$	$1+3\mathbb{Z}$

Figure 6.54 Quotient Group of $Z/3Z$.

Important Note   The reason G/H is called a "quotient group" comes from division of integers. The quotient $6/2 = 3$ means one can subdivide six objects into three subcollections, each containing two objects. The quotient G/H group is the same idea, although here we end up with a group rather than a number. However, the *order* of the quotient group (for finite groups) *is* the order of the group G divided by the order of the subgroup H, that is $|G/H| = |G|/|H|$.

Problems

1. **True or False**
   a) The order of any subgroup divides the order of the group.
   b) Every subgroup of a group contains an identity element.
   c) Some groups do not have any subgroups.
   d) $\mathbb{Z}$ is a subgroup of $\mathbb{R}$ under the operation of addition.
   e) The symmetric group S_2 has two subgroups.
   f) There are some groups where every subset is a subgroup.
   g) The set $\{e, h\}$ is a subgroup of the group of symmetries of a square, where e denotes the identity map, and h is the horizontal flip.
   h) There are five subgroups of order 2 of the group of symmetries of a square.

2. **Subgroups of $\mathbb{Z}_6$**
   List the subgroups of $\mathbb{Z}_6 = \{0, 1, 2, 3, 4, 5\}$ generated by the different elements of the group. What is the order of each of these generated groups?

3. **Cayley Table**
   Find the Cayley table for the subgroup $\{e, R_{180}, v, h\}$ of symmetries of a square.

4. **Cayley Table**
   Show that the group defined by the following Cayley table in Figure 6.55 is a subgroup of S_3.

*	( )	(123)	(132)
( )	( )	(123)	(132)
(123)	(123)	(132)	( )
(132)	(132)	( )	(123)

Figure 6.55 Subgroup S_3.

5. Subgroup Generated by R_{240}
Find the subgroup of the dihedral group D_3 of symmetries of an equilateral triangle generated by R_{240}.

6. Generated Groups of Symmetries of a Rectangle
In the Klein Four-group $\{e, R_{180}, v, h\}$ of symmetries of a rectangle, find the subgroups generated by each element in the group. What is the order of each member?

7. Center of a Group
The center $Z(G)$ of a group G consists of all elements of the group that commute with all elements of the group. That is

$$Z(G) = \{g \in G : gx = xg \text{ for all } x \in G\}$$

It can be shown that the center of any group is a subgroup of the group. Find the center of the group of symmetries of a rectangle. Note: The center of a group is never empty since the identity element of a group always commutes with every element of the group. The question is, are there other elements that commute with every element of the group?

8. Hasse Diagram
Draw the Hasse diagram for the subgroups of symmetries of a rectangle.

9. Subgroups of $\mathbb{Z}_8$
Find the subgroups of the cyclic group $\mathbb{Z}_8$.

10. Subgroups of $\mathbb{Z}_{11}$
Find the subgroups of $\mathbb{Z}_{11}$.

11. Matrix Subgroup
The set G of all 2×2 invertible matrices with real entries forms a group under matrix multiplication. Show that the subset of matrices of the form

$$H = \left\{ \begin{bmatrix} 1 & a \\ 0 & 1 \end{bmatrix} ; a \in \mathbb{Z} \right\}$$

is a subgroup of G.

12. Cosets

The set $4\mathbb{Z} = \{..., -12, -8, -4, 0, 4, 8, 12, ...\}$ is a subgroup of the integers $\mathbb{Z}$ with the addition operation. What are the cosets of $4\mathbb{Z}$ and show them along with $4\mathbb{Z}$ partition $\mathbb{Z}$ into disjoint sets.

13. Cosets in the Plane

The Cartesian plane

$$\mathbb{R}^2 = \{(x,y) : x \in \mathbb{R}, y \in \mathbb{R}\}$$

with the usual operation pointwise addition is a group and the line $H = \{(x, y) : y = x\}$ passing through the origin is a subgroup.

a) What are the cosets of this subgroup?
b) What is the sum of the two cosets

$$(0,1) + \{(x,y) : y = x\}$$
$$(2,3) + \{(x,y) : y = x\}$$

14. Internet Research

There is a wealth of information related to topics introduced in this section just waiting for curious minds. Try aiming your favorite search engine toward *why are subgroups important, number of subgroups,* and *groups with no proper list of small groups.*

6.5

Rings and Fields

> **Purpose of Section** To introduce the concept of an algebraic ring and an important type of ring called a field.

6.5.1 Introduction to Rings

Although the algebraic group has one binary operation, the algebraic system we studied in grade school has two binary operations: addition and multiplication. This leads us to the study of rings and fields. We begin with one of the most important abstract systems with two binary operations called a ring.[1] A ring is one of the basic structures of abstract algebra, which generalizes the common arithmetic operations of the integers, polynomials, matrices, and so on. Ring theory is used today to understand basic physical laws, such as those underlying such things as symmetry phenomena in molecular chemistry.

You have seen examples of rings before. The integers $\mathbb{Z}$ with ordinary addition (+) and multiplication (×) are an example of an algebraic ring. In this regard, you might think of a ring as "generalized integers." The study of rings was initiated (in part) by the German mathematician Richard Dedekind (1831–1916) in the late 1800s, and the axiomatic foundations were laid down in the 1920s, and most of them by the German mathematician Emmy Noether.

Definition A set $\{R, +, \times\}$ with two (closed) binary operations of +(addition) and ×(multiplication) is called a **ring** if:

a) The system with operation + forms a **commutative group**.
b) The operation × is **associative**. In the language of predicate logic:

$$(\forall a,b,c \in R)[(a \times b) \times c = a \times (b \times c)].$$

---

[1] The word "ring" was coined by the German mathematician David Hilbert (1862–1943).

Advanced Mathematics: A Transitional Reference, First Edition. Stanley J. Farlow.
© 2020 John Wiley & Sons, Inc. Published 2020 by John Wiley & Sons, Inc.
Companion website: www.wiley.com/go/farlow/advanced-mathematics

c) The operation × **distributive** over + both on the left and right. In other words

- $(\forall a, b, c \in R)[a \times (b + c) = (a \times b) + (a \times c)]$
- $(\forall a, b, c \in R)[(b + c) \times a = (b \times a) + (c \times a)]$

The first thing one notes about a ring is that practically no conditions are imposed on the multiplication, only that it is associative and obeys distributive laws. The action in a ring lies with the addition.

We denote the ring operation of addition by + and multiplication by ×, although they do not necessarily denote addition and multiplication of numbers. We often denote ring multiplication by ab for shorthand. We also call the additive identity in the ring, the **zero** (or **additive identity**) of the ring, and denote it by 0. A ring need not have a multiplicative identity, but when it does, we say the ring has a **multiplicative identity** (or **unity**) and is denoted by, you guessed it, 1.

Important Note Roughly, a "ring" is a set of elements having two operations, normally called addition and multiplication, which behave in many ways like the integers. You can add, subtract, and multiply elements in a ring, but not in general divide them. We must wait until we get to the general structure of a field before we can add, subtract, multiply, and divide.

Special Kinds of Rings

- **Commutative Rings:** If multiplication commutes, i.e. $ab = ba$, then the ring is called a **commutative ring**. We do not mention addition since addition always commutates in a ring.
- **Rings with Multiplicative Identity:** A ring with a multiplicative identity is called a **ring with identity**. Again, we do not mention the additive identity since rings always have an additive identity.
- **Ring with Zero Divisors:** Nonzero elements $a, b \in R$ in a ring are called **zero divisors** if their product is zero; that is $ab = 0$ or $ba = 0$ (0 being the additive identity in the ring). This condition may appear strange to the reader since the familiar rings of integers, rational, and real numbers with ordinary addition and multiplication do not have zero divisors. But some rings, such as rings of matrices with ordinary addition and multiplication, have nonzero matrices whose product is the zero matrix.

6.5.2 Common Rings

- **Example 1** The integers $\mathbb{Z} = \{0, \pm 1, \pm 2, ...\}$ with usual addition and multiplication are a commutative ring with the multiplicative identity of 1.
- **Example 2** The set $\mathbb{Z}[x]$ of all polynomials in x with integer coefficients and usual addition and multiplication is a commutative ring with multiplicative identity $f(x) = 1$.
- **Example 3** The set $2\mathbb{Z} = \{0, \pm 2, \pm 4, ...\}$ of even integers with usual addition and multiplication is a commutative ring without a multiplicative identity.
- **Example 4** The set $\mathbb{Z}_n = \{0, 1, 2, ..., n-1\}$ with addition and multiplication modulo n is a commutative ring with multiplicative identity 1. This ring is called the ring of integers modulo n.
- **Example 5** The set $M_2(\mathbb{Z})$ of all 2×2 matrices with integer entries is a noncommutative ring with multiplicative identity

$$\begin{bmatrix} 1 & 0 \\ 0 & 1 \end{bmatrix}.$$

- **Example 6** The sets $\mathbb{Z}, \mathbb{Q}, \mathbb{R}$ with the usual addition and multiplication are all rings. The additive identity in each of these rings is 0 and the multiplicative identity is 1.
- **Example 7** An important ring is the set of polynomials of the form

$$a_n x^n + a_{n-1} x^{n-1} + \cdots + a_1 x + a_0$$

for a given natural number n and real coefficients with normal addition and multiplication of polynomials. Note that addition, subtraction, and multiplication are closed operations, but division is not.

> **Important Note** You cannot always solve simple linear algebraic equations in rings. In the ring of integers $\mathbb{Z}$ with ordinary addition and multiplication, one cannot solve $2x = 1$ since 1/2 does not belong in the ring. Rings describe mathematical objects that can be added, subtracted, and multiplied, but not divided.

Example 7 Special Ring
Draw the addition and multiplication tables for the ring $\mathbb{Z}_3 = \{0, 1, 2\}$ with addition and multiplication modulo the prime number 3 (see Figure 6.56).

Solution
Carrying out these operations, we find.

Note that the addition table forms a commutative group, and if we only look at the nonzero members $\{1, 2\}$ of the multiplication table, they form a commutative group of order two. This is special kind of ring called a field. But not all rings have this much structure, as the following example illustrates.

+	0	1	2
0	0	1	2
1	1	2	0
2	2	0	1

×	0	1	2
0	0	0	0
1	0	1	2
2	0	2	1

Addition modulo 3   Multiplication modulo 3

Figure 6.56 Special ring.

Example 8   The Cyclic Group Ring
Draw the addition and multiplication tables for the ring

$$\mathbb{Z}_6 = \{0,1,2,3,4,5\}$$

with addition and multiplication modulo 6.

Solution
Carrying out these operations, we find the tables in Figure 6.57.

Although the addition table for $\mathbb{Z}_6$ forms a commutative group with additive identity 0, the nonzero members of the multiplication table do not form a group for a variety of reasons. First off, the numbers 2, 3, and 4 do not have

+	0	1	2	3	4	5
0	0	1	2	3	4	5
1	1	2	3	4	5	0
2	2	3	4	5	0	1
3	3	4	5	0	1	2
4	4	5	0	1	2	3
5	5	0	1	2	3	4

×	0	1	2	3	4	5
0	0	0	0	0	0	0
1	0	1	2	3	4	5
2	0	2	4	0	2	4
3	0	3	0	3	0	3
4	0	4	2	0	4	2
5	0	5	4	3	2	1

Addition modulo 6   Multiplication modulo 6

Figure 6.57 Cyclic ring.

multiplicative inverses since there is no member of the ring that satisfies the equations $2 \times a = 1$, $3 \times a = 1$, $4 \times a = 1$.

Historical Note   Emmy Noether (1882–1935) was a German mathematician who made groundbreaking contributions to ring theory and theoretical physics. She was described by Albert Einstein as the most important woman in the history of mathematics. *Noether's theorem* has been called one of the most significant results in theoretical physics and gives a fundamental connection between symmetry and conservation laws.

Although rings are important algebraic structures in many areas of mathematics,[2] they sometimes are too restrictive. For example, in the ring of integers $\mathbb{Z}$ under ordinary addition and multiplication, we cannot divide 3 by 5 to obtain a member of the ring. This leads us to the study of the algebraic field.[3]

6.5.3   Algebraic Fields

Rings allow for addition, subtraction, and multiplication, but not division. The equation $3x = 7$ has no solution in the ring of integers but does have a solution in the *field* of rational numbers. An algebraic field is the domain for many areas of mathematics, including real and complex analysis. We saw in Section 4.2 that the real numbers were a complete ordered field.

Definition   A **field** is a set F with at least two elements with two closed binary operations $+$ and $\times$, such that

- F is a commutative group under the operation $+$.
- The nonzero elements of F form a commutative group under $\times$.
- Multiplication distributive over addition.

When one thinks of a field, one thinks of a structure with two operations resembling addition and multiplication, where one can add, subtract, multiply, and divide. The reader is well aware of three common fields from analysis, the rational numbers $\mathbb{Q}$, the real numbers $\mathbb{R}$, and the complex numbers $\mathbb{C}$. There are "hypercomplex" number systems past the complex numbers, like

---

2 Ring theory is fundamental in algebraic geometry where rings of polynomials are important.
3 Someone once said students learn about rings in the third grade when they learn to multiply, and all about fields in the fourth grade when they learn to divide.

four-dimensional quaternions and the eight-dimensional octonians, neither of those number systems are fields, but more general abstract systems. Quaternion multiplication does not commute, and octonian multiplication not only does not commute but also is not even associative.

6.5.4 Finite Fields

Although we think of the rational numbers $\mathbb{Q}$, the real numbers $\mathbb{R}$, and the complex numbers $\mathbb{C}$ as the major infinite fields, there are many important finite fields, which contain only a finite number of elements. Finite fields play an important role in many areas of mathematics, including algebraic geometry and number theory, as well as applied areas like coding theory and cryptography.

An interesting property of finite fields is that they exist only for certain orders. For example there are finite fields of order 2, 3, 4, and 5, but none of order 6. There are finite fields of order 7, 8, 9, but none of order 10. To be specific, there are finite fields of order p^n, $n = 1, 2, \ldots$ where p is a prime number, but no others. These finite fields are called **Galois fields** and denoted by $GF(p^n)$.[4]

There are two main classifications of finite fields. There are the finite fields $GF(p)$ of prime order and then the more involved fields of order p^n when $n > 1$. The prime order fields $GF(p)$, when $n = 1$, are the field of permutations $\mathbb{Z}_p = \{0, 1, 2, \ldots, p-1\}$ of integers with addition and multiplication carried out modulo p. However, we saw in Example 8 that $\mathbb{Z}_6 = \{0, 1, 2, 3, 4, 5\}$ is not a field since 6 is not a prime number.

Example 9 Galois Field
Draw the addition and multiplication table for the Galois field

$$GF(7) = \mathbb{Z}_7 = \{0, 1, 2, 3, 4, 5, 6\}.$$

and find the additive and multiplicative inverses of each element 0, 1, 2, ..., 6.

Solution
Performing addition and multiplication modulo 7, we arrive at the tables in Figure 6.58. The additive inverse of a number is found by moving across the number's row until reaching 0, where the additive inverse is the column number. A similar principle holds for multiplicative inverses.

---

4 $GF(p^n)$ stands for Galois field in honor of the French mathematician Evariste Galois (1811–1832) who first studied them.

+	0	1	2	3	4	5	6
0	0	1	2	3	4	5	6
1	1	2	3	4	5	6	0
2	2	3	4	5	6	0	1
3	3	4	5	6	0	1	2
4	4	5	6	0	1	2	3
5	5	6	0	1	2	3	4
6	6	0	1	2	3	4	5

Addition modulo 7

×	0	1	2	3	4	5	6
0	0	0	0	0	0	0	0
1	0	1	2	3	4	5	6
2	0	2	4	6	1	3	5
3	0	3	6	2	5	1	4
4	0	4	1	5	2	6	3
5	0	5	3	1	6	4	2
6	0	6	5	4	3	2	1

Multiplication modulo 7

a	$-a$	a^{-1}
0	0	–
1	6	1
2	5	4
3	4	5
4	3	2
5	2	3
6	1	6

Additive and multiplicative inverses modulo 7

Figure 6.58 Arithmetic operations for the field GF(7) = $\mathbb{Z}_7$.

Example 10 Subtraction and Division Modulo 7

Since GF(7) = $\mathbb{Z}_7$ is a field, we should be able to carry out all four arithmetic operations: addition, subtraction, multiplication, and division. Since Figure 6.58 shows us how to add and multiply, how do we subtract and divide?

a) Find 2 – 5
b) Find 5/3

Solution

a) To find $x = 2 - 5$, we find x that satisfies $x + 5 = 2$. Figure 6.58 gives $x = 4$.

b) To find $4/3 = 4 \times 3^{-1}$, we first find the inverse 3^{-1} by finding the value of x that satisfies $3x = 1$. From Figure 6.58, we find $x = 5$ and so

$$4/3 = 4 \times 3^{-1} = 4 \times 5 = 6.$$

Problems

1. **True or False**
   a) A ring can be finite or infinite.
   b) In a ring $\{R, +, \times\}$, the set R with $\times$ is a group.
   c) In a ring $\{R, +, \times\}$, the set R with $+$ is a group.
   d) The ring $\mathbb{Z}_{11}$ is also a field.
   e) The ring $\mathbb{Z}_8$ is also a field.
   f) There are fields where $a \times b = 0$, but neither a, b are zero.

2. **Multiplicative Identity**
   For each of the following rings, tell if the ring is commutative and if there exists a multiplicative identity. If a multiplicative identity exists, what is it?
   a) The ring of integers $\mathbb{Z}$ with usual addition and multiplication.
   b) The ring of even integers $2\mathbb{Z}$ with usual addition and multiplication.
   c) The ring $C(\mathbb{R})$ of real-valued continuous functions with usual addition and multiplication.
   d) The ring consisting of the set

   $$\mathbb{Z}\left[\sqrt{2}\right] = \left\{ m + n\sqrt{2} : m, n \in \mathbb{Z} \right\}$$

   with usual addition and multiplication.

   e) The ring $\mathbb{Z}[x]$ of all polynomials in x whose coefficients are integer with ordinary addition and multiplication.
   f) The ring $\mathbb{Q}$ of rational numbers with ordinary addition and multiplication.
   g) The ring consisting of the set $\mathbb{Z}_3 = \{0, 1, 2\}$ where addition and multiplication are defined modulo 3.

3. **Ring of Matrices**
   Show that the set of all 2×2 matrices

   $$R = \left\{ \begin{bmatrix} a & b \\ c & d \end{bmatrix} : a, b, c, d \in \mathbb{Z} \right\}$$

   is a ring under matrix addition and matrix multiplication, but not a field.

4. **Rings that are not Fields**
   Why do the following rings fail to be fields?
   a) The ring of polynomials with real coefficients with the usual addition and multiplication.
   b) The ring of $n \times n$ matrices with the usual matrix addition and multiplication.
   c) The set $\mathbb{Z}_n = \{0, 1, 2, \ldots, n-1\}$, where the operations of addition and multiplication are performed mod n, where n is a composite natural number.

5. **Modulo 3 Field**
   The addition and multiplication tables for $\mathbb{Z}_3$ are shown in Figure 6.59. What are the additive and multiplicative inverses for each member of the field?

+	0	1	2	×	0	1	2
0	0	1	2	0	0	0	0
1	1	2	0	1	0	1	2
2	2	0	1	2	0	2	1

Figure 6.59 Modulo 3 field.

6. Construct the addition and multiplication tables for $\mathbb{Z}_5$.

7. **Arithmetic in $\mathbb{Z}_3$**
   In the field GF(3) = $\mathbb{Z}_3$, compute the following.
   a) $1 + 2$
   b) $1 - 2$
   c) 2×2
   d) $1/2$

8. **Modular Algebra**
   Find values of x that satisfies the following equations.
   a) $2x = 1 (\mathrm{mod}\, 3)$, $x \in \mathbb{Z}_3$
   b) $3x = 2 (\mathrm{mod}\, 5)$, $x \in \mathbb{Z}_5$
   c) $4x = 3 (\mathrm{mod}\, 7)$, $x \in \mathbb{Z}_7$

9. **Multiplicative Inverse**
   The integers $\mathbb{Z}$ under usual addition and multiplication form a commutative ring with unity 1. Do any members of this ring have multiplicative inverses? If so, what are they?

10. **More Modular Algebra**

Solve the equation $7x - 3 = 5$ in $\mathbb{Z}_{24}$.

11. **Type of Ring**

The set $\{0, a, b, c\}$ with operations of addition and multiplication, defined by the tables in Figure 6.60, forms a ring. Is this group commutative and does it have a multiplicative identity?

$\oplus$	0	a	b	c
0	0	a	b	c
a	a	0	c	b
b	b	c	0	a
c	c	b	a	0

$\otimes$	0	a	b	c
0	0	0	0	0
a	0	a	b	c
b	0	b	c	a
c	0	c	a	b

Figure 6.60 Special ring.

Zero Divisors

In some rings, things do not obey the arithmetic you learned in grade school. For example in the ring $\mathbb{Z}_4 = \{0, 1, 2, 3\}$ modulo 4 arithmetic, we found $2 \times 2 = 0$. In this case, we say that 2 is a zero divisor for this ring. In general, an element $a \in R$ in a ring is a **zero divisor** if there is a nonzero element $b \in R$ in the ring such that $ab = 0$. Matrix rings also have zero divisors.

12. **Zero Divisors**

Find a zero divisor in the ring of 3×3 matrices with integer entries using the usual operations of addition and multiplication?

13. **Internet Research**

There is a wealth of information related to topics introduced in this section just waiting for curious minds. Try aiming your favorite search engine toward *importance of ring theory in mathematics, importance of field theory in mathematics, important rings and fields in mathematics, history of ring theory in mathematics,* and *history of field theory in mathematics.*

Index

Advanced Mathematics: A Transitional Reference, First Edition. Stanley J. Farlow.
© 2020 John Wiley & Sons, Inc. Published 2020 by John Wiley & Sons, Inc.
Companion website: www.wiley.com/go/farlow/advanced-mathematics